Springer Series on Atomic, Optical, and Plasma Physics

Volume 112

The Springer Series on Atomic, Optical, and Plasma Physics covers in a comprehensive manner theory and experiment in the entire field of atoms and molecules and their interaction with electromagnetic radiation. Books in the series provide a rich source of new ideas and techniques with wide applications in fields such as chemistry, materials science, astrophysics, surface science, plasma technology, advanced optics, aeronomy, and engineering. Laser physics is a particular connecting theme that has provided much of the continuing impetus for new developments in the field, such as quantum computation and Bose-Einstein condensation. The purpose of the series is to cover the gap between standard undergraduate textbooks and the research literature with emphasis on the fundamental ideas, methods, techniques, and results in the field.

More information about this series at http://www.springer.com/series/411

Anders Kastberg

Structure of Multielectron Atoms

Springer

Anders Kastberg
Université Côte d'Azur
Nice, France

ISSN 1615-5653 ISSN 2197-6791 (electronic)
Springer Series on Atomic, Optical, and Plasma Physics
ISBN 978-3-030-36418-2 ISBN 978-3-030-36420-5 (eBook)
https://doi.org/10.1007/978-3-030-36420-5

This Springer imprint is published by the registered company Springer Nature Switzerland AG.
The registered company address is: Gewerbestrasse 11, 6330 Cham, Switzerland

This book is dedicated to three gentlemen who all were my mentors, thesis advisors, scientific role models and good friends:

Professors Arne Arnesen, Reinhold Hallin and Carl Nordling.

Preface

The field of atomic physics is, arguably, about a century and a half old. During that time, the field has constantly evolved. The most important paradigm shift happened around the 1920s, after the breakthrough in quantum mechanics. In terms of textbooks, this was manifested by *The Theory of Atomic Spectra* by E. U. Condon and G. H. Shortley, first published in 1935. This book is a classic, and it remains a useful reference to active scientists.

Many other books on atomic physics have been authored through the years, and their organisations and contents reflect how the field has evolved. To an increasing extent, these books have been divided into either being specialised and aimed at experts, or being rather a textbook written for advanced undergraduate or early postgraduate students.

The motivation for writing this book was a perceived need for an updated book specifically detailing *atomic structure*, with the intended principal audience of students at master's or PhD-level, answering to the present state of atomic physics. In a traditional university course, or textbook, in atomic physics, the theory of atomic structure is taught in parallel with spectroscopy. This is pertinent, since knowledge about atomic structure has been unravelled by spectroscopy, and a lot can be learned from the study of spectra.

However, with the great progress in theoretical and experimental research during the last decades, atomic physics as a research field has changed character. There is no longer a clear demarcation line between, for example, atomic physics and quantum optics. A top-level conference in atomic physics in 2019 will typically see rather little structure being discussed. Instead, the focus is often on subjects such as quantum gases, multi-body physics, fundamental metrology and quantum information. With this development, the traditional way of teaching atomic physics is no longer such an obvious choice as it used to be.

There are now textbooks, and academic courses, in atomic physics that begin by treating a two-level atom interacting with an electromagnetic field. From there, it goes on to prepare the students for the most pressing needs of contemporary atomic physics, and this does not always include the theory about atomic structure much more than in passing. My idea with writing this book is certainly not to challenge this viewpoint, but rather to present a complement. Learning about quantum optics dominated atomic physics, spectroscopy, experimental techniques *and* the theory of atomic structure does not have to be mutually exclusive.

Atomic structure theory provides an excellent case study of quantum mechanics, beneficial for anyone who wants to strive towards better grasp of the subject. Also for a deeper understanding of subjects such as molecular physics, condensed matter, nuclear physics and indeed modern experimental atomic physics, a solid background in atomic structure should be beneficial.

The book is intended for students who already have a grasp on some of the basics of quantum mechanics. This should include the non-relativistic solution to the Schrödinger equationfor hydrogen, the fundamentals of angular momentum, and perturbation theory. What I mean by *structure* in this book is essentially the stationary states — the solutions to the time-independent Schrödinger equation. This does not include transitions between states (radiative or otherwise). The latter is sometimes referred to as *dynamical structure*, but I do not deal with it in this book.

I have made the choice not to discuss the history of atomic structure theory. This is interesting, but that story is told elsewhere. Neither have I included spectroscopy, analysis of spectra, experimental techniques or interactions with oscillating electromagnetic fields. The book is purely about structure, and it involves all atoms in the periodic system, *except* hydrogen.

Hydrogen is special. With only one electron bound to a nucleus, analytic solutions are possible — also relativistic ones. It is even feasible to make very precise analytical predictions about hydrogen with quantum electrodynamics. When one moves from hydrogen to the rest of the elements, however, the philosophy must change. The presence of two (or more) interacting electrons essentially means that the exact methods that can be deployed for hydrogen become unavailable. Instead, one must turn to approximation methods, and the central aspects of these methods are, more often than not, couplings of angular momenta.

Having said that, the first chapter of the book *is* about one-electron atoms. This is partly intended as a repetition, but more as a way to establish some notation and nomenclature, and important ground rules. A central aspect in the book is the *central-field approximation* (introduced in chapter 5). In this, individual electrons are approximated to belonging to certain individual *orbitals* and, in that formalism, the notations developed for the hydrogen atom become relevant.

In many ways, the heart of this book are the angular momentum coupling schemes described in chapters 6–9. The preceding chapters lay the groundwork for this, while chapters 10–12 add interactions with external static fields and higher electromagnetic moments of the nucleus.

The book is about the *basics* of the structure of multielectron atoms. Complications may be added ad infinitum and, while they are all interesting, they are subjects for other works. A short exposé is given in chapter 13. Contemporary works in multielectron atomic structure theory almost always involve numerical methods. There are excellent books that introduce such methodologies, for example, *Atomic Many-Body Theory* by I. Lindgren and J. Morrison and *The Theory of Atomic Structure and Spectra* by R. D. Cowan. In the last chapter of this book, I have stopped short of explaining numerical methods in any detail, but have simply commented on them and recommend the reader to go on to read works such as the ones mentioned above.

In the ambition of achieving some fluidity, some longer derivations and explanations have been placed in appendices. A notable example is angular momentum theory, which is central for the book. This appendix could be read as a chapter, for example, after chapter 1.

Throughout the book, I have tried to use symbols and nomenclatures standardised by the International Union of Pure and Applied Physics (IUPAP), as far as it is practical. This is further explained in the relevant sections in the forematters of the book. As far as units are concerned, I predominantly use atomic units, and when I do not, it is always SI-units.

Energy levels are in the book frequently calculated with perturbation theory. To declutter the notation, I have chosen not to call energy perturbations ΔE, but rather just E with a subscript relevant for the specific problem. The text clarifies that this 'energy' really pertains to an *energy correction*.

This is the first edition of this book. I have tried to eliminate errors as much as possible, but there will be ones lingering. Please do report errors in tables, equations, figures or in the text, either to me directly or via the editors. In a potential second edition, there will be fewer inaccuracies.

While writing, I have made extensive use of other works. From my lists of recommended readings, and from lists of citations, it will be fairly obvious which my main sources have been. However, a couple that ought to be specifically highlighted, apart from those previously mentioned in this preface, are *Quantum theory of atomic structure* by J. C. Slater and *Physics of Atoms and Molecules* by B. H. Bransden and C. J. Joachain.

This book would never have been written without the constant support from my editors at *Springer*: Chris Caron, Sam Harrison and Nirmal Selvaraj, and their respective co-workers. Chris was the one who first presented me with the idea of writing a book.

During the years that I have worked on the manuscript, I have often pestered my colleagues with questions ranging from fundamental physics to very specific problems. Particular victims of this have been my friends and colleagues at the Raman Research Institute in Bengaluru and at Université Côte d'Azur in Nice. The concerned individuals are too many to list, but I still want to particularly acknowledge help from Olivier Alibart, Virginia D'Auria, Olivier Legrand, Vasant Natarajan, Bijaya Sahoo, Joseph Samuel and Sagar Sutradhar. During some intense weeks in 2019, Mariama Mattoir proofread all long tables, checked many figures and a lot of the maths.

English is not my native language, and I have been greatly helped by Ken Coey, who read the entire manuscript, corrected mistakes and inconsistencies, and gave valuable advice about style.

Finally, I want to thank my wife Vanessa for all her patience while I have been busy with this manuscript.

Nice, France Anders Kastberg
September 2019

Contents

Notation, Used Symbols

As far as possible, the notations in this book will follow ones standardised by the International Union of Pure and Applied Physics (IUPAP). The description in the table below will predominantly be consistently followed, with some exceptions called for by pedagogical considerations. For spectroscopic notation and quantum numbers, we refer to section 1.5. See also the list of constants and the introduction to atomic units in appendix A.

Symbol	Description
$\mathbf{r}_i$	Coordinate vector of an individual bound electron
$\mathbf{r} = (\mathbf{r}_1, \ldots, \mathbf{r}_N)$	Ensemble of spatial coordinates for N electrons
θ	Zenith angle (latitude)
φ	Azimuthal angle (longitude)
$\mathbf{q}_i$	Compact notation for all (continuous) spatial coordinates *and* (discrete) spin quantum numbers, for an individual electron
$\mathbf{q} = (\mathbf{q}_1, \ldots, \mathbf{q}_N)$	Ensemble of combined coordinates for N electrons
$\psi_i(\mathbf{r}_i)$	Spatial wave function of *one* bound electron
$\psi(\mathbf{r})$	Spatial electronic wave function of an atom, as function of electron spatial coordinates, with spin excluded
$\zeta_i(m_{si})$	Spin wave function of *one* bound electron, as function of the discrete variable m_{si}
$\mathscr{X}(m_{s1}, \ldots, m_{sN})$	Total electronic spin wave function of an atom
$\Psi_i(\mathbf{q}_i)$	Total wave function of one bound electron, including spatial and spin variables
$\Psi(\mathbf{q})$	Total atomic electronic wave function
$\mathbb{S}$	Slater determinant
$Y_{k,q}(\theta, \varphi)$	Spherical harmonics

Continued on next page

continued from previous page

Symbol	Description
$\mathrm{L}_n^{(k)}(x)$	Associated Laguerre polynomial
$\mathrm{P}_n(x)$	Legendre polynomial
$\mathrm{P}_l^{(m)}(x)$	Associated Legendre polynomial
∇^2	Laplacian operator
$\forall$	Universal quantification ('given any')
$\delta(\alpha,\beta)$	Discrete Dirac delta function
$\mathbb{1}$	Unity matrix
μ	Reduced mass
Z	Nuclear charge state
E	Energy
V	Potential

Acronyms

Some acronyms used consistently in the book are listed below. Some others will be explained in the text.

a.u.	atomic units
c.c.	complex conjugate
CoM	centre-of-mass
CFA	central-field approximation
CGC	Clebsch–Gordan coefficient
hfs	hyperfine structure
QED	quantum electrodynamics
SCF	self-consistent field
SI	international system of units

Constants

The table below shows definitions of symbols used throughout the text for physical constants, and their accepted values at the time of writing (2019). Some data are taken from published CODATA values from 2014 [1]. Revisions from that have been made, taking into account the redefinition of the SI-system, which came into effect 20th May 2019 [2]. As the main source for the table, we have used [3]. For the current error bars, we refer to the sources cited above.

Symbol	Numerical value	Description
c	$= 299\,792\,458\ \mathrm{m\,s^{-1}}$ **exact**	speed of light in vacuum
μ_0	$= 4\pi \times 1.000\,000\,0008 \times 10^{-7}\ \mathrm{N\,A^{-2}}$ $= 12.566\,370\,6212 \times 10^{-7}\ \mathrm{N\,A^{-2}}$	magnetic constant
ε_0	$= (\mu_0 c^2)^{-1}$ $= 8.854\,187\,8128 \times 10^{-12}\ \mathrm{F\,m^{-1}}$	electric constant
h	$= 6.626\,070\,15 \times 10^{-34}\ \mathrm{J\,s}$ **exact** $= 4.135\,667\,696 \ldots \times 10^{-15}\ \mathrm{eV\,s}$	Planck constant
$\hbar$	$= h/(2\pi)$ **exact** $= 1.054\,571\,817 \ldots \times 10^{-34}\ \mathrm{J\,s}$ $= 6.582\,119\,569 \ldots \times 10^{-16}\ \mathrm{eV\,s}$	reduced Planck constant
e	$= 1.602\,176\,634 \times 10^{-19}\ \mathrm{C}$ **exact**	elementary charge
μ_B	$= e\hbar/(2m_e)$ $= 9.274\,010\,0783 \times 10^{-24}\ \mathrm{J\,T^{-1}}$ $= 5.788\,381\,8060 \times 10^{-5}\ \mathrm{eV\,T^{-1}}$	Bohr magneton

Continued on next page

continued from previous page

Symbol	Numerical value	Description
μ_N	$= e\hbar/(2m_p)$	nuclear magneton
	$= 5.050\,783\,7461 \times 10^{-27}$ J T^{-1}	
	$= 3.152\,451\,258\,44 \times 10^{-8}$ eV T^{-1}	
α	$= e^2/(4\pi\varepsilon_0 \hbar c)$	fine-structure constant
	$= 7.297\,352\,5693 \times 10^{-3}$	
α^{-1}	$= 137.035\,999\,084$	inverse fine-structure constant
R_∞	$= \alpha^2 m_e c/(2h)$	Rydberg constant
	$= 10\,973\,731.568\,160$ m^{-1}	
$R_\infty hc$	$= 2.179\,872\,361\,1035 \times 10^{-18}$ J	(energy equivalent)
	$= 13.605\,693\,122\,994$ eV	
a_0	$= \alpha/(4\pi R_\infty) = 4\pi\varepsilon_0 \hbar^2/(m_e e^2)$	Bohr radius
	$= 0.529\,177\,210\,903 \times 10^{-10}$ m	
E_h	$= e^2/(4\pi\varepsilon_0 a_0)$	Hartree energy
	$= 2R_\infty hc = \alpha^2 m_e c^2$	
	$= 4.359\,744\,722\,2071 \times 10^{-18}$ J	
	$= 27.211\,386\,245\,988$ eV	
m_e	$= 9.109\,383\,7015 \times 10^{-31}$ kg	electron mass
	$= 5.485\,799\,090\,65 \times 10^{-4}\, m_u$	
$m_e c^2$	$= 8.187\,105\,7769 \times 10^{-14}$ J	(energy equivalent)
	$= 0.510\,998\,950\,00$ MeV	
m_p	$= 1.672\,621\,923\,69 \times 10^{-27}$ kg	proton mass
	$= 1.007\,276\,466\,621\, m_u$	
$m_p c^2$	$= 1.503\,277\,615\,98 \times 10^{-10}$ J	(energy equivalent)
	$= 938.272\,088\,16$ MeV	
m_n	$= 1.674\,927\,498\,04 \times 10^{-27}$ kg	neutron mass
	$= 1.008\,664\,915\,95\, m_u$	
$m_n c^2$	$= 1.505\,349\,762\,87 \times 10^{-10}$ J	(energy equivalent)
	$= 939.565\,420\,52$ MeV	

Continued on next page

continued from previous page

Symbol	Numerical value	Description
m_p/m_e	$= 1836.152\,673\,43$	proton-electron mass ratio
m_n/m_e	$= 1838.683\,661\,73$	neutron-electron mass ratio
m_n/m_p	$= 1.001\,378\,419\,31$	neutron-proton mass ratio
N_A	$= 6.022\,140\,76 \times 10^{23}$ mol^{-1} **exact**	Avogadro constant
m_u	$= 1.660\,539\,066\,60 \times 10^{-27}$ kg	atomic mass constant
k_B	$= 1.380\,649 \times 10^{-23}$ J K^{-1} **exact**	Boltzmann constant

References

1. P.J. Mohr, D.B. Newell, B.N. Taylor, Journal of Physical and Chemical Reference Data **45**, 043102 (2016)
2. D.B. Newell, F. Cabiati, J. Fischer, K. Fujii, S.G. Karshenboim, H.S. Margolis, E. de Mirandés, P.J. Mohr, F. Nez, K. Pachucki, T.J. Quinn, B.N. Taylor, M. Wang, B.M. Wood, Z. Zhang, Metrologia **55**, L13 (2018)
3. Physical Measurement Laboratory of NIST. The NIST reference on constants, units and uncertainties. [Online]. Available: https://physics.nist.gov/cuu/Constants/index.html (2019). Accessed: 2019-05-22

Chapter 1
The One-Electron Atom

There is a fundamental difference between the standard theoretical approaches for describing atoms with at least two electrons, and for atomic systems with just a single electron. For any quantum mechanical problem, a preferred way to approach a problem is to formulate the Schrödinger equation, with the proper Hamiltonian, and to try to solve it analytically. In the case of hydrogen, or a hydrogen-like ion, this is feasible. When including interactions with external fields, relativistic effects, such as the electron spin, and possibly also QED effects, an analytical solution may turn out to be very difficult, albeit generally possible. As soon as we go to helium, or any other system with a nucleus and two or more electrons, we have a multibody problem, and a direct assault on the Schrödinger equation, without serious approximations, will in most cases not lead very far.

For multielectron atoms, we will in this book principally resort to perturbation theory as the approximation tool of choice. When doing this, we will most of the time use the central-field approximation (see chapter 5), where the zeroth order states are products of individual electron wave functions. The latter are in turn found as solutions to single electrons moving in time-averaged *central-field potentials*, a little different for each electron. The exact shape of these average central fields is different from that for one sole electron and an unscreened nucleus, and thus the wave functions will also be different. However, their general forms are similar, and even without exact knowledge of the wave functions, we are able to understand a great deal about the energetic structure of the atoms if we start with basis functions that are similar to the solutions for the hydrogen atom, and in particular ones that can be functionally separated and individually specified with quantum numbers, in a manner analogous to that used for hydrogen.

This aspect makes the one-electron atom formalism highly relevant as a first stepping stone for the study of multielectron atoms, and therefore we begin this book by a brief recapitulation of the hydrogen problem. This chapter also serves the purpose of establishing important nomenclature, which will be extensively used throughout the book. In section 1.1, we formulate the Schrödinger equation and factorise it in radial and angular parts. The solutions to these are summarily reviewed in the two following sections, with a more detailed coverage provided in appendices B and C.

A. Kastberg, *Structure of Multielectron Atoms*, Springer Series on Atomic, Optical, and Plasma Physics 112, https://doi.org/10.1007/978-3-030-36420-5_1

Sections 1.4 and 1.5 summarise the insights in one-electron atoms that we need for the remainder of the book, including notations for angular momenta and quantum numbers.

More detailed accounts of the solution to the hydrogen problem can be found in many basic quantum mechanics texts and books on general atomic physics, see, for example, the suggested list at the end of this chapter.

1.1 Formulation of the Schrödinger equation for the Hydrogen Atom

In this overview, we will make some practical approximations and simplifications. We are, for the moment, only trying to establish the general form of the hydrogenic wave functions, and therefore this is accurate enough for our purposes. To start with, we assume that the nucleus has zero extension, we ignore the centre-of-mass (CoM) motion and we place the origin at the CoM. Since the nucleus is at least 1800 times heavier than the electron, the problem essentially reduces to that of a single particle, the electron, moving in a central-field potential in this first approximation.

To take the finite mass of the nucleus into account, we initially replace the electron mass with the reduced mass, μ, of the two-body problem. We will ignore the influence on the wave function of relativistic effects in this chapter, which automatically implies that we ignore the spins of the electron and of the nucleus. This makes us ready to formulate the Hamiltonian and the Schrödinger equation of the problem.

The potential is the classical Coulomb attraction between two particles of opposite charges. Using spherical coordinates, and taking r as the radial distance between the particles, this is:

$$V(r) = -\frac{Ze^2}{4\pi\varepsilon_0 r} \quad , \tag{1.1}$$

with Z being the charge state of the positively charged nucleus. The time-independent and non-relativistic Schrödinger equation, as function of the displacement vector $\mathbf{r}$, is:

$$-\frac{\hbar^2}{2\mu}\nabla^2\psi(\mathbf{r}) + V(r)\,\psi(\mathbf{r}) = E\,\psi(\mathbf{r}) \quad , \tag{1.2}$$

where the Laplacian in spherical coordinates is:

$$\nabla^2 = \frac{1}{r^2}\frac{\partial}{\partial r}\left(r^2\frac{\partial}{\partial r}\right) + \frac{1}{r^2\sin\theta}\frac{\partial}{\partial\theta}\left(\sin\theta\frac{\partial}{\partial\theta}\right) + \frac{1}{r^2\sin^2\theta}\frac{\partial^2}{\partial\varphi^2} \quad . \tag{1.3}$$

Since the potential is purely central, the solution to (1.2) can be factorised into one radial and one angular part:

$$\psi(r,\theta,\varphi) = R(r)\,Y(\theta,\varphi) \; . \tag{1.4}$$

Substituting this into (1.2), the Schrödinger equation becomes:

$$\frac{1}{R(r)}\frac{\partial}{\partial r}\left[r^2\frac{\partial R(r)}{\partial r}\right]+\frac{2\mu r^2}{\hbar^2}\left[\frac{Ze^2}{4\pi\varepsilon_0 r}+E\right]$$
$$=-\frac{1}{Y(\theta,\varphi)}\left[\frac{1}{\sin\theta}\frac{\partial}{\partial\theta}\left(\sin\theta\frac{\partial}{\partial\theta}\right)+\frac{1}{\sin^2\theta}\frac{\partial^2}{\partial\varphi^2}\right]Y(\theta,\varphi)\ . \quad (1.5)$$

The left side of the equation only depends on r, and the right side only on the angular parameters — the latitude and the longitude. For the equality to hold, this means that both sides of (1.5) must be constant.

Before proceeding, we will simplify the notation by introducing atomic units and the angular momentum operator. The motivation for using atomic units is that when performing long derivations, the inclusion of many symbols and constants makes the work cumbersome. To circumvent that, some key natural constants are set to unity [1, 2]:

$$e=m_e=\hbar=\frac{1}{4\pi\varepsilon_0}=1\ . \quad (1.6)$$

The units for the involved physical quantities then have to be adapted accordingly, whenever quantified answers are sought. An introduction to, and a list of, atomic units is given in appendix A. In the continuation of this book, we will use atomic units, when we do not explicitly state otherwise.

The expression within the square brackets on the right side of (1.5), with the preceding minus sign included, is identical to the quantum mechanical operator for the square of the orbital angular momentum, $\mathbf{L}^2$, except for a factor $\hbar^2$. This means that $Y(\theta,\varphi)$ is eigenfunction to $\mathbf{L}^2$ — see (C.28). A more thorough discussion on angular momentum is presented in appendix C. Therein, it is also shown that the eigenvalue equation for $\mathbf{L}^2$ (in a.u.) is:

$$\mathbf{L}^2\,Y(\theta,\varphi)=l(l+1)\,Y(\theta,\varphi)\ , \quad (1.7)$$

where the introduced quantum number l has to be a positive integer or zero.

From (1.7), we get the constant that equals both sides of (1.5). In atomic units, we can write the Schrödinger equation as:

$$\frac{1}{R(r)}\frac{\partial}{\partial r}\left[r^2\frac{\partial R(r)}{\partial r}\right]+2Zr+2E\,r^2=\frac{1}{Y(\theta,\varphi)}\mathbf{L}^2\,Y(\theta,\varphi)=l(l+1)\ . \quad (1.8)$$

Here, we have set $\mu\approx m_e$. This approximation amounts to taking the nuclear mass as infinite, which places the origin firmly at the nuclear position and it makes $\mathbf{r}$ the position vector of the electron.

Since the radial and angular parameters are separated, and the wave function is factorised, we are left with two uncoupled differential equations, which can be solved independently. The angular part of (1.8) is independent of the potential, as it is for all central potentials. The absolute value of the angular momentum, as well

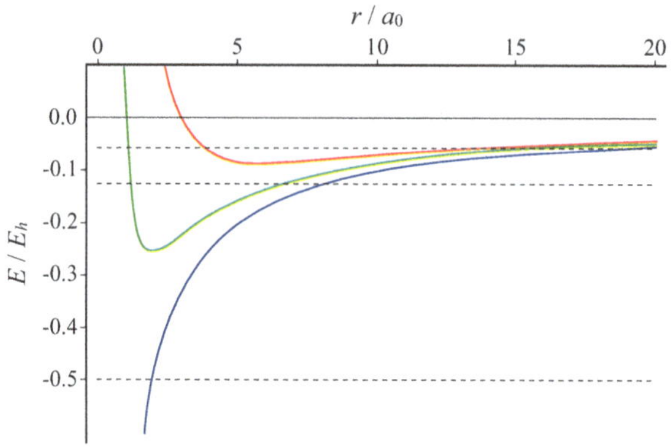

Fig. 1.1 Effective potential, $V_{\rm eff}(r)$, for the radial part of the Schrödinger equation for a hydrogenic system (1.9), with $Z=1$. Potentials are shown for three different values of the angular momentum quantum number: $l=0$ (blue), $l=1$ (green) and $l=2$ (red). The axes are in atomic units and zero energy corresponds to an electron infinitely distant from the nucleus. The three dashed horizontal lines show the energies of the three lowest energy eigenstates (see section 1.4.1).

as its components, are constants of the motion in the absence of an angular force component. The solutions to the angular equation are in the form of the standard spherical harmonics, see appendix C.

The energy eigenvalue appears solely in the radial part of (1.8), and therefore the energies will, at this level of approximation, be independent of the angular coordinates. In the two upcoming sections, we will treat the radial and angular problems separately.

1.2 Solutions to the Radial Equation

The radial equation, extracted from (1.8), reads:

$$\frac{1}{r^2}\frac{\mathrm{d}}{\mathrm{d}r}\left[r^2\frac{\mathrm{d}}{\mathrm{d}r}R(r)\right]+\left[\frac{2Z}{r}+2E-\frac{l(l+1)}{r^2}\right]R(r)=0\,. \tag{1.9}$$

This represents a one-dimensional problem of a particle moving in an effective potential, $V_{\rm eff}(r)$. This has two parts: a central Coulomb attraction term and an additional term with opposite sign corresponding to a centrifugal barrier. The effective potential is:

$$V_{\rm eff}(r)=-\frac{Z}{r}+\frac{l(l+1)}{2\,r^2}\,. \tag{1.10}$$

This is illustrated in figure 1.1 (for $Z=1$), in which we can clearly see the dependence on the quantum number l — the angular momentum. For $l>0$, the centrifugal force will prevent the electron from getting too close to $r=0$.

Positive energy solutions to (1.9) give diffusive solutions that are relevant for scattering phenomena, but which will not be dealt with here. In the current treatment, we explicitly look for bound states, that is, solutions for which $E<0$. In the following, we will give an outline of the solution. For a more detailed account, see appendix B.

A first step is to introduce the substitution $U(r) \equiv rR(r)$. This leaves us with the equation:

$$-\frac{1}{2}\frac{\mathrm{d}^2}{\mathrm{d}r^2}U(r)+V_{\mathrm{eff}}(r)\,U(r)=E\,U(r)\ . \tag{1.11}$$

Note that since we have formulated the Schrödinger equation in atomic units, the energy in (1.11) will be in E_{h} and r has to be given in a_0 (see appendix A).

The solutions to (1.11) are periodic and discretised into two quantum numbers n and l, respectively, the *principal quantum number* and the *orbital angular momentum quantum number*, both introduced in section 1.1. From the solution of the radial equation follows constraints for these two quantum numbers (see appendix B). They must both be integers, and fulfil the inequality:

$$0 \leq l < n\ . \tag{1.12}$$

1.2.1 Eigenstates

The eigenstates found from a solution of the radial function (1.11) are in the form of associated Laguerre polynomials, $\mathrm{L}_n^{(k)}(x)$ — see, for example, [3]. It is practical to write the general solution in terms of a rescaled radial parameter:

$$\rho=\frac{2Zr}{n}\ . \tag{1.13}$$

The solutions, derived in (B.13), are:

$$R_{nl}(r)=-\frac{2Z^{3/2}}{n^2}\sqrt{\frac{(n-l-1)!}{[(n+l)!]^3}}\ \rho^l\,\mathrm{e}^{-\rho/2}\ \mathrm{L}_{n+l}^{(2l+1)}(\rho)\ . \tag{1.14}$$

This equation is normalised such that:

$$\int_0^\infty R_{nl}^*(r)\,R_{nl}(r)\,r^2\,\mathrm{d}r=1\ , \tag{1.15}$$

and the functions $R_{nl}(r)$ are mutually orthogonal.

In (1.16), we present some of the lowest order normalised radial functions for one-electron systems, in the format $R_{nl}(r)$, compiled from (1.14).

$$
\begin{aligned}
R_{10}(r) &= R_{1s}(r) = 2Z^{3/2}\,\mathrm{e}^{-Zr}\\
R_{20}(r) &= R_{2s}(r) = \frac{Z^{3/2}}{\sqrt{2}}\left[1-\frac{Zr}{2}\right]\mathrm{e}^{-Zr/2}\\
R_{21}(r) &= R_{2p}(r) = \frac{Z^{5/2}}{2\sqrt{6}}\,r\,\mathrm{e}^{-Zr/2}\\
R_{30}(r) &= R_{3s}(r) = \frac{2Z^{3/2}}{3\sqrt{3}}\left[1-\frac{2Zr}{3}+\frac{2(Zr)^2}{27}\right]\mathrm{e}^{-Zr/3}\\
R_{31}(r) &= R_{3p}(r) = \frac{8Z^{5/2}}{27\sqrt{6}}\,r\left[1-\frac{Zr}{6}\right]\mathrm{e}^{-Zr/3}\\
R_{32}(r) &= R_{3d}(r) = \frac{4Z^{7/2}}{81\sqrt{30}}\,r^2\,\mathrm{e}^{-Zr/3}\\
R_{40}(r) &= R_{4s}(r) = \frac{Z^{3/2}}{4}\left[1-\frac{3Zr}{4}+\frac{(Zr)^2}{8}-\frac{(Zr)^3}{192}\right]\mathrm{e}^{-Zr/4}\\
R_{41}(r) &= R_{4p}(r) = \frac{\sqrt{5}Z^{5/2}}{16\sqrt{3}}\,r\left[1-\frac{Zr}{4}+\frac{(Zr)^2}{80}\right]\mathrm{e}^{-Zr/4}\\
R_{42}(r) &= R_{4d}(r) = \frac{Z^{7/2}}{64\sqrt{5}}\,r^2\left[1-\frac{Zr}{12}\right]\mathrm{e}^{-Zr/4}\\
R_{43}(r) &= R_{4f}(r) = \frac{Z^{9/2}}{768\sqrt{35}}\,r^3\,\mathrm{e}^{-Zr/4}\,.
\end{aligned}
\tag{1.16}
$$

In the second part of the equations in (1.16) we have used the standard spectroscopic symbols for different values of l. This will be further explained in section 1.5.

In figure 1.2, we plot radial wave functions from (1.16), for principal quantum numbers from $n=1$ to $n=4$. One conclusion is that $R_{nl}(r)$ can be non-zero at $r=0$ only for s-states ($l=0$). That means that such states are the only ones to have a finite probability amplitude at the origin. This is expected, since the centrifugal term cancels only for these. Another thing to note is that each function $R_{nl}(r)$ crosses zero $(n-l-1)$ times.

1.2.2 Radial Probability Distribution

The probability of finding the electron inside a spherical shell between r and $\mathrm{d}r$ can be found from:

$$
\begin{aligned}
P(r, r+\mathrm{d}r) &= \int_0^{2\pi}\!\!\int_0^{\pi} |\psi(r,\theta,\varphi)|^2\, r^2 \sin\theta\,\mathrm{d}r\,\mathrm{d}\theta\,\mathrm{d}\varphi\\
&= |R_{nl}(r)|^2 r^2\,\mathrm{d}r \int_0^{2\pi}\!\!\int_0^{\pi} |Y(\theta,\varphi)|^2 \sin\theta\,\mathrm{d}\theta\,\mathrm{d}\varphi = |U_{nl}(r)|^2\,\mathrm{d}r\,.
\end{aligned}
\tag{1.17}
$$

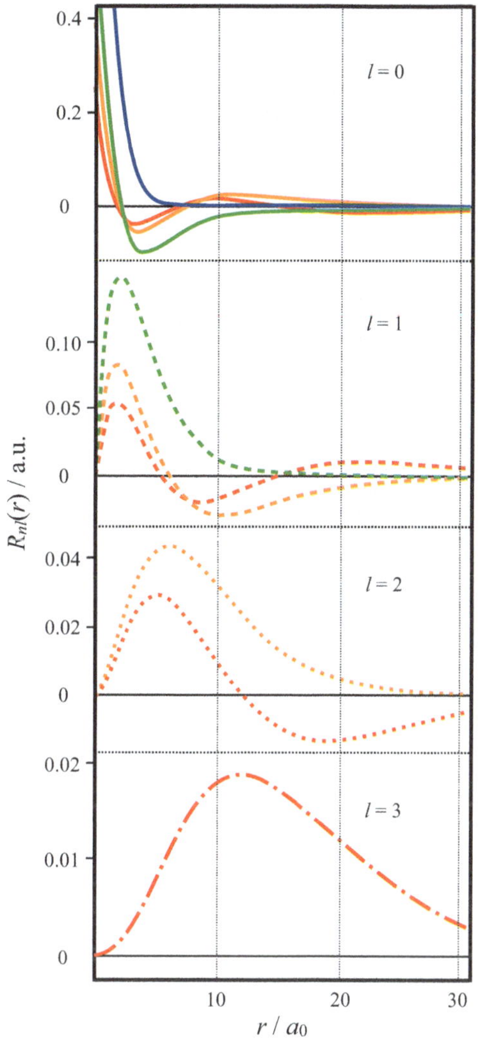

Fig. 1.2 Radial wave functions — $R_{nl}(r)$, from (1.16) — for $Z=1$ (in a.u.). The blue curve is for $n=1$, the green ones for $n=2$, the orange for $n=3$ and the red for $n=4$. The top graph, with full lines, is for $l=0$; the second graph, with dashed lines, for $l=1$; the third graph, with dotted lines, for $l=2$; and the bottom graph, with a dash-dotted line, for $l=3$. Note the different scales on the abscissas.

Here we have assumed that the angular functions can be taken as normalised (see the following section). The factor r^2 is a reflection of the fact that the volume of a spherical shell at a radial distance of r must be proportional to $r^2 \mathrm{d}r$. In figure 1.3, we have plotted such radial probability distributions for the lowest principal quantum numbers, for different angular momenta l and for $Z=1$. This should be compared to figure 1.2. Note that the number of nodes in each distribution is given by $n-l-1$, and the number of anti-nodes by $n-l$. These anti-nodes will manifest themselves as concentric spherical shells with zero charge density. For increasing n, the electron will on the average be further away from the nucleus. Another important observation is that for a given n, the maximum of the charge density lies closer to the nucleus for a larger l, and this holds true also for the expectation value of r.

Fig. 1.3 Normalised radial probability distributions — $[rR(r)]^2$, see (1.17) — corresponding to the radial wave functions in (1.16), with $Z = 1$. The blue curve is for $n = 1$, the green ones for $n = 2$, the orange for $n = 3$ and the red for $n = 4$. The top graph, with full lines, is for $l = 0$; the second graph from the top, with dashed lines, for $l = 1$; the third graph, with dotted lines, for $l = 2$; and the bottom graph, with a dash-dotted line, for $l = 3$. Note the different scales on the abscissas. The vertical coloured lines show the expectation values of the radial parameter — see (1.18) and table 1.1

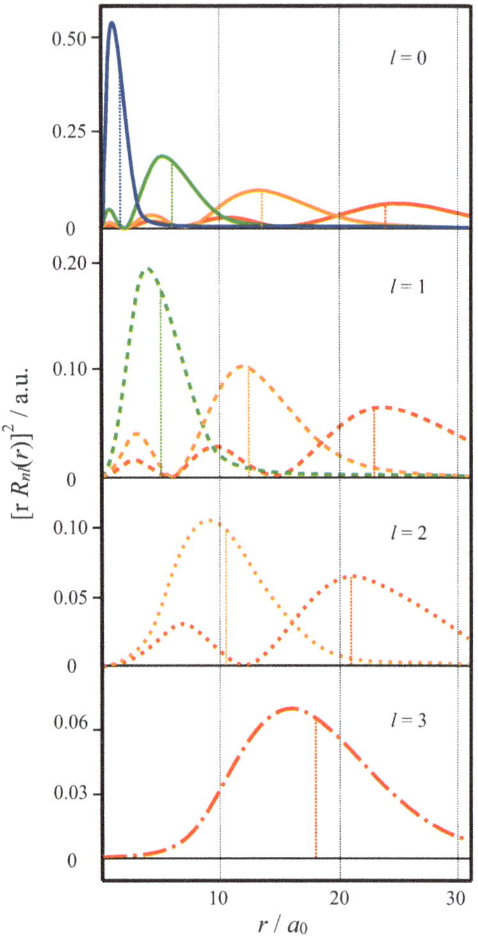

The expectation value of the radial parameter is found through:

$$\langle r \rangle_{nl} = \int_0^{2\pi} \int_0^{\pi} \int_0^{\infty} \psi_{nl}^*(r,\theta,\varphi)\, r\, \psi_{nl}(r,\theta,\varphi)\, r^2 \sin\theta \,\mathrm{d}r\,\mathrm{d}\theta\,\mathrm{d}\varphi$$

$$= \int_0^{\infty} |R_{nl}(r)|^2 r^3 \,\mathrm{d}r = \frac{1}{2Z}\left[3n^2 - l(l+1)\right] , \qquad (1.18)$$

in the atomic unit a_0. The solution to the integral is taken from (B.42) in appendix B. In table 1.1, we show the maxima of the respective probability distributions, as well as the expectation values for the radial parameter for all the curves plotted in figure 1.3.

Table 1.1 Radii for maximal probabilities and expectation values for the radial probability distributions in figure 1.3 (with $Z = 1$). The numerical values are given in atomic units (a_0) and are rounded off to two decimal places.

	$n=1$		$n=2$		$n=3$		$n=4$		$n=5$	
	r_{max}	$\langle r \rangle$	r_{max}	$\langle r \rangle$	r_{max}	$\langle r \rangle$	r_{max}	$\langle r \rangle$	r_{max}	$\langle r \rangle$
$l=0$	1.00	1.50	5.24	6.00	13.07	13.50	24.62	24.00	39.92	37.50
$l=1$			4.00	5.00	12.00	12.50	23.58	23.00	38.90	36.50
$l=2$					9.00	10.50	21.21	21.00	36.69	34.50
$l=3$							16.00	18.00	32.81	31.50
$l=4$									25.00	27.50

1.3 Solution of the Angular Equation

Since both the radial and the angular sides of (1.8) equal $l(l+1)$, for a given eigenfunction, the angular part of the equation is:

$$-\left[\frac{1}{\sin\theta}\frac{\partial}{\partial\theta}\left(\sin\theta\frac{\partial}{\partial\theta}\right)+\frac{1}{\sin^2\theta}\frac{\partial^2}{\partial\varphi^2}\right]Y(\theta,\varphi) = l(l+1)\,Y(\theta,\varphi)\,. \qquad (1.19)$$

As stated in section 1.1, the operator on the left side of (1.19) is the operator for the square of the orbital angular momentum divided by $\hbar^2$ (see appendix C), which justifies the identification of the constant as $l(l+1)$, and the formulation of the eigenvalue equation in (1.7). For the projection of $\mathbf{L}$ along a quantisation axis $\hat{e}_z$, we will use the quantum number m_l (eigenvalue to L_z), and the solutions to (1.19) will be quantised in l and m_l.

With (1.19) being in the form of the standard generator for the spherical harmonics, the solution to this differential equation is valid for any central potential. The angular functions are (see appendix C):

$$Y_{l,m_l}(\theta,\varphi) = [-1]^{(m_l+|m_l|)/2}\sqrt{\frac{(2l+1)\,(l-|m_l|)!}{4\pi\,(l+|m_l|)!}}\;\mathrm{P}_l^{(|m_l|)}(\cos\theta)\;\mathrm{e}^{\mathrm{i}\varphi m_l}\,. \qquad (1.20)$$

The functions $\mathrm{P}_l^{(m)}(\cos\theta)$ are l^{th} degree, m^{th} order associated Legendre polynomials [3] — see (C.48). The functions $Y_{l,m_l}(\theta,\varphi)$ are mutually orthogonal and normalised as:

$$\int_0^{2\pi}\!\!\int_0^{\pi} Y^*_{l,m_l}(\theta,\varphi)\,Y_{l,m_l}(\theta,\varphi)\,\sin\theta\,\mathrm{d}\theta\,\mathrm{d}\varphi = 1\;. \qquad (1.21)$$

The explicit forms of some of the lowest order solutions are:

$$\begin{aligned}
Y_{0,0}(\theta,\varphi) &= \frac{1}{2}\frac{1}{\sqrt{\pi}} \\
Y_{1,0}(\theta,\varphi) &= \frac{1}{2}\sqrt{\frac{3}{\pi}}\cos\theta \\
Y_{1,\pm1}(\theta,\varphi) &= \mp\frac{1}{2}\sqrt{\frac{3}{2\pi}}\sin\theta\,\mathrm{e}^{\pm\mathrm{i}\varphi} \\
Y_{2,0}(\theta,\varphi) &= \frac{1}{4}\sqrt{\frac{5}{\pi}}(3\cos^2\theta-1) \\
Y_{2,\pm1}(\theta,\varphi) &= \mp\frac{1}{2}\sqrt{\frac{15}{2\pi}}\sin\theta\cos\theta\,\mathrm{e}^{\pm\mathrm{i}\varphi} \\
Y_{2,\pm2}(\theta,\varphi) &= \frac{1}{4}\sqrt{\frac{15}{2\pi}}\sin^2\theta\,\mathrm{e}^{\pm2\mathrm{i}\varphi} \\
Y_{3,0}(\theta,\varphi) &= \frac{1}{4}\sqrt{\frac{7}{\pi}}(5\cos^3\theta-3\cos\theta) \\
Y_{3,\pm1}(\theta,\varphi) &= \mp\frac{1}{8}\sqrt{\frac{21}{\pi}}\sin\theta\,(5\cos^2\theta-1)\,\mathrm{e}^{\pm\mathrm{i}\varphi} \\
Y_{3,\pm2}(\theta,\varphi) &= \frac{1}{4}\sqrt{\frac{105}{2\pi}}\sin^2\theta\cos\theta\,\mathrm{e}^{\pm2\mathrm{i}\varphi} \\
Y_{3,\pm3}(\theta,\varphi) &= \mp\frac{1}{8}\sqrt{\frac{35}{\pi}}\sin^3\theta\,\mathrm{e}^{\pm3\mathrm{i}\varphi}\,.
\end{aligned} \tag{1.22}$$

In figure 1.4, we illustrate the behaviour of $Y_{l,m_l}(\theta,\varphi)$ as function of latitude, (angle θ), for values $l \leq 3$. Note that the respective curves cross zero $l-|m_l|$ times. For the values of θ that correspond to these zero-crossings, the overall probability distribution will have a node. The real part of the functions $\mathrm{e}^{\pm\mathrm{i}\varphi m_l}$ has $2m$ zero points between 0 and 2π, which corresponds to m nodal meridians. The total angular distribution will thus have l nodal lines. Apart from a normalisation factor, the graphs are identical to the respective associated Legendre functions.

The angular probability distribution for an electron with a specific angular eigenfunction $Y_{l,m_l}(\theta,\varphi)$ is obtained by taking the modulus squared of the spherical harmonics for the different combinations of l and m_l. The lowest orders of these, corresponding to (1.22), are illustrated graphically in table 1.2.

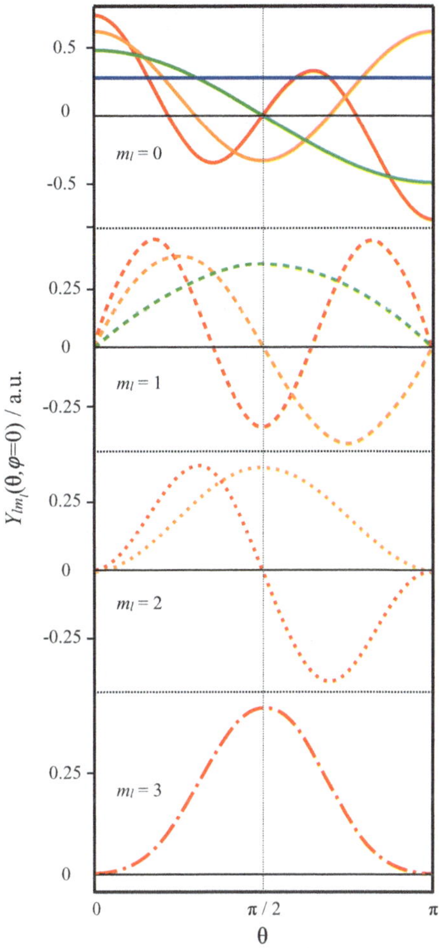

Fig. 1.4 The angular wave functions, $Y_{l,m_l}(\theta,\varphi)$, — the spherical harmonics, from (1.22) — as functions of the zenith angle θ. The azimuthal angle φ has been set to zero in these plots. The blue curve is for $l=0$, the green ones for $l=1$, the orange for $l=2$ and the red for $l=3$. The top graph, with full lines, is for $m_l=0$; the second graph from the top, with dashed lines, for $|m_l|=1$; the third graph, with dotted lines, for $|m_l|=2$; and the bottom graph, with a dash-dotted line, for $|m_l|=3$. Note the different scales on the abscissas.

1.4 The Total Hydrogenic Wave Function

Since the total wave function is the product of $R(r)$ and $Y(\theta,\varphi)$, the complete hydrogenic spatial wave function — from (1.14) and (1.20) — is:

$$\psi_{nlm_l}(r,\theta,\varphi) = (-1)^{\frac{m_l+|m_l|}{2}} \sqrt{\frac{(n-l-1)!\,(2l+1)\,(l-|m_l|)!}{4\pi n^2\,[(n+l)!]^3\,(l+|m_l|)!}}$$
$$\times \left(\frac{2Z}{n}\right)^{l+1} r^l\, \mathrm{e}^{-Zr/n}\, \mathrm{L}_{n+l}^{(2l+1)}(\rho)\, \mathrm{P}_l^{(|m_l|)}(\cos\theta)\, \mathrm{e}^{\mathrm{i}\varphi m_l}\,. \quad (1.23)$$

Table 1.2 Angular probability distributions for a one-electron atom — corresponding to the modulus squared of the spherical harmonics — for different values of the quantum numbers l and m_l. The figures correspond to the equations in (1.22), and to the plots as function of θ in figure 1.4).

	$l = 0$	$l = 1$	$l = 2$	$l = 3$
$m_l = 0$:		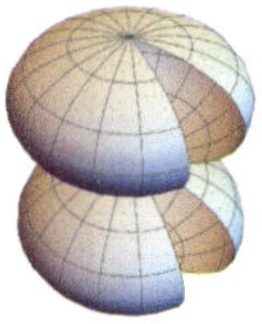		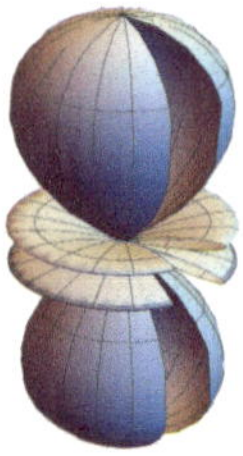
$m_l = \pm 1$:		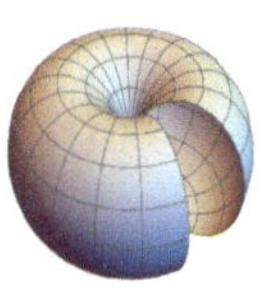	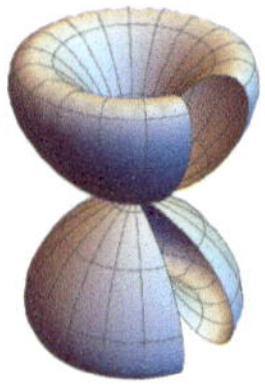	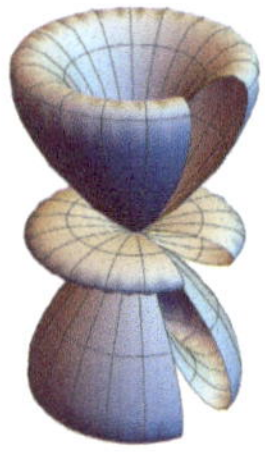
$m_l = \pm 2$:			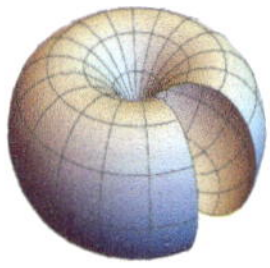	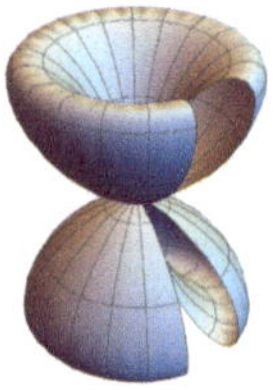
$m_l = \pm 3$:				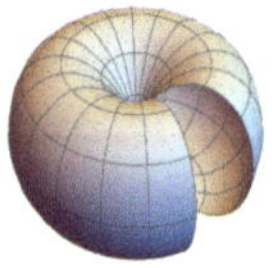

The argument to the associated Laguerre polynomial, $\mathrm{L}_{n+l}^{(2l+1)}(\rho)$, is as in (1.13), and the associated Legendre polynomial, $\mathrm{P}_l^{(|m_l|)}(\cos\theta)$, is the same as in (1.20). Equation (1.23) is the wave function in atomic units. This means that for any quantitative calculations, r must be given in a_0. Bound state solutions are discrete in the quantum numbers n, l and m_l.

1.4.1 Energy Levels

The eigenenergies corresponding to the solutions in (1.23), with the ionisation limit taken as zero energy, are (see appendix B):

$$E_n = -\frac{Z^2}{2n^2} . \tag{1.24}$$

In SI-units this is:

$$E_n = -\frac{Z^2 m_e e^4}{2(4\pi\varepsilon_0)^2 \hbar^2 n^2} = -\frac{Z^2 \alpha^2 m_e c^2}{2n^2} . \tag{1.25}$$

The equation is known as the *Balmer formula*, and in the last step, we have introduced the dimensionless *fine-structure constant* α (see also appendix A). The electron mass in (1.25) can be replaced by the reduced mass, if higher precision is wanted. The energies depend only on the principal quantum number, and the eigenstates are energy degenerate in l and m_l. For every value of l, there are $2l+1$ values of m_l, and for every value of n, there are values of l from 0 up to $n-1$. This means that the degeneracy for a certain principal quantum number n is:

$$g_n = \sum_{l=0}^{n-1} (2l+1) = n^2 , \tag{1.26}$$

(keep in mind that we still ignore the electron spin). The degeneracy in m_l is a consequence of spherical symmetry and the absence of any external fields, and it will remain also for atoms with more electrons. The degeneracy for l, however, is unique to hydrogen-like atoms. As we will see later (chapter 12), this latter degeneracy will make hydrogen uniquely sensitive to external electric fields. The progression in principal quantum numbers continues to infinity, which is a consequence of the asymptotic approach to zero of the Coulomb potential (and in contrast to a finite square potential, which has a finite number of bound states).

The energies in (1.25) are identical to the ones found from the Bohr model, which is not surprising given that the Bohr model was adapted to fit experimental data. The energy levels have been included in the diagram in figure 1.1. This provides a graphical illustration to the constraint in (1.12). For example, for $l=1$, the centrifugal barrier inhibits energies lower than about $-0.25\,E_h$, excluding the ground state. For the latter, $E_1 = -0.5\,E_h$ and the potential for $l=0$ is the only one possible. For $E_2 = -0.125\,E_h$, the only possible orbital angular momentum quantum numbers are $l=0$ and $l=1$ and so on.

1.4.2 Spin

In the treatment of the hydrogen atom in this chapter, we have thus far ignored the fact that both the electron and the nucleons have spins, and we have also ignored all other relativistic effects. We will still not revert to the relativistic version of the one-electron Schrödinger equation (also known as the Dirac equation). For that, we refer to appendix E and general volumes (for example, [4, 5] or [6]). We also defer the treatment of the interaction between the spin and orbital angular momentum of the electron, to chapter 4 and appendix F.

With the crude approximation that the electron spin, **S**, neither interacts with the orbital angular momentum, **L**, nor with the nucleus, its part of the total wave function can be factored out, and it will have no influence on the energy, in the case of the hydrogen atom. Even so, the inclusion of the spin part of the wave function has an effect on the symmetry properties of the wave function, which is important for larger atoms, and therefore we include it already here. The total electronic wave function can then be written as:

$$\Psi(\mathbf{q}) = \psi_{nlm_l}(r,\theta,\varphi)\,\zeta(s,m_s) \ , \tag{1.27}$$

with the spatial wave function taken from (1.23), and $\zeta(s,m_s)$ as the spin wave function. The discrete variables s and m_s are the quantum numbers corresponding to the spin angular momentum operators $\mathbf{S}^2$ and S_z. In (1.27), $\mathbf{q}$ is used as a compound notation for the spatial coordinates and the parameters specifying the spin. Formally, this is inaccurate since it means that we mix continuous and discrete variables. However, we introduce this way of writing in order to keep the notation compact, and we then keep in mind that for any subsequent integration over $\mathbf{q}$, the part of it which is over the spin variables should be discrete summations.

The spin is an angular momentum, and the general properties will be the same as those outlined in appendix C for general angular momenta. Some specificities are presented in appendix E. The eigenvalue equations for the operators $\mathbf{S}^2$ and S_z are (in atomic units):

$$\begin{aligned} \mathbf{S}^2\,\zeta(s,m_s) &= s(s+1)\,\zeta(s,m_s) \\ S_z\,\zeta(s,m_s) &= m_s\,\zeta(s,m_s) \ . \end{aligned} \tag{1.28}$$

In the case of an electron (as well as for a proton and a neutron), we always have $s = 1/2$. This makes the parameter s in $\zeta(s,m_s)$ redundant, and for the most part, we will leave it out. The lack of flexibility for the absolute value of the electron spin leaves us with just two possibilities for its orientation: $m_s = \pm 1/2$. These different spins are typically referred to as spin-up and spin-down, and we will occasionally annotate the respective single-electron wave functions by ζ_+ and ζ_-.

1.4.3 Parity

The parity of a wave function describes how its value changes during an inversion of spatial parameters, that is:

$$\begin{aligned} x &\to -x \\ y &\to -y \\ x &\to -z \end{aligned} \qquad \text{or} \qquad \begin{aligned} r &\to r \\ \theta &\to \pi - \theta \\ \varphi &\to \pi + \varphi \end{aligned} \quad , \tag{1.29}$$

in Cartesian and spherical coordinates, respectively. When $\mathbf{r}$ changes to $-\mathbf{r}$, the momentum changes from $\mathbf{p}$ to $-\mathbf{p}$. The hydrogen Hamiltonian remains unchanged since neither the Coulomb potential nor the kinetic energy depends on the signs of $\mathbf{r}$ or $\mathbf{p}$. Thus, we can write the Schrödinger equation for the two cases:

$$\begin{aligned} H\,\psi(\mathbf{r}) &= E\,\psi(\mathbf{r}) \\ H\,\psi(-\mathbf{r}) &= E\,\psi(-\mathbf{r}) \ . \end{aligned} \tag{1.30}$$

Since $\psi(\mathbf{r})$ and $\psi(-\mathbf{r})$ are solutions to the same Schrödinger equation, and have the same energy, they can only differ by a scalar factor:

$$\psi(-\mathbf{r}) = c\,\psi(\mathbf{r}) \, . \tag{1.31}$$

This must hold for all position vectors $\mathbf{r}$, and for $-\mathbf{r}$ we get:

$$\psi(\mathbf{r}) = c\,\psi(-\mathbf{r}) = c^2\,\psi(\mathbf{r}) \ . \tag{1.32}$$

Thus, the only possible values for the constant c are $+1$ and -1, and wave functions with these respective properties are referred to as having *even* or *odd* parity.

From (1.23), and by applying the rules in (1.29), one can conclude that a parity transformation of the hydrogen wave function gives:

$$\psi_{nlm_l}(\mathbf{r}) = R_{nl}(r)\,Y_{lm_l}(\theta,\varphi) \ \to \ \psi_{nlm_l}(-\mathbf{r}) = (-1)^l\,R_{nl}(r)\,Y_{lm_l}(\theta,\varphi) \ . \tag{1.33}$$

As a consequence, the parity of a hydrogen wave function is:

$$P = (-1)^l \, . \tag{1.34}$$

We see that wave functions with an even value of l have even parity, and those with odd l have odd parity.

1.5 Spectroscopic Notation

In section 1.2, we saw that radial hydrogenic wave functions are quantised into two quantum numbers, n and l, corresponding to the radial energy and to the orbital angular momentum (the operator $\mathbf{L}^2$). In the following chapters, we will need increasingly more quantum numbers in order to specify atomic states, and therefore we here dedicate a section to describing the notation that we will adopt.

Many of the conventions for how to annotate atomic states have historical origins that no longer are topically pertinent, but they are still relatively universally used and they remain practical. Many of the effects that the quantum numbers in the following table are referring to are yet to be explained in the upcoming chapters. However, we introduce the standard spectroscopic notations already now, in table 1.3, for use in the rest of the book.

Table 1.3: Standard spectroscopic notation used for quantum numbers in this book

Corresponds to:	Represented by:	Constraints:
Principal quantum number, n :		
Main quantisation of a central-field Hamiltonian	Numerical value	$n \geq 1$
Single-electron angular momentum, l (or l_i) :		
Quantisation of the operator $\mathbf{L}_i^2$ for a single electron	The letters s, p, d, f, g, h, i, k ...; corresponding to respectively $l = 0, 1, 2, 3, 4, 5, 6, 7 \ldots$	$0 \leq l < n$
Projection of the single-electron angular momentum, m_l (or m_{li}) :		
Quantisation of the operator L_{zi} for a single electron	Numerical value	$-l \leq m_l \leq l$
Single-electron spin, s (or s_i) :		
Quantisation of the operator $\mathbf{S}_i^2$ for a single electron	Numerical value	$s = 1/2$

Continued on next page

continued from previous page

Corresponds to:	Represented by:	Constraints:
Projection of the single-electron spin, m_s (or m_{si}) :		
Quantisation of the operator S_{zi} for a single electron	Numerical value	$m_s = \pm 1/2$
Total single-electron angular momentum, j (or j_i) :		
Quantisation of the operator $\mathbf{J}_i^2$, with $\mathbf{J}_i \equiv \mathbf{L}_i + \mathbf{S}_i$, for a single electron	Numerical value	$\lvert l - 1/2 \rvert \leq j \leq l + 1/2$
Projection of the total single-electron angular momentum, m_j (or m_{ji}) :		
Quantisation of the operator J_z for a single electron	Numerical value	$-j \leq m_j \leq j$
Single-electron orbital, $(nl)^w$, with occupation number w :		
The values of n and l	The numerical value of n and the letter symbol for l	$w \leq 2(2l+1)$
Total electronic angular momentum for an N-electron atom, L :		
Quantisation of the operator $\mathbf{L}^2$ $\left(\mathbf{L} = \sum_{i=1}^{N} \mathbf{L}_i\right)$	The letters S, P, D, F, G, H, I, K . . . ; corresponding to respectively $L = 0, 1, 2, 3, 4, 5, 6, 7 \ldots$	$\min\left[\sum_i^N \pm l_i\right] \leq L \leq \sum_i^N l_i$
Projection of the total electronic angular momentum for an N-electron atom, M_L :		
Quantisation of the operator L_z	Numerical value	$-L \leq M_L \leq L$

Continued on next page

continued from previous page

Corresponds to:	Represented by:	Constraints:
Total electronic spin for an N-electron atom, S :		
Quantisation of the operator $\mathbf{S}^2$ $\left(\mathbf{S}=\sum_{i=1}^{N}\mathbf{S}_i\right)$	Numerical value and, in connection with an atomic term (see below), by a numerical value for $2S+1$	$\min\left[\sum_i^N \pm s_i\right] \leq S \leq \sum_i^N s_i$
Projection of the total electronic spin for an N-electron atom, M_S :		
Quantisation of the operator S_z	Numerical value	$-S \leq M_S \leq S$
Total electronic angular momentum for an N-electron atom, J :		
Quantisation of the operator $\mathbf{J}^2$, with $\left(\mathbf{J}=\sum_{i=1}^{N}\mathbf{L}_i+\sum_{i=1}^{N}\mathbf{S}_i\right)$ The order of the summation depends on the coupling scheme – see chapter 6	Numerical value	
Atomic term in the LS-coupling scheme (see chapter 6), $^{2S+1}L_J$:		
The values of L, S and J for a state that is well approximated by LS-coupling	The letter symbol for L preceded by the numerical value of $2S+1$ as a superscript and followed by the numerical value of J as a subscript	
Nuclear spin, I :		
Quantisation of the operator $\mathbf{I}^2$	Numerical value	$I=0,\ 1/2,\ 1, 3/2,\ 2,\ \ldots$
Projection of the nuclear spin, M_I :		
Quantisation of the operator I_z	Numerical value	$-I \leq M_I \leq I$
Total angular momentum for an N-electron atom, including the nucleus, F :		
Quantisation of the operator $\mathbf{F}^2$, with $\mathbf{F}\equiv\mathbf{J}+\mathbf{I}$	Numerical value	$\lvert I-J\rvert \leq F \leq I+J$

Continued on next page

continued from previous page

Corresponds to:	Represented by:	Constraints:
Projection of the total angular momentum for an N-electron atom including the nucleus, M_F :		
Quantisation of the operator F_z	Numerical value	$-F \leq M_F \leq F$

1.6 Central-Field Wave Functions in Multielectron Atoms

A major part of what has been described in this chapter is valid for any purely central potential. The separation of radial and angular parameters and the treatment of the angular part will be the same. A potential that only depends on the radial parameter r is not able to produce any torque on a system and, without any other influence, the angular momentum is decoupled from the potential and stays constant (it is a constant of the motion). Only the solution of the radial function will differ for a different central potential.

In a substantial portion of this book, when we treat the multielectron case, we will as a first approximation consider the purely central part of the multielectron potential. That enables us to factorise the individual electron wave functions. From the thus obtained set of basis functions, we proceed with a perturbative approach. The angular properties of the separable single-electron basis functions are identical to those depicted in table 1.2. The radial functions are not the same as the ones in section 1.2, since the central part of the total potential includes interelectron interactions, such as charge screening. However, many of the physical features of the here derived hydrogenic wave functions will still hold fairly well also for multielectron atoms.

Further Reading

The theory of atomic spectra, by Condon & Shortley [7]
Quantum Mechanics of One- and Two-Electron Atoms, by Bethe & Salpeter [4]
Quantum theory of atomic structure, by Slater [8]
The theory of atomic structure and spectra, by Cowan [9]
Quantum Mechanics – Course of th. physics, v. 3, by Landau & Lifshitz [10]
Atomic Many-Body Theory, by Lindgren & Morrison [11]
Physics of Atoms and Molecules, by Bransden & Joachain [5]

References

1. P.J. Mohr, D.B. Newell, B.N. Taylor, Journal of Physical and Chemical Reference Data **45**, 043102 (2016)
2. W.E. Baylis, G.W.F. Drake, in *Springer Handbook of Atomic, Molecular, and Optical Physics*, ed. by G.W.F. Drake (Springer-Verlag, New York, 2006), p. 1
3. G.B. Arfken, H.J. Weber, F.E. Harris, *Mathematical Methods for Physicists: A Comprehensive Guide*, 7th edn. (Academic Press, Amsterdam, 2012)
4. H.A. Bethe, E.E. Salpeter, *Quantum Mechanics of One- and Two-Electron Atoms* (Springer-Verlag, Berlin, 1957)
5. B.H. Bransden, C.J. Joachain, *Physics of Atoms and Molecules*, 2nd edn. (Prentice Hall, Harlow, England, 2003)
6. E.D. Commins, *Quantum mechanics: an experimentalist's approach* (Cambridge University Press, New York, 2014)
7. E.U. Condon, G.H. Shortley, *The theory of atomic spectra* (Cambridge University Press, Cambridge, 1935)
8. J.C. Slater, *Quantum theory of atomic structure* (McGraw-Hill, New York, 1960)
9. R.D. Cowan, *The theory of atomic structure and spectra* (University of California press, Berkeley, 1981)
10. L.D. Landau, E.M. Lifshitz, *Quantum Mechanics — Course of Theoretical Physics, volume 3*, 3rd edn. (Butterworth-Heinemann, Amsterdam, 1981)
11. I. Lindgren, J. Morrison, *Atomic Many-Body Theory*, 2nd edn. (Springer Verlag, Berlin, 1986)

Chapter 2
Atoms with Two or More Electrons – Symmetries

Following the revisit of the lowest order, non-relativistic solutions for one-electron systems in chapter 1, it is time to turn to the principal subject of the book — atoms with more than one electron. In this chapter, we will formulate the Schrödinger equation for the helium atom, and then generalise it for larger systems. Already with a second electron added, we see that we are no longer able to solve the Schrödinger equation without approximations. If we want to compute wave functions, or if we want to determine eigenenergies, we either have to use methods such as perturbation theory or the variational method (see chapter 3), do a numerical calculation (see chapter 14) or a combination of these. Either way, we need a deeper understanding of the impact of the departure from a two-body system on the symmetry properties of the wave function, and on internal interactions that are not spherically symmetric.

With more than one electron, we need to take into account that we are dealing with ensembles of identical fermionic particles, which imposes symmetry constraints. After the formulation of the Schrödinger equation for an N-electron atom, we introduce the importance of exchange symmetries for atoms with several electrons and apply it to a heuristic perturbative approach for the helium and lithium atoms, initiated here and brought to quantitative results in chapter 3. The concept of the so-called Slater determinants and their importance for antisymmetric wave functions is then introduced, as is an introduction into proper usage of determinantal wave functions in quantum mechanics.

2.1 The Multielectron Atom Schrödinger equation

Generalising the Schrödinger equation from the hydrogen atom to helium, or to any other two-electron system, will involve not just adding the kinetic energy and the Coulomb attraction to the nucleus for the second electron, but also the electrostatic repulsion between the two electrons. The presence of an exchange term in the overall potential means that we cannot treat the system as two isolated two-body problems.

A. Kastberg, *Structure of Multielectron Atoms*, Springer Series on Atomic, Optical, and Plasma Physics 112, https://doi.org/10.1007/978-3-030-36420-5_2

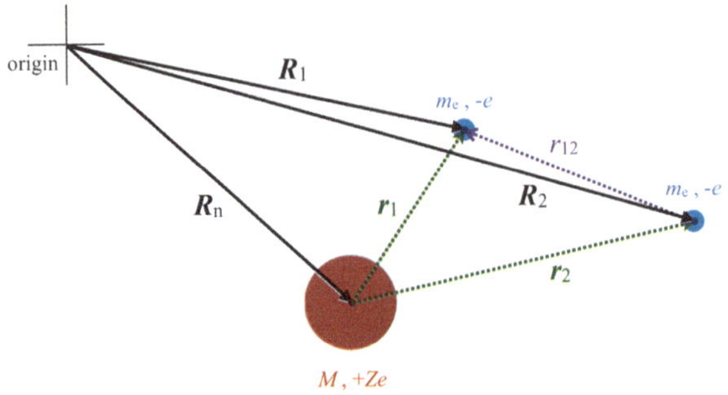

Fig. 2.1 Schematic illustration of the three-body problem of a two-electron atom. The red ball is the nucleus with mass M and charge $+Ze$. In the laboratory frame of reference, it has spatial coordinate $\mathbf{R}_\mathrm{n}$. The two electrons (the blue balls) have coordinates $\mathbf{R}_1$ and $\mathbf{R}_2$, masses m_e and charges $-e$. The vectors $\mathbf{r}_1$ and $\mathbf{r}_2$, shown in green, are the position vectors of the electrons relative to the nucleus, and r_{12} is the separation between the electrons.

A first step towards an understanding of this more complicated situation should be to generalise the Schrödinger equation in (1.2) to a three-body system.

2.1.1 The Three-Body Hamiltonian

We consider a system of a nucleus of mass M and charge $+Ze$, and two much lighter electrons of masses m_e and charges $-e$, as illustrated in figure 2.1 In a fixed laboratory frame of reference, the vectors $\mathbf{R}_\mathrm{n}$, $\mathbf{R}_1$ and $\mathbf{R}_2$ represent the dynamic positions of the nucleus and the two electrons, respectively. In this frame, the total Hamiltonian is (in SI-units):

$$\begin{aligned} H = & -\frac{\hbar^2}{2M}\nabla^2_{\mathbf{R}_\mathrm{n}} - \frac{\hbar^2}{2m_\mathrm{e}}\left(\nabla^2_{\mathbf{R}_1} + \nabla^2_{\mathbf{R}_2}\right) \\ & -\frac{Ze^2}{4\pi\varepsilon_0}\left(\frac{1}{|\mathbf{R}_\mathrm{n}-\mathbf{R}_1|} + \frac{1}{|\mathbf{R}_\mathrm{n}-\mathbf{R}_2|}\right) + \frac{e^2}{4\pi\varepsilon_0\,|\mathbf{R}_1-\mathbf{R}_2|} \,. \end{aligned} \tag{2.1}$$

The expression in (2.1) is complete but not practical. We want to know the structure of the atom, which means that the thing that interests us is the internal degrees of freedom. To get to the core of the problem, we have to separate out the centre-of-mass motion. We introduce the CoM position coordinate:

$$\mathbf{R} \equiv \frac{M\,\mathbf{R}_\mathrm{n} + m_\mathrm{e}\,(\mathbf{R}_1 + \mathbf{R}_2)}{M + 2m_\mathrm{e}} \,, \tag{2.2}$$

and we define the position vectors of the electrons relative to the nucleus:

$$\begin{aligned}\mathbf{r}_1 &\equiv \mathbf{R}_1 - \mathbf{R}_\mathrm{n} \\ \mathbf{r}_2 &\equiv \mathbf{R}_2 - \mathbf{R}_\mathrm{n}\,.\end{aligned} \tag{2.3}$$

From this, the Laplacians in (2.1) can be converted to ones in the coordinates $\mathbf{R}$, $\mathbf{r}_1$ and $\mathbf{r}_2$:

$$\begin{aligned}\nabla^2_{\mathbf{R}_\mathrm{n}} &= \left(\frac{M}{M+2m_\mathrm{e}}\right)^2 \nabla^2_{\mathbf{R}} - \frac{2M}{M+2m_\mathrm{e}}\left(\nabla_{\mathbf{R}}\cdot\nabla_{\mathbf{r}_1} + \nabla_{\mathbf{R}}\cdot\nabla_{\mathbf{r}_2}\right) + \left(\nabla_{\mathbf{r}_1} + \nabla_{\mathbf{r}_2}\right)^2 \\ \nabla^2_{\mathbf{R}_1} &= \left(\frac{m_\mathrm{e}}{M+2m_\mathrm{e}}\right)^2 \nabla^2_{\mathbf{R}} + \frac{2m_\mathrm{e}}{M+2m_\mathrm{e}}\nabla_{\mathbf{R}}\cdot\nabla_{\mathbf{r}_1} + \nabla^2_{\mathbf{r}_1} \\ \nabla^2_{\mathbf{R}_2} &= \left(\frac{m_\mathrm{e}}{M+2m_\mathrm{e}}\right)^2 \nabla^2_{\mathbf{R}} + \frac{2m_\mathrm{e}}{M+2m_\mathrm{e}}\nabla_{\mathbf{R}}\cdot\nabla_{\mathbf{r}_2} + \nabla^2_{\mathbf{r}_2}\,.\end{aligned} \tag{2.4}$$

With the change of reference frame in (2.4), we can now rewrite the Hamiltonian in (2.1) in CoM coordinates:

$$\begin{aligned}H = &-\frac{\hbar^2}{2(M+2m_\mathrm{e})}\nabla^2_{\mathbf{R}} - \frac{\hbar^2}{2}\left(\frac{1}{M}+\frac{1}{m_\mathrm{e}}\right)\left(\nabla^2_{\mathbf{r}_1}+\nabla^2_{\mathbf{r}_2}\right) - \frac{\hbar^2}{M}\left(\nabla_{\mathbf{r}_1}\cdot\nabla_{\mathbf{r}_2}\right) \\ &- \frac{Ze^2}{4\pi\varepsilon_0}\left(\frac{1}{r_1}+\frac{1}{r_2}\right) + \frac{e^2}{4\pi\varepsilon_0 r_{12}}\,.\end{aligned} \tag{2.5}$$

Here, the definitions of the scalars r_1, r_2 and r_{12} are, respectively, the separations of the electrons from the nucleus and the separation between the two electrons (see figure 2.1).

In (2.5), the CoM coordinate is only present in one term. This means that if we only want to know the internal atomic structure, this term can be ignored. This brings us to the *structural* Schrödinger equation for a two-electron atom:

$$\begin{aligned}&\left[-\frac{\hbar^2}{2}\left(\frac{1}{M}+\frac{1}{m_\mathrm{e}}\right)\left(\nabla^2_{\mathbf{r}_1}+\nabla^2_{\mathbf{r}_2}\right) - \frac{\hbar^2}{M}\left(\nabla_{\mathbf{r}_1}\cdot\nabla_{\mathbf{r}_2}\right)\right]\psi(\mathbf{r}_1,\mathbf{r}_2) \\ &-\left[\frac{Ze^2}{4\pi\varepsilon_0}\left(\frac{1}{r_1}+\frac{1}{r_2}\right) - \frac{e^2}{4\pi\varepsilon_0 r_{12}}\right]\psi(\mathbf{r}_1,\mathbf{r}_2) = E\,\psi(\mathbf{r}_1,\mathbf{r}_2)\,.\end{aligned} \tag{2.6}$$

The term with the scalar product $\nabla_{\mathbf{r}_1}\cdot\nabla_{\mathbf{r}_2}$ is called the *mass polarisation term* and it appears as a consequence of the motion of the nucleus relative to the CoM. In view of the fact that $M \gg m_\mathrm{e}$, we can in an initial treatment omit this term — we will reconsider it in section 10.1. Ignoring the nuclear motion in this way also simplifies the prefactor for the kinetic energy terms for the individual electrons. We take all this into account, and convert to atomic units, and the Schrödinger equation becomes:

$$\left(-\frac{\nabla^2_{\mathbf{r}_1}}{2} - \frac{\nabla^2_{\mathbf{r}_2}}{2} - \frac{Z}{r_1} - \frac{Z}{r_2} + \frac{1}{r_{12}}\right)\psi(\mathbf{r}_1,\mathbf{r}_2) = E\,\psi(\mathbf{r}_1,\mathbf{r}_2)\,. \tag{2.7}$$

2.1.2 Generalisation to an N-Electron Atom

Generalising the analysis of the two-electron atom and (2.7) to three or more electrons is straightforward. We get a sum of single-electron kinetic energy terms, a sum of Coulomb interactions with the nucleus and a number of exchange terms. For the latter, we just have to take care to avoid double counting. The N-electron Schrödinger equation becomes:

$$\left(-\sum_{i=1}^{N}\frac{\nabla_{\mathbf{r}_i}^2}{2}-\sum_{i=1}^{N}\frac{Z}{r_i}+\sum_{i=1}^{N}\sum_{j>i}^{N}\frac{1}{r_{ij}}\right)\psi(\mathbf{r}_1,\mathbf{r}_2,\ldots,\mathbf{r}_N)=E\,\psi(\mathbf{r}_1,\mathbf{r}_2,\ldots,\mathbf{r}_N)\ . \quad (2.8)$$

If we could solve the Schrödinger equation in (2.8), we would get functional expressions for wave functions and energies, just as for hydrogen in chapter 1. However, the presence of the cross terms means that we cannot factorise the wave function. Rather, the eigenstates will always be intrinsically entangled states, and finding an analytical solution to (2.8) will be difficult, even in this simplified example, where electron spins and the nuclear motion have been ignored.

In chapter 5, and thereafter following chapters, we will show that by splitting up the cross term into approximate central and angular parts, we are able to analyse a vast number of the structural and spectroscopic properties of the atom, even without knowing the exact functional form of the wave functions, or the exact atomic energies. Before doing that, however, we demonstrate what we may learn from treating the entire cross term as a perturbation, and in chapter 3 we also look at other approximation methods. An added value of the simplified perturbative analysis is that it forces us to consider the impact of exchange symmetries.

2.2 Exchange Symmetries

The only qualities that may differ between two electrons are their spins and their trajectories. In quantum mechanics, the latter may not be precisely defined, and thus the correct way to describe a system with many electrons is by an ensemble wave function:

$$\psi(\mathbf{r}_1,\mathbf{r}_2,\ldots,\mathbf{r}_N)\ . \quad (2.9)$$

We define a *pair-exchange operator*, $P_{\text{ex}}(i,j)$ as the exchange of spatial coordinates between electrons i and j. The eigenvalues of this operator must be either 1 or -1, since the operator applied twice must bring us back to the initial wave function, and any observable must be invariant with an exchange of electron coordinates.

More generally, an ensemble wave function for identical particles must always be either symmetric or antisymmetric with respect to an exchange of any two particle coordinates. In the case of atomic structure, we deal with ensembles of electrons, which are fermions. A compound wave function of a multifermion system must always be antisymmetric (see, for example, [1] or [2]), which leads to the Pauli principle. For example, consider the following two symmetrised wave functions for a two-particle system, which for simplicity we have assumed to be factorisable:

$$\begin{aligned}\psi_S &= \frac{1}{\sqrt{2}}\left[\psi_a(\mathbf{r}_1)\psi_b(\mathbf{r}_2) + \psi_a(\mathbf{r}_2)\psi_b(\mathbf{r}_1)\right] \\ \psi_A &= \frac{1}{\sqrt{2}}\left[\psi_a(\mathbf{r}_1)\psi_b(\mathbf{r}_2) - \psi_a(\mathbf{r}_2)\psi_b(\mathbf{r}_1)\right] . \end{aligned} \tag{2.10}$$

Here S and A stand for symmetric and antisymmetric. For fermions, an exchange of particles leads to a change of sign, and thus ψ_S above would be rendered zero.

For our present purposes, a consequence is that in a multielectron atom, two individual electrons may not have identical individual electronic wave functions (they can never be in the same state). We have thus far not included the electron spin into the atomic Hamiltonians that we consider, but the symmetry constraints and the Pauli principle mean that we must nevertheless somehow add the spin to the analysis of the wave function. The modulus of the spin for a single electron is always the same (the expectation value of $|\mathbf{S}|$ is $\sqrt{3}\hbar/2$, corresponding to the quantum number $s = 1/2$) and it has one of two possible spin-projections ($\pm\hbar/2$). The existence of two different possible spins and the different ways in which we can combine them in a multielectron atom will matter for the compound wave functions and for the associated energies.

The absence of the spin in the Hamiltonian means that we can factorise the spatial and spin parts of the wave functions, as was done in section 1.4.2. Following the nomenclature introduced there, and in (1.27), we write the total wave function for an *individual* electron as:

$$\Psi_i(\mathbf{q}_i) = \psi_i(\mathbf{r}_i)\,\zeta_i(m_{si}) . \tag{2.11}$$

We then look at the effect of exchange symmetry for the spin and spatial parts separately, before we combine it all to a compound multielectron wave function.

Our goal is to establish general symmetry properties, rather than achieving exact results. Therefore, we will adhere to the crude approximation already hinted at previously, namely that the cross term in (2.8) is left to a perturbative calculation. We can then temporarily proceed by writing a zero-order total atomic wave function as a product of the functions for the individual electrons, and see what the consequences of Fermi–Dirac statistics are. For an initial approximation of the energy, this can then be followed up by perturbation theory, with the cross term giving an energy offset and possibly a splitting of levels. We will here begin the perturbative treatment for a couple of small atoms, and in chapter 3 we will carry this further quantitatively.

2.2.1 The Compound Spin Wave Function

We commence by studying the compound spin wave function, and by analysing its possible symmetries. Note that this part of the wave function does not necessarily need to be exchange antisymmetric. That constraint will apply only to the *total* wave function. However, it does have to be either symmetric or antisymmetric with respect to the swapping of two electron coordinates.

To illustrate this, we begin with an example with a two-electron atom, or to be even more specific, we look at the spin part of the wave function for the ground state of neutral He. We write a total zero-order wave function as:

$$\begin{aligned}\Psi^{(0)}(\mathbf{r}_1, m_{s1}, \mathbf{r}_2, m_{s2}) &= \Psi^{(0)}(\mathbf{q}_1, \mathbf{q}_2) \\ &= \psi^{(0)}(\mathbf{r}_1, \mathbf{r}_2)\, \mathscr{X}(m_{s1}, m_{s2}) \\ &= \psi_{n_1 l_1}(\mathbf{r}_1)\, \psi_{n_2 l_2}(\mathbf{r}_2)\, \zeta_1(m_{s1})\, \zeta_2(m_{s1})\,. \end{aligned} \tag{2.12}$$

In the first line, we have used the vector $\mathbf{q}_i$ — the shorthand notation introduced in section 1.4.2. We have also eliminated the redundant quantum numbers s_1 and s_2. For each of the two spin functions, there are only two possible eigenstates, namely spin-up and spin-down, which we will write as ζ_+ or ζ_-, corresponding to the eigenvalues $\pm 1/2$ of S_z, as introduced in section 1.4.2. The task now is to make sure that if $\psi^{(0)}(\mathbf{r}_1, \mathbf{r}_2)$ in (2.12) is symmetric, $\mathscr{X}(m_{s1}, m_{s2})$ is antisymmetric or vice versa.

We start with the combined spin wave function, $\mathscr{X}(m_{s1}, m_{s2})$. With the two possibilities for the two electrons, we have a four-dimensional Hilbert space. The most intuitive tentative choice for the basis is:

$$\begin{aligned}\mathscr{X}_\alpha(m_{s1}, m_{s2}) &= \zeta_{1+}\,\zeta_{2+} \\ \mathscr{X}_\beta(m_{s1}, m_{s2}) &= \zeta_{1+}\,\zeta_{2-} \\ \mathscr{X}_\gamma(m_{s1}, m_{s2}) &= \zeta_{1-}\,\zeta_{2+} \\ \mathscr{X}_\delta(m_{s1}, m_{s2}) &= \zeta_{1-}\,\zeta_{2-}\,. \end{aligned} \tag{2.13}$$

Of these functions, $\mathscr{X}_\alpha(m_{s1}, m_{s2})$ and $\mathscr{X}_\delta(m_{s1}, m_{s2})$ are exchange symmetric. However, we have no justification for assuming that $\mathscr{X}_\beta(m_{s1}, m_{s2}) = \pm\mathscr{X}_\beta(m_{s2}, m_{s1})$ or $\mathscr{X}_\gamma(m_{s1}, m_{s2}) = \pm\mathscr{X}_\gamma(m_{s2}, m_{s1})$, and thus $\mathscr{X}_\beta(m_{s1}, m_{s2})$ and $\mathscr{X}_\gamma(m_{s1}, m_{s2})$ are neither symmetric nor antisymmetric. We do need a basis of functions that all have either of the two symmetries, and to get that we must instead form normalised superpositions of $\mathscr{X}_\beta(m_{s1}, m_{s2})$ and $\mathscr{X}_\gamma(m_{s1}, m_{s2})$, in the spirit of (2.10).

Together with $\mathscr{X}_\alpha(m_{s1}, m_{s2})$ and $\mathscr{X}_\delta(m_{s1}, m_{s2})$, and if the order is rearranged in a convenient way, we choose as basis functions:

$$\begin{cases}\mathscr{X}_\mathrm{a}(m_{s1}, m_{s2}) = \zeta_{1+}\,\zeta_{2+} \\ \mathscr{X}_\mathrm{b}(m_{s1}, m_{s2}) = \dfrac{1}{\sqrt{2}}(\zeta_{1+}\,\zeta_{2-} + \zeta_{1-}\,\zeta_{2+}) \\ \mathscr{X}_\mathrm{c}(m_{s1}, m_{s2}) = \zeta_{1-}\,\zeta_{2-}\end{cases} \tag{2.14}$$

$$\mathscr{X}_\mathrm{d}(m_{s1}, m_{s2}) = \frac{1}{\sqrt{2}}(\zeta_{1+}\,\zeta_{2-} - \zeta_{1-}\,\zeta_{2+})\,. \tag{2.15}$$

Here the top three functions — those in (2.14) — are all exchange symmetric. Together, they constitute a *triplet*. In contrast, the last function, the *singlet* (2.15), is exchange antisymmetric.

2.2.2 The Helium Ground State

We have not yet looked into the possible symmetries of the spatial wave function, but since we are for the moment just considering the ground state, we are left without choice. The energy for the factorised approximation will be determined solely by the spatial wave function, which in turn will be a product of hydrogen-like orbitals. For the ground state, we have to have

$$\psi^{(0)}(\mathbf{r}_1,\mathbf{r}_2) = \psi_{1s}(\mathbf{r}_1)\,\psi_{1s}(\mathbf{r}_2)\ , \tag{2.16}$$

in order to minimise the energy.

The function in (2.16) may seem to violate the Pauli principle, but it does not. The two electrons can have (and indeed must have) opposite spins, and thus they are in different states. What the function does, however, is that it fixes the spin function. In (2.16), $\psi(\mathbf{r}_1,\mathbf{r}_2)$ is exchange symmetric and therefore the spin part of the ground state wave function for He must be the antisymmetric $\mathscr{X}_d(\mathbf{S}_1,\mathbf{S}_2)$ in (2.15). The zero-order wave function for the helium ground state is thus found to be the product state:

$$\Psi_0^{(0)}(\mathbf{q}_1,\mathbf{q}_2) = \frac{1}{\sqrt{2}}(\zeta_{1+}\,\zeta_{2-} - \zeta_{1-}\,\zeta_{2+})\,\psi_{1s}(\mathbf{r}_1)\,\psi_{1s}(\mathbf{r}_2)\,. \tag{2.17}$$

2.2.3 The Spatial Wave Function

For excited two-electron atoms, we have to take a more general point of view and consider also symmetry constraints for the spatial part of the wave function. Since we are still only investigating symmetry properties, it will suffice with an example for which just one of the electrons has been excited from 1s to an arbitrary nl-orbital.

The factorised zeroth order spatial wave function, ignoring symmetry, is:

$$\psi_a^{(0)}(\mathbf{r}_1,\mathbf{r}_2) = \psi_{1s}(\mathbf{r}_1)\,\psi_{nl}(\mathbf{r}_2)\,, \tag{2.18}$$

which has the zero-order energy:

$$E_a^{(0)} = E_{1s} + E_{nl}\,. \tag{2.19}$$

Here we encounter *exchange degeneracy*. We can just as well construct a wave function where the indices have been exchanged:

$$\psi_b^{(0)}(\mathbf{r}_1,\mathbf{r}_2) = \psi_{nl}(\mathbf{r}_1)\,\psi_{1s}(\mathbf{r}_2)\,, \tag{2.20}$$

with the same energy as $\psi_{\mathrm{a}}^{(0)}(\mathbf{r}_1,\mathbf{r}_2)$. This has two consequences. One is that if we want to proceed to calculate the energy with perturbation theory, this must be of the degenerate kind, where the basis functions used are superpositions of the degenerate states. The second is that these superpositions provide us with a natural way to form symmetric and antisymmetric spatial wave functions.

The normalised symmetric and antisymmetric superpositions are:

$$
\begin{aligned}
\psi_{\mathrm{S}}^{(0)}(\mathbf{r}_1,\mathbf{r}_2) &= \frac{1}{\sqrt{2}}\left[\psi_{1\mathrm{s}}(\mathbf{r}_1)\,\psi_{nl}(\mathbf{r}_2)+\psi_{nl}(\mathbf{r}_1)\,\psi_{1\mathrm{s}}(\mathbf{r}_2)\right] \\
\psi_{\mathrm{A}}^{(0)}(\mathbf{r}_1,\mathbf{r}_2) &= \frac{1}{\sqrt{2}}\left[\psi_{1\mathrm{s}}(\mathbf{r}_1)\,\psi_{nl}(\mathbf{r}_2)-\psi_{nl}(\mathbf{r}_1)\,\psi_{1\mathrm{s}}(\mathbf{r}_2)\right] .
\end{aligned} \tag{2.21}
$$

From the discussion in the preceding section on spin wave functions, and taking (2.14) and (2.15) into account, we can conclude that the symmetric spatial function, ψ_{S} must always be combined with the singlet spin function in (2.15), whereas ψ_{A} can be combined with any of the spin functions in the triplet (2.14).

We could now write the four possible total compound zero-order wave functions for the $1snl$-excited state in He, by taking the proper products between the equations in (2.21), and the spin singlet and spin triplet, respectively. We will, however, leave that to the next section — see equation (2.24) — where we will also generalise the symmetry considerations to bigger atoms.

2.3 Slater Determinants

We have seen that the total wave function, including spatial and spin parameters, must be antisymmetric, and that achieving this needs care. In the preceding section, this was limited to an atom with two electrons. In the following, we shall generalise it to an N-electron system, but in order to get to that, we start by slightly reformulating what was established in the previous section for helium.

For three or more electrons, exchange degeneracy is even more of an issue then it is for helium. Degenerate perturbation theory is essentially the procedure of first creating a basis consisting of superpositions of all product states of the degenerate zero-order states, in all possible permutations. This basis will provide a non-diagonal Hamiltonian when the perturbation is added, and its diagonalisation will provide the new eigenstates. For the two-electron problem, the dimension of the secular equation is just two. For the N-electron case, we will get a Hilbert space of dimension $N!$.

2.3.1 Two-Electron Antisymmetric Wave Functions

When we dealt with the He antisymmetric wave function in section 2.2, we did it in two steps. We first analysed the spin wave function, which led to a triplet of symmetric states and one singlet state that is antisymmetric. We then took on the spatial function, and while considering the exchange degeneracy, we prepared for degenerate perturbation theory — with the interaction term as the perturbation — by formulating the zeroth order superposition basis states (2.21). After that, we could combine symmetric spatial functions with antisymmetric spin functions or vice versa. This approach is fine as long as we only have two electrons. With many electrons it becomes unmanageable, and instead, it is better to take on the combined wave function directly, with both spatial and spin parts, before forming the basis states.

Instead of looking at the spin and spatial parts separably, we thus start by formulating *spin-orbitals* — one-electron product functions of spin and spatial functions — as in (1.27). We form product states of the type:

$$\Psi_{\mathrm{a}}^{(0)}(\mathbf{q}_1,\mathbf{q}_2) = \psi_{n_1 l_1}(\mathbf{r}_1)\,\zeta_1(m_{s1})\,\psi_{n_2 l_2}(\mathbf{r}_2)\,\zeta_2(m_{s2})\,. \tag{2.22}$$

We assume that the electrons are in different spatial orbitals, that is, $n_1 l_1 \neq n_2 l_2$, thereby avoiding conflicts with the Pauli principle. Taking the exchange degeneracy into account, we know that $\Psi_{\mathrm{a}}^{(0)}$ is accompanied by the function:

$$\Psi_{\mathrm{b}}^{(0)}(\mathbf{q}_1,\mathbf{q}_2) = \psi_{n_2 l_2}(\mathbf{r}_1)\,\zeta_1(m_{s1})\,\psi_{n_1 l_1}(\mathbf{r}_2)\,\zeta_2(m_{s2})\,, \tag{2.23}$$

which will have the same energy. Furthermore, the individual electron spins are as always either 'up' or 'down', so with all possible spin permutations taken into account, this will leave us with eight degenerate functions.

In the Hilbert space formed by those eight functions, we can diagonalise the Hamiltonian and find the eigenstates. This is easier than it appears since orthogonality assures that most antidiagonal elements are zero, and we get away with four secular equations of dimension two. The outcome of the diagonalisation is eight new eigenfunctions: four of them exchange symmetric, and the other four exchange antisymmetric. Again relying on the Pauli exclusion principle, we can exclude the symmetric ones, and we end up with four functions:

$$\begin{aligned}
\Psi_{\alpha}^{(0)}(\mathbf{q}_1,\mathbf{q}_2) &= \frac{1}{\sqrt{2}}\left[\psi_{n_1 l_1}(\mathbf{r}_1)\,\psi_{n_2 l_2}(\mathbf{r}_2) - \psi_{n_2 l_2}(\mathbf{r}_1)\,\psi_{n_1 l_1}(\mathbf{r}_2)\right]\,\zeta_{1+}\,\zeta_{2+}\\
\Psi_{\beta}^{(0)}(\mathbf{q}_1,\mathbf{q}_2) &= \frac{1}{2}\left[\psi_{n_1 l_1}(\mathbf{r}_1)\,\psi_{n_2 l_2}(\mathbf{r}_2)\right.\\
&\qquad \left. -\psi_{n_2 l_2}(\mathbf{r}_1)\,\psi_{n_1 l_1}(\mathbf{r}_2)\right]\,(\zeta_{1+}\,\zeta_{2-} + \zeta_{1-}\,\zeta_{2+})\\
\Psi_{\gamma}^{(0)}(\mathbf{q}_1,\mathbf{q}_2) &= \frac{1}{\sqrt{2}}\left[\psi_{n_1 l_1}(\mathbf{r}_1)\,\psi_{n_2 l_2}(\mathbf{r}_2) - \psi_{n_2 l_2}(\mathbf{r}_1)\,\psi_{n_1 l_1}(\mathbf{r}_2)\right]\,\zeta_{1-}\,\zeta_{2-}\\
\Psi_{\delta}^{(0)}(\mathbf{q}_1,\mathbf{q}_2) &= \frac{1}{2}\left[\psi_{n_1 l_1}(\mathbf{r}_1)\,\psi_{n_2 l_2}(\mathbf{r}_2)\right.\\
&\qquad \left. +\psi_{n_2 l_2}(\mathbf{r}_1)\,\psi_{n_1 l_1}(\mathbf{r}_2)\right]\,(\zeta_{1+}\,\zeta_{2-} - \zeta_{1-}\,\zeta_{2+})\,.
\end{aligned} \tag{2.24}$$

This is fully consistent with what was found in section 2.2.

2.3.2 N-Electron Atom Antisymmetric Functions

We can now follow the procedure used for deriving (2.24), but for an atom with three or more electrons. We will have product states of the type:

$$\begin{aligned}\Psi_{\mathrm{a}}^{(0)}(\mathbf{q}_1,\mathbf{q}_2,\ldots,\mathbf{q}_n) &= \Psi_{\mathrm{a}}^{(0)}(\mathbf{q}) \\ &= \psi_{n_1 l_1}(\mathbf{r}_1)\,\zeta_1(m_{s1})\;\psi_{n_2 l_2}(\mathbf{r}_2)\,\zeta_2(m_{s2})\ldots\;\psi_{n_N l_N}(\mathbf{r}_N)\,\zeta_N(m_{sN}) \\ &= \Psi_1(\mathbf{q}_1)\,\Psi_2(\mathbf{q}_2)\ldots\,\Psi_N(\mathbf{q}_N)\,. \end{aligned} \tag{2.25}$$

The vector $\mathbf{q}$ is here used as a shorthand for all coordinates (spatial and spin) for all N electrons (and as before, we take the mathematical liberty to mix discrete and continuous variables in one single symbol). With all possible permutations of the indices, we will all in all have $N!$ degenerate product functions. In the Dirac vector notation, we write the state vector corresponding to (2.25) as the convolution:

$$|1\rangle_1 * |2\rangle_2 * \ldots * |N\rangle_N\,, \tag{2.26}$$

where the numbers inside the kets typically represent a set of particular quantum numbers, and the subscript refers to the respective particle coordinate.

When we construct normalised superpositions of the different permutations, we will find a variety of symmetries. Only one is fully symmetric with respect to exchange of any two electron coordinates (the one where all functions are summed with equal weight), and one other is antisymmetric. The rest have more complicated types of symmetry. The only one we need to retain when we look for a suitable zero-order function is the fully exchange antisymmetric one.

With many electrons, the superposition of $N!$ product states becomes a long expression. A convenient way to write this, and also to find the antisymmetric state, is to write the superposition in the form of a determinant [3]:

$$\psi^{(0)}(\mathbf{q}) = \frac{1}{\sqrt{N!}}\begin{vmatrix}\Psi_1(\mathbf{q}_1) & \Psi_1(\mathbf{q}_2) & \ldots & \Psi_1(\mathbf{q}_N)\\ \Psi_2(\mathbf{q}_1) & \Psi_2(\mathbf{q}_2) & \ldots & \Psi_2(\mathbf{q}_N)\\ \vdots & \vdots & \ddots & \vdots \\ \Psi_N(\mathbf{q}_1) & \Psi_N(\mathbf{q}_2) & \ldots & \Psi_N(\mathbf{q}_N)\end{vmatrix} \equiv \mathbb{S}_{\Psi}\,. \tag{2.27}$$

This is known as a *Slater determinant*. Due to the symmetry properties of a determinant (the interchange of two columns, or two rows, changes the sign of the determinant), such functions will always be exchange antisymmetric. In the last part of (2.27), we have introduced a notation for Slater determinants that will be used also in the upcoming chapters of this book.

Slater determinant superposition functions are necessary starting points when attempts are being made to form multielectron wave functions. They are necessary for many numerical solutions to the Schrödinger equation (see chapter 14). We also need them when we calculate matrix elements for perturbation Hamiltonians; the wave functions going into such expectation values must for a multielectron atom be Slater determinant functions. How to proceed with integrals over Slater determinants will be explained in section 2.3.4.

2.3.3 Antisymmetric Wave Functions for Li

As an example, we will formulate a wave function for the ground state of the three-electron atom lithium. In the approximation where we neglect the electron–electron interaction term, two electrons will be in the 1s-orbital, just as for helium, and due to the Pauli principle, they must have opposite spins. The same principle prevents the third electron to go into 1s, so this must be in the lowest energy orbital that remains, that is, 2s (the degeneracy between 2s and 2p is broken for multielectron atoms, see chapter 4). The spin direction of the third electron is irrelevant.

The normalised antisymmetric zero-order wave function is the Slater determinant:

$$\Psi_0^{(0)}(\mathbf{q}_1,\mathbf{q}_2,\mathbf{q}_3) = \frac{1}{\sqrt{6}} \begin{vmatrix} \psi_{1s}(\mathbf{r}_1)\,\zeta_{1+} & \psi_{1s}(\mathbf{r}_2)\,\zeta_{2+} & \psi_{1s}(\mathbf{r}_3)\,\zeta_{3+} \\ \psi_{1s}(\mathbf{r}_1)\,\zeta_{1-} & \psi_{1s}(\mathbf{r}_2)\,\zeta_{2-} & \psi_{1s}(\mathbf{r}_3)\,\zeta_{3-} \\ \psi_{2s}(\mathbf{r}_1)\,\zeta_{1+} & \psi_{2s}(\mathbf{r}_2)\,\zeta_{2+} & \psi_{2s}(\mathbf{r}_3)\,\zeta_{3+} \end{vmatrix} . \tag{2.28}$$

Each term in the superposition (2.28) is a solution, with the same energy, to a separable Schrödinger equation. Therefore, their linear combination is also a solution. An exchange of any two electron labels in (2.28) will change the sign of the function, but nothing else. That is, the determinantal wave function in (2.28) is exchange antisymmetric.

With (2.28) as a starting point, we can directly apply perturbation theory in order to find the energy of the ground state of Li. We will perform that quantitative analysis in section 3.1.3, where we shall calculate the matrix element of the summed perturbation operator $(r_{12}^{-1}+r_{13}^{-1}+r_{23}^{-1})$ for (2.28).

2.3.4 Matrix Elements for Slater Determinants

When we describe multielectron atoms with Slater determinant wave functions, we need to know how to calculate the corresponding matrix elements (see [4]). We investigate integrals involving an operator Q of the type:

$$\begin{aligned} I(Q) &\equiv \langle \mathbb{S}_\Psi \,|\, Q \,|\, \mathbb{S}_\Phi \rangle \\ &\equiv \left(\frac{1}{\sqrt{N!}} \begin{vmatrix} \langle \Psi_1 |_1 & \cdots & \langle \Psi_1 |_N \\ \vdots & \ddots & \vdots \\ \langle \Psi_N |_1 & \cdots & \langle \Psi_N |_N \end{vmatrix} \right) Q \left(\frac{1}{\sqrt{N!}} \begin{vmatrix} | \Phi_1 \rangle_1 & \cdots & | \Phi_1 \rangle_N \\ \vdots & \ddots & \vdots \\ | \Phi_N \rangle_1 & \cdots & | \Phi_N \rangle_N) \end{vmatrix} \right) \\ &= \frac{1}{N!} \int \begin{vmatrix} \Psi_1^*(\mathbf{q}_1) & \cdots & \Psi_1^*(\mathbf{q}_N) \\ \vdots & \ddots & \vdots \\ \Psi_N^*(\mathbf{q}_1) & \cdots & \Psi_N^*(\mathbf{q}_N) \end{vmatrix} Q \begin{vmatrix} \Phi_1(\mathbf{q}_1) & \cdots & \Phi_1(\mathbf{q}_N) \\ \vdots & \ddots & \vdots \\ \Phi_N(\mathbf{q}_1) & \cdots & \Phi_N(\mathbf{q}_N) \end{vmatrix} \mathrm{d}\mathbf{q} . \end{aligned} \tag{2.29}$$

The determinants of state vectors in the middle line are defined as superpositions of convolutions, such as in (2.26). The differential is $\mathbf{dq} \equiv \mathbf{dq_1} \dots \mathbf{dq_N}$. That is, the integration goes over all spatial variables and all spins (in this case, 'integration' over a spin parameter will simply mean summing states with spin-up and spin-down). The functions Ψ_i and Φ_i belong to the same basis, and they are products of spatial functions and spin functions. The diagonal matrix element is the one for which all $\Phi_i = \Psi_i$. We assume all wave functions to be orthonormal. The operator Q could be any operator which is a function of the variables $\mathbf{q}_i$, but we shall limit this study to three special cases, where Q is either a constant, a single-electron operator or an electron pair operator.

Each of the two determinants, when developed, will have $N!$ terms, all being product functions with N components. To begin with, we shall consider two smaller integrals that each contain just one such term for the *first* determinant. The first product term is that for which the Ψ-determinant in (2.29) is replaced by the product of the functions on the principal diagonal:

$$t_a = \frac{1}{N!} \int \left[\Psi_1^*(\mathbf{q}_1)\, \Psi_2^*(\mathbf{q}_2)\, \Psi_3^*(\mathbf{q}_3) \dots \Psi_N^*(\mathbf{q}_N) \right] \\ \times Q \begin{vmatrix} \Phi_1(\mathbf{q}_1) & \dots & \Phi_1(\mathbf{q}_N) \\ \vdots & \ddots & \vdots \\ \Phi_N(\mathbf{q}_1) & \dots & \Phi_N(\mathbf{q}_N) \end{vmatrix} \mathbf{dq}\,. \tag{2.30}$$

Next we take another product term from the first determinant, and we choose the one for which only the coordinates for the first two functions have been interchanged. This must have the opposite sign, so the integral is:

$$t_b = -\frac{1}{N!} \int \left[\Psi_1^*(\mathbf{q}_2)\, \Psi_2^*(\mathbf{q}_1)\, \Psi_3^*(\mathbf{q}_3) \dots \Psi_N^*(\mathbf{q}_N) \right] \\ \times Q \begin{vmatrix} \Phi_1(\mathbf{q}_1) & \dots & \Phi_1(\mathbf{q}_N) \\ \vdots & \ddots & \vdots \\ \Phi_N(\mathbf{q}_1) & \dots & \Phi_N(\mathbf{q}_N) \end{vmatrix} \mathbf{dq}\,. \tag{2.31}$$

For a determinant, interchanging two column changes the sign. This means that we can also write t_b as:

$$t_b = \frac{1}{N!} \int \left[\Psi_1^*(\mathbf{q}_2)\, \Psi_2^*(\mathbf{q}_1)\, \Psi_3^*(\mathbf{q}_3) \dots \Psi_N^*(\mathbf{q}_N) \right] \\ \times Q \begin{vmatrix} \Phi_1(\mathbf{q}_2) & \Phi_1(\mathbf{q}_1) & \dots & \Phi_1(\mathbf{q}_N) \\ \vdots & \vdots & \ddots & \vdots \\ \Phi_N(\mathbf{q}_2) & \Phi_N(\mathbf{q}_1) & \dots & \Phi_N(\mathbf{q}_N) \end{vmatrix} \mathbf{dq}\,, \tag{2.32}$$

where columns 1 and 2 have changed place in the second determinant. If we now compare (2.32) with (2.30), we see that the only difference is that the coordinates

$\mathbf{q}_1$ and $\mathbf{q}_1$ have swapped places, both in the determinant and in the product of Ψ^*'s. Since these are just integration variables, this does not change the values of the integrals, and thus $t_b = t_a$.

The same analysis can be made by taking any two terms of the first determinant — there will be $N!$ such terms and they will all be the same. This means that we can simplify (2.29) considerably:

$$I(Q) = \int \left[\Psi_1^*(\mathbf{q}_1) \ldots \Psi_N^*(\mathbf{q}_N)\right] Q \begin{vmatrix} \Phi_1(\mathbf{q}_1) & \ldots & \Phi_1(\mathbf{q}_N) \\ \vdots & \ddots & \vdots \\ \Phi_N(\mathbf{q}_1) & \ldots & \Phi_N(\mathbf{q}_N) \end{vmatrix} \mathrm{d}\mathbf{q} . \tag{2.33}$$

In this way, we have eliminated not just the first of the two determinants, but also the prefactor $N!$.

If the operator Q is a constant (that is, it does not act on any of the coordinates $\mathbf{q}_i$), the orthogonality of the functions Φ_i and Ψ_i means that $I(Q)$ will only be non-zero if there is one term in the determinant of the Φ's that exactly equals the only term of the Ψ's that has been retained in (2.33). For this eventuality, this diagonal matrix element will exactly equal the constant itself. If no term to the right equals the one to the left, the integral returns zero. This requires that all the functions Ψ_i and Φ_i are orthogonal, but we do limit this discussion to that situation.

If Q is a single-electron operator — for clarity called $Q^{(1)}$ — such as the kinetic energy operator or the Coulomb attraction to the nucleus, we get two possible kinds of non-zero matrix elements. If the functions in the Φ-determinant are exactly the same as those in the Ψ-term, we get the diagonal element:

$$\text{if} \quad \Phi_i = \Psi_i \quad \forall\ i$$

$$I(Q^{(1)}) = \left\langle \mathbb{S}_\Psi \middle| Q^{(1)} \middle| \mathbb{S}_\Phi \right\rangle = \sum_{i=1}^{N} \int \Psi_i^*(\mathbf{q}_i)\, Q^{(1)}\, \Psi_i(\mathbf{q}_i)\, \mathrm{d}\mathbf{q}_i . \tag{2.34}$$

The matrix element will be the sum of all diagonal single-function integrals of the entire basis.

The other option is if one permutation among the Φ's differs from the Ψ-product by just *one* function. An additional requirement is that all other functions remain in the same order. Then we have:

$$\text{if all} \quad \Phi_i = \Psi_i \quad \forall\ i, \ \text{except for}\ i = a$$

$$I(Q^{(1)}) = \left\langle \mathbb{S}_\Psi \middle| Q^{(1)} \middle| \mathbb{S}_\Phi \right\rangle = \int \Psi_a^*(\mathbf{q}_a)\, Q^{(1)}\, \Phi_a(\mathbf{q}_a)\, \mathrm{d}\mathbf{q}_a . \tag{2.35}$$

The total matrix element will here equal that of just *one* single-electron integral.

Finally, we look at the possibility of a two-electron operator — now referred to as $Q^{(2)}$ — such as the electron–electron interaction Hamiltonian for one pair of electrons, $1/r_{ij}$. For the matrix element of such an operator, each partial integral has to run over two electronic wave functions on each side of $Q^{(2)}$, and all integrals have to be taken over the full coordinate and spin spaces of two electrons. This means that maximum *two* functions among Φ's and Ψ's may differ. The diagonal case, where the set of functions on both sides of the operator are the same, is:

$$
\begin{aligned}
&\text{if} \quad \Phi_i = \Psi_i \quad \forall\ i \\
I(Q^{(2)}) &= \left\langle \mathbb{S}_\Psi \middle| Q^{(2)} \middle| \mathbb{S}_\Phi \right\rangle \\
&= \sum_{i=1}^{N} \sum_{j>i}^{N} \left[\iint \Psi_i^*(\mathbf{q}_i)\, \Psi_j^*(\mathbf{q}_j)\, Q^{(2)}\, \Psi_i(\mathbf{q}_i)\, \Psi_j(\mathbf{q}_j)\, \mathrm{d}\mathbf{q}_i \mathrm{d}\mathbf{q}_j \right. \\
&\left. - \iint \Psi_i^*(\mathbf{q}_i)\, \Psi_j^*(\mathbf{q}_j)\, Q^{(2)}\, \Psi_i(\mathbf{q}_j)\, \Psi_j(\mathbf{q}_i)\, \mathrm{d}\mathbf{q}_i \mathrm{d}\mathbf{q}_j \right] .
\end{aligned} \tag{2.36}
$$

The integration variables above are just that, and the indices do not really have a meaning. The integrals just have to be taken over two sets of complete electron coordinate spaces. Equation (2.36) will be indispensable when we have to calculate expectation values of electron–electron interaction Hamiltonians in chapters 7 and 8.

For the two-electron operator, we have two non-diagonal possibilities, because now one or two electron coordinates may be different. The former of these eventualities gives the result:

$$
\begin{aligned}
&\text{if all} \quad \Phi_i = \Psi_i \quad \forall\ i, \ \text{except for } i = a \\
I(Q^{(2)}) &= \left\langle \mathbb{S}_\Psi \middle| Q^{(2)} \middle| \mathbb{S}_\Phi \right\rangle \\
&= \sum_{i \neq a}^{N} \left[\iint \Psi_a^*(\mathbf{q}_a)\, \Psi_i^*(\mathbf{q}_i)\, Q^{(2)}\, \Phi_a(\mathbf{q}_a)\, \Psi_i(\mathbf{q}_i)\, \mathrm{d}\mathbf{q}_i \mathrm{d}\mathbf{q}_a \right. \\
&\left. - \iint \Psi_a^*(\mathbf{q}_a)\, \Psi_i^*(\mathbf{q}_i)\, Q^{(2)}\, \Psi_i(\mathbf{q}_a)\, \Phi_a(\mathbf{q}_i)\, \mathrm{d}\mathbf{q}_i \mathrm{d}\mathbf{q}_a \right] .
\end{aligned} \tag{2.37}
$$

For the latter circumstance we have:

$$
\begin{aligned}
&\text{if all} \quad \Phi_i = \Psi_i \quad \forall\ i, \ \text{except for } i = a \ \text{and for } i = b \\
I(Q^{(2)}) &= \left\langle \mathbb{S}_\Psi \middle| Q^{(2)} \middle| \mathbb{S}_\Phi \right\rangle \\
&= \iint \Psi_a^*(\mathbf{q}_a)\, \Psi_b^*(\mathbf{q}_b)\, Q^{(2)}\, \Phi_a(\mathbf{q}_a)\, \Phi_b(\mathbf{q}_b)\, \mathrm{d}\mathbf{q}_a \mathrm{d}\mathbf{q}_b \\
&- \iint \Psi_a^*(\mathbf{q}_a)\, \Psi_b^*(\mathbf{q}_b)\, Q^{(2)}\, \Phi_b(\mathbf{q}_a)\, \Phi_a(\mathbf{q}_b)\, \mathrm{d}\mathbf{q}_a \mathrm{d}\mathbf{q}_b .
\end{aligned} \tag{2.38}
$$

2.4 Symmetry in Atomic Structure

The Schrödinger equation in (2.8) is general. We can use it for any atomic system. We have seen, however, that the exact formulation is of limited use since exact solutions will be wanting. We will attempt to form product states as a zeroth order approximations in the continuation of this book, and we have already begun that treatment in this chapter. However, in a strict meaning, it will *always* be wrong to assign certain different labels for each individual electron. The correct wave function is an entangled state, where each electron in the atom has a finite probability to be at any point in space where the probability density does not have a node.

Besides the entanglement, the presence of two or more electrons — identical fermions — means that the multielectron wave function has to be antisymmetric with regard to an exchange of any two electron parameters. For many of the upcoming approximative approaches, we have to begin by a trial solution, or a zero-order solution, that has the form of a Slater determinant, as in (2.27).

We have initiated a perturbative calculation of energies in this chapter, and we will carry it further to quantitative results in chapter 3 for a few examples. Even in the simplest approximation, where the entire exchange term in (2.8) is treated as a perturbation, the exchange degeneracy has the consequence that we have to take antisymmetric linear combinations of degenerate solutions to the zero-order Hamiltonian as the starting point for the calculation. Another way to stress the importance of the symmetry properties is that symmetry imposes constraints on how different spins can be organised and how they combine with spatial functions. In the next chapter, this will be shown to produce considerable energy shifts, even when the spin is left out of the Hamiltonian.

Further Reading

The theory of atomic spectra, by Condon & Shortley [5]
Quantum theory of atomic structure, by Slater [4]
Introduction to Modern Statistical Mechanics, by Chandler [1]
Physics of Atoms and Molecules, by Bransden & Joachain [6]

References

1. D. Chandler, *Introduction to Modern Statistical Mechanics* (Oxford University Press, New York, 1987)
2. E.D. Commins, *Quantum mechanics: an experimentalist's approach* (Cambridge University Press, New York, 2014)
3. J.C. Slater, Phys. Rev. **34**, 1293 (1929)
4. J.C. Slater, *Quantum theory of atomic structure* (McGraw-Hill, New York, 1960)
5. E.U. Condon, G.H. Shortley, *The theory of atomic spectra* (Cambridge University Press, Cambridge, 1935)
6. B.H. Bransden, C.J. Joachain, *Physics of Atoms and Molecules*, 2nd edn. (Prentice Hall, Harlow, England, 2003)

Chapter 3
Approximation Methods

Since we will not easily find exact analytical solutions to the Schrödinger equation for atoms with more than one electron, good approximative approaches are called for. The most obvious one to try is time-independent perturbation theory, and in chapter 2 we initiated perturbative treatments for the atoms helium and lithium while taking the entire electron–electron interaction term in the Hamiltonian as the perturbation. In this chapter, we continue this and we derive quantitative results. The perturbative approach will be modified and enhanced in chapters 5 and 6, where the interelectronic term is split up in a central part and an angular one — a strategy that works well for almost all atomic systems.

Following the review of the perturbative method, we continue with an application of the variational principle to the smallest multielectron atoms, and we will see that this can give impressively accurate results. Finally, we introduce the Thomas-Fermi model. This is a way to use a general application of Fermi-Dirac statistics in order to construct approximative compound wave functions, which, for example, can be used as starting points in numerical calculations. We continue to postpone relativistic effects to later chapters.

The objective of this chapter is not so much to find good quantitative agreements between models and empirical results, but rather to gain an improved qualitative understanding of atomic structure. The insights gained will then be applied in the subsequent chapters.

3.1 Perturbative Treatment of the Exchange Term

The strategy in this section is to start by forming zero-order wave functions from a simplified Hamiltonian, without the electron–electron interaction contributions:

$$H^{(0)} = -\sum_{i=1}^{N} \frac{1}{2}\nabla_{r_i}^2 - \sum_{i=1}^{N} \frac{Z}{r_i} \,. \tag{3.1}$$

A. Kastberg, *Structure of Multielectron Atoms*, Springer Series on Atomic, Optical, and Plasma Physics 112, https://doi.org/10.1007/978-3-030-36420-5_3

This provides product states of single-electron orbitals and zero-order energies. To these we then apply the perturbation:

$$H_{ee} = \sum_{i=1}^{N} \sum_{j>i}^{N} \frac{1}{r_{ij}} . \tag{3.2}$$

This treatment was already begun in chapter 2, but here we derive quantitative results for energy eigenvalues, and these will be compared to experimental data. That will tell us something about the quality of the approximation, and before that is done, there is no need to calculate the functional form of the perturbed wave functions. We limit the study to first-order perturbation theory and to two- and three-electron atoms.

3.1.1 The Energy of the Helium Ground State

For helium, the lowest energy zero-order solution is the product state in (2.16). We do not have to include the spin in the calculation of the energy since we still ignore the contribution of this to the Hamiltonian. As was seen in chapter 2, the two possible orientations of each spin do, however, allow us to have two electrons in the 1s orbital and still be consistent with the Pauli principle.

The zero-order energy, in atomic units, is given by the hydrogenic solution in (1.24), with $Z=2$:

$$E_0^{(0)'} = 2E_{1s} = -\frac{Z^2}{n^2} = -4\,E_h . \tag{3.3}$$

The reason for the prime in (3.3) is that we have not yet taken into account that the definition of zero energy might need to be modified for He. This will be further explained shortly.

The first-order perturbative contribution, coming from the Coulomb repulsion between the electrons, can be calculated from:

$$\begin{aligned} \Delta E^{(1)} &= \iint \psi_{1s}^*(\mathbf{r}_1)\,\psi_{1s}^*(\mathbf{r}_2)\,\frac{1}{r_{12}}\,\psi_{1s}(\mathbf{r}_1)\,\psi_{1s}(\mathbf{r}_2)\,d\mathbf{r}_1 d\mathbf{r}_2 \\ &= \iint |\psi_{1s}(\mathbf{r}_1)|^2\,|\psi_{1s}(\mathbf{r}_2)|^2\,\frac{1}{r_{12}}\,d\mathbf{r}_1 d\mathbf{r}_2 \equiv J_{1s:1s} . \end{aligned} \tag{3.4}$$

At the end of (3.4), we have defined the *Coulomb integral*, $J_{n_i l_i : n_j l_j}$, in the electron–electron interaction. In this specific example, it pertains to the He ground state with the two electrons both in the 1s orbital.

The Coulomb integral has a straightforward physical interpretation. The square moduli of the wave functions are the probability densities of the respective electrons, and when multiplied by $-e$ this will be the same as the charge distribution. A charge in the volume $d\mathbf{r}_1$ multiplied by that in $d\mathbf{r}_2$, divided by the distance between the charges r_{12}, is the classical expression for the Coulomb repulsion between two charge clouds. After integration over the six spatial coordinates, the resulting value of $J_{n_il_i:n_jl_j}$ will correspond to the total classical energy contribution due to the electrostatic repulsion between the electrons in orbitals n_il_i and n_jl_j. Since the interaction is repulsive, we would expect this addition to the total energy to be positive (a decrease of the binding energy).

In appendix D, we show how such electron overlap integrals can be computed analytically, and we give some examples. From table D.2 we find that for $J_{1\mathrm{s}:1\mathrm{s}}$, the energy contribution is $5Z/8$. For neutral He, this means an energy shift of $1.25\,E_\mathrm{h}$, corresponding to approximately 34.0 eV. The adjusted first-order value of the ground state energy then becomes:

$$E_0^{(1)'} = E_0^{(0)'} + \Delta E^{(1)} = -2.75\,E_\mathrm{h} \approx -74.8\,\mathrm{eV}\ . \tag{3.5}$$

In order to relate the energy in (3.5) to an experimental value, we have to be careful with what we define as zero energy. The way we have formulated the problem, the energy -74.8 eV corresponds to the amount needed in order to remove *both* electrons from a ground state helium atom, and then end up with a bare nucleus. For a comparison with a tabulated value, we must keep in mind that the standard way to define the zero energy for an atom is the *first ionisation limit*. This is the energy needed to remove *one* (typically the most loosely bound) electron.

This means that the value calculated above is a measure of the ionisation energy for neutral He *plus* the energy needed to go from He^+ to He^{2+}. The latter we can compute with the one-electron formula (1.24), and it will be just half of the energy in (3.3), that is, $2\,E_\mathrm{h}$. Our adjusted value of the He ground state energy is thus:

$$E_0^{(1)} = -0.75\,E_\mathrm{h} \approx -20.4\,\mathrm{eV}\ . \tag{3.6}$$

The experimentally measured number for the He ionisation energy is about 24.6 eV [1]. The discrepancy is smaller than it appears, due to the fact that we have redefined the energy scale. If we keep the zero energy defined as it is for the Schrödinger equation in (2.7), the calculated value differs from the measured one by about 5%. For two-electron ions, the error decreases with increasing Z, and it is really bad only for H^- (see, for example, [2] or [3]).

Besides giving us the correct order of magnitude for the He ground state energy, this calculation has shown us how the overlap of the charge clouds can enter into calculations, and we have introduced the Coulomb integral. We could continue to calculate also the perturbed wave functions in the first order, but that is of limited use. A direct second-order calculation of the energy can be made, but the improvement is small, while it gets mathematically heavy. Even with higher order calculations, the convergence is slow, and the calculations typically have to be combined with the variational method (see [4]).

3.1.2 Energies of the Lowest Excited States in Helium

We now turn to a perturbative calculation of the energy of the first few excited states in helium. Again we will take as a zero-order wave function a product state of single-electron orbitals. For the first excited state, this means that one of the electrons is left in the 1s-orbital, while the other is excited to $n=2$. In contrast to the analysis of the ground state, we are now faced with a number of different degeneracies.

To begin with, we have the exchange degeneracy, as was explained in chapter 2. Pure product states will not do, but instead, we have to form a basis that consists of superposition states, as in (2.21):

$$\psi^{(0)}_{1s2l_+}(\mathbf{r}_1,\mathbf{r}_2) = \frac{1}{\sqrt{2}}\left[\psi_{1s}(\mathbf{r}_1)\psi_{2l}(\mathbf{r}_2)+\psi_{2l}(\mathbf{r}_1)\psi_{1s}(\mathbf{r}_2)\right]$$
$$\psi^{(0)}_{1s2l_-}(\mathbf{r}_1,\mathbf{r}_2) = \frac{1}{\sqrt{2}}\left[\psi_{1s}(\mathbf{r}_1)\psi_{2l}(\mathbf{r}_2)-\psi_{2l}(\mathbf{r}_1)\psi_{1s}(\mathbf{r}_2)\right]\,. \tag{3.7}$$

We have substituted the subscripts 'S' and 'A' from (2.21) with '+' and '−'. An additional degeneracy is that for the excited electron, we can have either $l=0$ or $l=1$, which will give identical energies for the uncoupled wave functions. The zero-order energy is:

$$E^{(0)'}_{1s2l} = E_{1s}+E_{2l} = -\frac{5Z^2}{8}\,E_\mathrm{h} = -\frac{5}{2}\,E_\mathrm{h}\,. \tag{3.8}$$

The energy in (3.8) is that needed to dissociate both electrons from the nucleus, in analogy with the calculation for the ground state. However, this time the starting point is that of one electron being in $n=2$.

The functions in (3.7) have been formed in order to account for the exchange degeneracy. If we use these as two different zero-order wave functions, we can proceed with standard first-order perturbation theory in order to calculate the energy offsets. We saw in chapter 2 that spatial wave functions of particular symmetries, such as those in (3.7), must be combined with their proper spin counterparts. We have still not included the spin in the Hamiltonian, but nevertheless, we now see that the way that the spin wave functions are arranged may affect the energy. This is caused by exchange degeneracy and symmetry constraints, and the effect is substantial.

There are additional degeneracies, besides the exchange one. We work with hydrogenic, non-relativistic product states, which means that we have the already mentioned degeneracy in l, and also one in m_l. The single-electron wave functions $\psi_{2s_0}, \psi_{2p_{-1}}, \psi_{2p_0}$ and $\psi_{2p_{-1}}$ (with the value of m_l as the last subscript) all have the same zero-order energies. With these four functions, and the exchange degeneracy in (3.7) also taken into account, it would appear that we need to form a secular equation with an 8×8-matrix. This can, however, be circumvented by considering other symmetry constraints.

Most non-diagonal elements in the complete matrix will involve integrals over the perturbation $1/r_{12}$ and two wave functions of differing l or m_l. If the perturbation is expanded in spherical harmonics (see appendix C.3), one finds that all such integrals will cancel (an effect of $1/r_{12}$ being an even function). As a consequence, instead of one big matrix, we are left with four 2×2 ones along the main diagonal.

We thus apply zero-order functions in the form of (3.7), and with those, we can directly calculate the first-order perturbative corrections for all four exchange degenerate states. The respective energy offsets are:

$$\Delta E^{(1)}_{1s2l_m\pm} = \iint \left[\psi^{(0)}_{1s2l_m\pm}(\mathbf{r}_1,\mathbf{r}_2)\right]^* \frac{1}{r_{12}}\, \psi^{(0)}_{1s2l_m\pm}(\mathbf{r}_1,\mathbf{r}_2)\, \mathrm{d}\mathbf{r}_1 \mathrm{d}\mathbf{r}_2 \,. \tag{3.9}$$

When we expand this, we get two types of integrals:

$$\begin{aligned} J_{1s:2l_m} &\equiv \iint |\psi_{1s}(\mathbf{r}_1)|^2 \frac{1}{r_{12}} |\psi_{2l_m}(\mathbf{r}_2)|^2\, \mathrm{d}\mathbf{r}_1 \mathrm{d}\mathbf{r}_2 \\ K_{1s:2l_m} &\equiv \iint \psi^*_{1s}(\mathbf{r}_1)\, \psi^*_{2l_m}(\mathbf{r}_2) \frac{1}{r_{12}}\, \psi_{2l_m}(\mathbf{r}_1)\, \psi_{1s}(\mathbf{r}_2)\, \mathrm{d}\mathbf{r}_1 \mathrm{d}\mathbf{r}_2 \,. \end{aligned} \tag{3.10}$$

The integral $J_{1s:2l_m}$ in (3.10) is the same Coulomb integral as was introduced in section 3.1.1. It describes the electrostatic repulsion between the two charge clouds, and its energy contribution will always be positive. The other integral, $K_{1s:2l_m}$, is called the *exchange integral*. This is a quantum interference effect that has no classical analogue, and its contribution to the energy can be either positive or negative.

For each of the four possible combinations of values for l and m_l for the second electron, we get two energies:

$$E^{(1)'}_{1s2l_m} = E^{(0)'}_{1s2l_m} + \Delta E^{(1)}_{1s2l_m\pm} = E^{(0)'}_{1s2l_m} + J_{1s:2l_m} \pm K_{1s:2l_m} \,, \tag{3.11}$$

where we have not yet reset the energy scale as in section 3.1.1. Solutions to the Coulomb and exchange integrals are given in appendix D, in table D.2. For the present example, we find that the degeneracy in m_l is retained, and therefore we will drop the m_l-subscripts for the integrals $J_{n_1l_1:n_2l_2}$ and $K_{n_1l_1:n_2l_2}$. We end up with two different energies for the 1s2s configuration, and two others that are the same for all 1s2p states. In atomic units, the values of the integrals are (from table D.2 and with $Z=2$)

$$\begin{aligned} J_{1s:2s} &= \frac{34}{81}\, E_h \\ K_{1s:2s} &= \frac{32}{729}\, E_h \\ J_{1s:2p} &= \frac{118}{243}\, E_h \\ K_{1s:2p} &= \frac{224}{6561}\, E_h \,. \end{aligned} \tag{3.12}$$

We note from table D.2 that in contrast to the zero-order energy, these energy contributions are linear in Z.

The energy levels are shown in table 3.1 and figure 3.1, together with empirical data for comparison. Note that the values in the table and in the figure are referenced to a zero-point energy corresponding to both electrons being at an infinite radial distance, leaving behind a bare nucleus.

Table 3.1 Energies of low lying excited states in neutral He, as calculated by first-order perturbation theory, compared with experimental values from [5]. The tabled values refer to the double ionisation energies, as in figure 3.1.

Energy level	First-order correction	Calculated energy	Experimental value
E_{1s2s_-}	$J_{1s:2s} - K_{1s:2s}$	-57.8 eV	-59.19 eV
E_{1s2s_+}	$J_{1s:2s} + K_{1s:2s}$	-55.4 eV	-58.39 eV
E_{1s2p_-}	$J_{1s:2p} - K_{1s:2p}$	-55.7 eV	-58.04 eV
E_{1s2p_+}	$J_{1s:2p} + K_{1s:2p}$	-53.9 eV	-57.79 eV

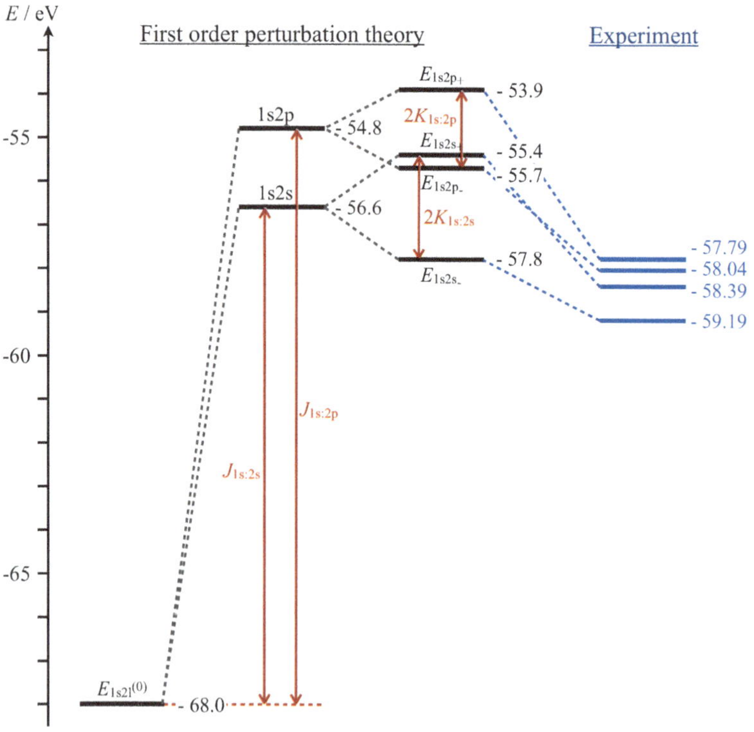

Fig. 3.1 Energies of the levels in the 1s2s and 1s2p configurations of neutral He, calculated with first-order perturbation theory, and compared with experimentally measured values [5]. Zero energy is taken as that for both electrons being removed to infinity, as in table 3.1.

The quantitative agreement with empirical data is similar to what it is for the ground state. An important error in the calculation that shows up for the excited state, however, is that the energy of the antisymmetric 1s2p-state is below that of the symmetric 1s2s-state. This is contrary to experimental results and the involved level crossing can hardly be explained with physical arguments. This is indeed an artefact of the approximations involved.

From a qualitative point of view, there are important lessons to be learned from this calculation. One is that due to the exchange degeneracy alone, energies of 1s*nl*-states split. The resulting levels correspond to spatial (and spin) eigenstates of different symmetries. Furthermore, the degeneracy in l is removed, and experimentally (as well as in higher order perturbative calculations) the splitting between 1s2s and 1s2p is greater than that between symmetric and antisymmetric spatial wave functions [6]. These observations can give us some physical insight.

The fact that we have a higher energy for the 1s2p states than for 1s2s is a phenomenon that will hold for essentially all multielectron atoms. Valence orbitals with higher l are more loosely bound. The reason for this is that orbitals with lower l will be less affected by the centrifugal barrier introduced in section 1.2. This means that they have a greater probability of penetrating the charge screen provided by inner electrons and feel a greater attractive force by the nucleus. Thus, such an electron will be more tightly bound and have a lower energy.

The difference in energy between symmetric and antisymmetric spatial states is more subtle but is likewise found for most atomic states. To explain this, we can first recall some results from section 2.2. There, we showed that in order to have a total wave function which is antisymmetric — a requirement for fermions — we must combine a symmetric spatial function with an antisymmetric spin function and vice versa. Keep in mind also that the exchange term in the Hamiltonian, $1/r_{12}$, gives a positive energy contribution.

The spin singlet in (2.15), which is antisymmetric, only has terms with the two electron spins pointing in opposite directions. This means that these electrons can be close to each other without violating the Pauli principle. As a consequence, r_{12} may be small and the exchange term large. Since the spin singlet has to be combined with a symmetric spatial wave function, the latter will have a higher energy.

This qualitative argument can be improved. A greater energy resulting from exchange degeneracy must indicate a larger contribution from the $1/r_{12}$ term. The symmetric spatial function in (3.7) will not cancel for $\mathbf{r}_2 = \mathbf{r}_1$, whereas the antisymmetric one will. The conclusion is the same as for the less stringent argument in the preceding paragraph: the antisymmetric spatial function, which is combined with a spin triplet in the total wave function, will have a lower energy.

3.1.3 The Lithium Ground State

We will end this perturbation theory analysis by pushing it one step further — to a calculation for the ground state of neutral Li. The purpose is yet again more to learn general qualitative lessons, rather than to gain precise numerical results, and we take up the thread where we left it in section 2.3.3.

The electron configuration for the lowest energy state is $1s^2 2s$, and the zero-order energy is:

$$E_0^{(0)} = -Z^2 \sum_{i=1}^{3} \frac{1}{2n_i^2} = -\frac{81}{8} E_h \approx -10.1\, E_h\,. \tag{3.13}$$

The zero-order antisymmetric ground state wave function is the Slater determinant in (2.28), and we can calculate the perturbation from:

$$\Delta E^{(1)} = \int \left|\Psi^{(0)}(\mathbf{q})\right|^2 \left(\frac{1}{r_{12}} + \frac{1}{r_{13}} + \frac{1}{r_{23}}\right) d\mathbf{q}\,, \tag{3.14}$$

with the integral taken over all nine electron coordinates, and over all the spins.

For a solution of this integral, we could apply (2.36) for the diagonal matrix element of the sum of the $1/r_{ij}$ operators. However, we would then have to take into account a range of symmetry considerations that we have deferred to chapter 7 and appendix D. Since a three-dimensional system is still quite manageable, we instead opt for developing the determinant and factorising out the spin functions.

We write:

$$\Psi^{(0)}(\mathbf{q}) = \frac{1}{\sqrt{6}} \left[\psi_a(\mathbf{r})\, \zeta_{1+}\zeta_{2-}\zeta_{3+} + \psi_b(\mathbf{r})\, \zeta_{1-}\zeta_{2+}\zeta_{3+} + \psi_c(\mathbf{r})\, \zeta_{1+}\zeta_{2+}\zeta_{3-}\right], \tag{3.15}$$

with the spatial functions defined as:

$$\begin{cases} \psi_a(\mathbf{r}) \equiv [\psi_{1s}(\mathbf{r}_1)\psi_{1s}(\mathbf{r}_2)\psi_{2s}(\mathbf{r}_3) - \psi_{2s}(\mathbf{r}_1)\psi_{1s}(\mathbf{r}_2)\psi_{1s}(\mathbf{r}_3)] \\ \psi_b(\mathbf{r}) \equiv [\psi_{1s}(\mathbf{r}_1)\psi_{2s}(\mathbf{r}_2)\psi_{1s}(\mathbf{r}_3) - \psi_{1s}(\mathbf{r}_1)\psi_{1s}(\mathbf{r}_2)\psi_{2s}(\mathbf{r}_3)] \\ \psi_c(\mathbf{r}) \equiv [\psi_{2s}(\mathbf{r}_1)\psi_{1s}(\mathbf{r}_2)\psi_{1s}(\mathbf{r}_3) - \psi_{1s}(\mathbf{r}_1)\psi_{2s}(\mathbf{r}_2)\psi_{1s}(\mathbf{r}_3)] \end{cases}. \tag{3.16}$$

This leaves us with a vast number of integrals when (3.14) is developed, but since the Hamiltonian is spin independent, and the spin wave functions are orthonormal, many of the integrals vanish. The perturbative energy correction can be simplified as:

$$\Delta E^{(1)} = \int \left(|\psi_a(\mathbf{r})|^2 + |\psi_b(\mathbf{r})|^2 + |\psi_c(\mathbf{r})|^2\right) \left(\frac{1}{r_{12}} + \frac{1}{r_{13}} + \frac{1}{r_{23}}\right) d\mathbf{r}\,. \tag{3.17}$$

At this point, we can take the hydrogenic solutions for the wave functions, and the result is a number of integrals of the same type as in section 3.1.2, calculated in appendix D. Many of the integrals return zero, due to the orthogonality between ψ_{1s} and ψ_{2s}, and all the others will be two-electron Coulomb and exchange integrals. The latter will be of exactly the same form as the ones we calculated for the He excited states, and even the numerical values will be the same, apart from the nuclear charge factor Z. The energy shift reduces to:

$$\begin{aligned} \Delta E^{(1)} &= J_{1s:1s} + 2J_{1s:2s} - K_{1s:2s} \\ &= Z\left(\frac{5}{8} + \frac{34}{81} - \frac{16}{729}\right) E_h = \frac{5965}{1944} E_h \approx 3.07\, E_h\,, \end{aligned} \tag{3.18}$$

with numerical values taken from table D.2, and using $Z = 3$.

The three contributions in (3.18) can be readily understood. The Coulomb integrals are the positive energy contribution from the overlap of the three charge clouds. There are two such 1s–2s interactions and one 1s–1s. There is only one term with an exchange integral. This is because the exchange integral terms cancel for electrons of different spins (see chapter 7).

The total energy for the Li ground state in this first-order perturbative calculation is:

$$E_0^{(1)\prime} = E_0^{(0)} + \Delta E^{(1)} = -\frac{6859}{972} E_\mathrm{h} \approx -7.06\, E_\mathrm{h} \,, \tag{3.19}$$

or about 192 eV. As before, this is the total binding energy of all three electrons. The experimental number for this is about 203 eV (with values from [7, 8] and [9]). The accuracy is rather decent, considering the approximations made. More important is that we have revealed some qualitative facts that we will be able to use further on in the book.

3.2 Atomic Wave Functions via the Variational Principle

In the following attempt to improve on the approximate results reached in the previous section, we will begin with an informed guess about the form of the ground state wave function, involving one or more adjustable parameters. With this, we will calculate the corresponding expectation value of the full Hamiltonian. If we find a global minimum for this, when we vary the wave function parameters, we will have found an approximate functional form for the ground state. This is the essence of Rayleigh-Ritz variational principle (see, for example, [10]), which we will here apply to small atomic systems. The results for the wave function, and the minimised energy, will still be approximations, but ones that give a better agreement between predicted quantities and experimental results than the perturbation calculation in section 3.1.

In section 3.2.1 we calculate the approximate ground state energy of helium, using the most rudimentary version of the variational principle. In section 3.2.2, we expand the survey to modern versions of the method, including also excited states. Finally, we briefly discuss larger atoms than He.

3.2.1 Basic Atomic Variational Analyses

The variational method, as applied to atoms, is described in many general textbooks on quantum mechanics or atomic physics, see, for example, [11] for an introduction. In [12], contemporary variational techniques for atomic calculations are reviewed. The method is applied for physical systems for which the Hamiltonian is known, while it has a form that makes the Schrödinger equation difficult or impossible to solve analytically. The basic idea is to formulate a trial function, $\psi(\alpha)$, as a function of one or more introduced variable parameters. This function should be normalised and must be bound for all space. It should also be of a form that allows a calculation of the expectation value of the Hamiltonian.

The expectation value integral will always yield a value that is greater than, or equal to, the ground state energy E_0:

$$\frac{\langle\, \psi(\alpha)\,|H|\,\psi(\alpha)\,\rangle}{\langle\, \psi(\alpha)\,|\,\psi(\alpha)\,\rangle} \geq E_0\,. \tag{3.20}$$

If the quantity to the left in (3.20) is minimised while varying the parameter α, an approximation to the ground state will have been found. The equality option in (3.20) will only be in play when the trial function exactly equals the ground state wave function, $\psi(\alpha)=\psi_0$. The quality of the approximation will depend strongly on the choice of trial function and its parameterisation.

When we treated the He ground state with perturbation theory in section 3.1.1, we stopped short of calculating the wave function. However, the zero-order ground state used in (3.4), taken from (1.16), is the spherically symmetric function:

$$\psi^{(0)}_{1\mathrm{s}^2}(\mathbf{r}_1,\mathbf{r}_2) = \psi_{1\mathrm{s}}(\mathbf{r}_1)\,\psi_{1\mathrm{s}}(\mathbf{r}_2) = \frac{Z^3}{\pi}\,\mathrm{e}^{-Z(r_1+r_2)}\,. \tag{3.21}$$

This is a wave function for an atom with two independent electrons, which both feel the full Coulomb attraction from the nucleus of charge Z while not interacting with each other.

We know from the preceding section, and by reasoning, that on average the effect of the interaction term is that the electrons will feel a smaller net nuclear attraction. Therefore, we begin the variational analysis with a normalised trial solution, which is a central Coulomb potential with an effective charge $Z_{\mathrm{eff}} < Z$:

$$\psi^{(Z_{\mathrm{eff}})}_{1\mathrm{s}^2}(\mathbf{r}_1,\mathbf{r}_2) = \frac{Z^3_{\mathrm{eff}}}{\pi}\,\mathrm{e}^{-Z_{\mathrm{eff}}(r_1+r_2)}\,, \tag{3.22}$$

and we are going to let Z_{eff} be a variable parameter.

The Hamiltonian is the one from (2.7), and the expectation value that we need to compute is:

$$\langle E\rangle^{(Z_{\mathrm{eff}})} = \left\langle \psi^{(Z_{\mathrm{eff}})}_{1\mathrm{s}^2}(\mathbf{r}_1,\mathbf{r}_2) \middle| -\frac{\nabla^2_{r_1}}{2} - \frac{\nabla^2_{r_2}}{2} - \frac{Z}{r_1} - \frac{Z}{r_2} + \frac{1}{r_{12}} \middle| \psi^{(Z_{\mathrm{eff}})}_{1\mathrm{s}^2}(\mathbf{r}_1,\mathbf{r}_2) \right\rangle\,. \tag{3.23}$$

The mean value of the electron–electron interaction term is identical with the Coulomb integral formulated in (3.4), and it is calculated analytically in appendix D, *except* that for neutral He the wave functions for the 1s-orbitals should here have $Z_{\mathrm{eff}} < 2$ instead of $Z=2$. The four first terms only include one electron at a time. They can be found analytically using (1.16), and some integrals of this type are given in appendix B.3.

The expectation values of the five terms, for the trial ground state function, are:

$$\left\langle -\frac{\nabla_{r_1}^2}{2} \right\rangle = \left\langle -\frac{\nabla_{r_2}^2}{2} \right\rangle = \frac{Z_{\mathrm{eff}}^2}{2}$$
$$\left\langle -\frac{Z}{r_1} \right\rangle = \left\langle -\frac{Z}{r_2} \right\rangle = -Z Z_{\mathrm{eff}}$$
$$\left\langle \frac{1}{r_{12}} \right\rangle = \frac{5 Z_{\mathrm{eff}}}{8} \; , \tag{3.24}$$

and the expression for the total mean energy becomes:

$$\langle E \rangle^{(Z_{\mathrm{eff}})} = Z_{\mathrm{eff}}^2 - 2 Z Z_{\mathrm{eff}} + \frac{5 Z_{\mathrm{eff}}}{8} \; . \tag{3.25}$$

The next step is to minimise (3.25) while varying Z_{eff}. This will yield an approximative functional form for the ground state wave function, and also a corresponding Z_{eff}. The result is:

$$Z_{\mathrm{eff}} = Z - \frac{5}{16} \; . \tag{3.26}$$

The energy is the minimised value:

$$\langle E \rangle_{\mathrm{min}} (Z_{\mathrm{eff}}) = -\left(Z - \frac{5}{16} \right)^2 E_{\mathrm{h}} \; . \tag{3.27}$$

For neutral He, the corresponding wave function is:

$$\psi_{1s^2}^{(Z_{\mathrm{eff}})}(\mathbf{r}_1, \mathbf{r}_2) = \frac{19683}{4096\,\pi} \, \mathrm{e}^{-\frac{27}{16}(r_1 + r_2)} \; . \tag{3.28}$$

As for the perturbative calculation in section 3.1.1, the energy in (3.27) corresponds to the one necessary for removing both electrons from a ground state two-electron atom. If we subtract the energy $-Z^2/2$ needed to remove the second electron, when one is already gone, and use $Z = 2$, the ionisation energy for the helium ground state is then found to be about -23.1 eV.

This is considerably closer to the experimental value (-24.6 eV [1]) than the one achieved with first-order perturbation theory. It is remarkably good considering the simplicity of the calculation, and that it is non-relativistic. This calculation also gives a qualitative understanding of the effective screening of the nuclear charge from one electron by the other. For a two-electron atomic system in the ground state, we have found a screening constant of $Z - Z_{\mathrm{eff}} \approx 5/16$.

3.2.2 High Precision Variational Computations

The analysis done in the preceding section can be made much more sophisticated. Variational calculation on small atoms is an active research subject and with the latest results, the wave function for He is now essentially exactly known (in a non-relativistic limit) [12, 13].

A two-electron atomic trial function can be written as a serial expansion of a chosen basis set:

$$\psi_{\text{tr}}^{(c_\mu)}(\mathbf{r}_1,\mathbf{r}_2) = \sum_{\mu=1}^{N} c_\mu \, \phi_\mu(\mathbf{r}_1,\mathbf{r}_2) \, . \tag{3.29}$$

The form of the basis functions ϕ_μ can, for example, be defined as:

$$\phi_\mu^{(\alpha,\beta)}(s,t,u) = s^i \, t^j \, u^k \, \mathrm{e}^{-\alpha r_1 - \beta r_2} \, . \tag{3.30}$$

This includes the spatial parameters:

$$\begin{aligned} s &\equiv r_1 + r_2 \\ t &\equiv r_1 - r_2 \\ u &\equiv r_{12} \, , \end{aligned} \tag{3.31}$$

known as *Hylleraas parameters* [14]. The indices μ should represent a set of the three integer powers $\{i, j, k\}$. N is the number of polynomial terms included in the calculation, with more terms giving better accuracy. The physical roles of α and β are similar to the effective charge Z_{eff}, used in the preceding section.

With the Hamiltonian (2.7), the variational analysis is now:

$$\frac{\partial}{\partial c_\mu} \left\langle \psi_{\text{tr}}^{(c_\mu)} \, | H | \, \psi_{\text{tr}}^{(c_\mu)} \right\rangle = 0 \, . \tag{3.32}$$

For μ ranging from one to N, this constitutes a secular equation of dimension N, wherein each eigenvalue ε_μ will be an approximation (or more accurately an upper bound) to a He energy level.

This method will give the energy for the ground state, E_1, but also excited states. The more orders that are included in the expansion in (3.29), the more eigenenergies are obtained and the better is the accuracy. The proof of this is known as the Hylleraas-Undheim-MacDonald theorem and for details about this, we refer to [12]. A diagram illustrating the progression of improved accuracy and increased numbers of excited state energies is shown in figure 3.2.

Modern variational calculations give vastly better approximations to the level energies than does an analysis where the exchange term is taken as a perturbation (see, for example, [15] or [12]). In a non-relativistic limit, determined energy levels for He can be regarded as exact, at least up to principal quantum number $n = 10$.

3.2.3 Lithium and Larger Atoms

Also for atoms with three or more electrons, there exist calculations that use the variational principle to calculate approximate wave functions, in particular for the

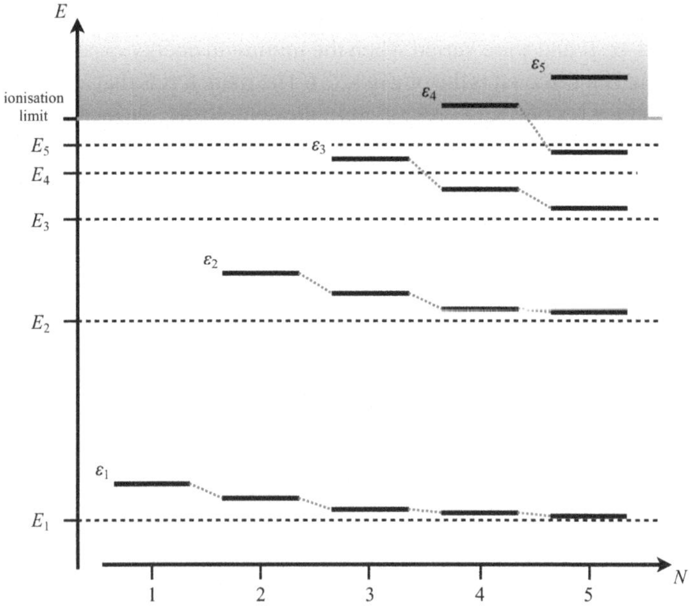

Fig. 3.2 Illustration of the progressively improved accuracy in variational approximations of energy levels, with increased size of the basis set (the figure is redrawn from [12]). Every eigenvalue found by a solution to (3.32) will give an upper bound to a true energy level. As more powers are added in the expansion (3.29), rows and columns are added to the Hamiltonian matrix in (3.32). This means that further energy levels are obtained, and the deviations from the true energies move asymptotically towards zero.

ground state, but also for excited states. The larger the atom becomes, the more cumbersome the calculation becomes and the poorer the result.

One main complication with such calculations is to properly account for the correlation between electrons. In order to reduce errors emanating from this, one typically introduces wave functions containing interelectronic separations, in the spirit of the Hylleraas method (see section 3.2.2 and (3.31)). This has proven to give a faster convergence and more accurate results, but a very large number of terms have to be included in order to achieve acceptable wave functions.

One example of how this can be done for a three-electron atom, see [15], is to make an expansion in a Hylleraas basis, as in (3.29):

$$\psi_{\text{tr}}^{(c_\mu)}(\mathbf{r}_1,\mathbf{r}_2,\mathbf{r}_3) = \sum_{\mu=1}^{N} c_\mu\, \phi_\mu(\mathbf{r}_1,\mathbf{r}_2,\mathbf{r}_3) \ , \tag{3.33}$$

with the basis functions:

$$\phi_\mu^{(\alpha,\beta,\gamma)}(\mathbf{r}_1,\mathbf{r}_2,\mathbf{r}_3) = r_1^i\, r_2^j\, r_3^k\, r_{12}^l\, r_{13}^m\, r_{23}^n\ \mathrm{e}^{-\alpha r_1-\beta r_2-\gamma r_3} \ . \tag{3.34}$$

Here, the summation index μ corresponds to all the six indices $\{i, j, k, l, m, n\}$, and the parameters α, β and γ are varied when the minimum energy expectation value is calculated. The Hamiltonian is the one in (2.8). The more terms that are included, the better the accuracy becomes. In order to limit the complexity, various simplifications exists, such as never having more than one interelectronic parameter per term.

The most direct way to assess the quality of a variational calculation is by a comparison with experimental values for energies. The precision of the calculations is such, however, that for such a comparison to be meaningful, relativistic and QED corrections must also be taken into account. The latter is usually done by adding layers of perturbation theory to a good zero-order eigenfunction found by the variational method. It is not always obvious that the energy provides the most sensitive test of the calculation. With good wave functions, expectation values of other quantities could also be calculated. Things that can be computed are, for example, ionisation potentials and the electron density at the nucleus (useful for hyperfine structure analyses). With even more refinements, excited states can be estimated, and with them also transition probabilities, which can be compared to experimental measurements of radiative lifetimes.

As the methods are applied to increasingly heavier atoms, the complexity escalates very fast. A partial exception is Be or ions with four electrons. For such systems, two electron orbitals are filled, and the resulting spherical symmetry enables some simplifications (see section 6.2). There exist in the literature ground state calculations at least up to Ne (with ten electrons) [16].

3.3 The Thomas-Fermi Model

There exist many methods for estimating electronic potentials $V(\mathbf{r}_i)$, which can be used as starting points for further calculations of atomic quantities. One such is the Thomas-Fermi method. This is a statistical model that is generally applicable on ensembles of Fermi-Dirac particles. When it is applied to atomic structural theory, one takes the specific system of a central attractive potential and assumes a spherically symmetric radial potential. The Coulomb repulsion between electrons is initially ignored, and by just using the ordering resulting from symmetry considerations and the Pauli principle, a remarkably good first guess of $V(\mathbf{r}_i)$ can be had.

3.3.1 The Thomas-Fermi Method in Atomic Structure

We assume a small volume, for example a cube of side L, at a radial distance of r from an atomic nucleus, and we consider the electronic charge distribution inside this. Electrons are spin one-half fermions and we assume them to interact electronically with the positively charged nucleus, but not with each other. The volume is chosen small enough for the potential inside to be taken as homogeneous, and since the zero point in energy can be chosen at will, this particular potential energy can be set to zero. On that scale, the sum of the kinetic energies of all electrons inside the volume will account for the latter's entire energy content.

To get a handle on the kinetic energy, we initially consider the cubic volume as having hard walls. This oversimplification will be remedied when we eventually integrate over all real space. In this imaginary small infinite well, the kinetic energy of each of the N electrons inside the volume can be described by three quantum numbers, all integers larger than zero (in SI-units):

$$E_{\text{kin}-i} = \frac{\pi^2 \hbar^2}{2 m_e L^2} \left(n_x^2 + n_y^2 + n_z^2\right) \ . \tag{3.35}$$

In a *state space* defined by n_x, n_y and n_z, a particular energy inside the cubic well, $E(r)$, will be represented by a spherical segment with radius:

$$\sqrt{n_x^2 + n_y^2 + n_z^2} = \left[\frac{2 m_e L^2 E(r)}{\pi^2 \hbar^2}\right]^{1/2} \ . \tag{3.36}$$

We only have positive values for the quantum numbers and thus it will suffice with one octant of the full sphere. If we take $E_{\max}(r)$ to be the maximum kinetic energy possible for an electron to stay confined in the volume (now forgetting the 'hard walls'), and we assume that all quantum states with energy lower than that are occupied by exactly two electrons (one per spin), the number of confined electrons must be:

$$N_{E(r)} = \frac{1}{3\pi^2} \left[\frac{L \sqrt{2 m_e E_{\max}(r)}}{\hbar}\right]^3 \ . \tag{3.37}$$

By multiplying this with $-e$, we get the total charge enclosed in the volume.

Since the actual physical system is electrons in a central attractive potential, this potential must perfectly balance the kinetic energy in (3.37), or else the electrons would not be in bound states. If we quantify the nuclear attraction by the electromagnetic scalar potential $\phi(r)$, the total energy of one electron in the volume is:

$$E_{\text{tot}-i} = E_{\max}(r) - e\,\phi(r) \ . \tag{3.38}$$

This together with (3.37) gives us the following expression for the electronic charge density:

$$\rho_q(r) = -\frac{e}{3\pi^2} \left\{\frac{\sqrt{2 m_e \left[E_{\text{tot}} + e\,\phi(r)\right]}}{\hbar}\right\}^3 \ . \tag{3.39}$$

Equation 3.39 can now be used together with Poisson's equation (see, for example, [17] or [18]) for the potential, and this will result in the differential equation:

$$\nabla^2 \phi(r) = -\frac{\rho_q(r)}{\varepsilon_0} \ . \tag{3.40}$$

We take (3.40) to represent the entire atom, and we consider limiting conditions. If the nuclear charge is Ze, the potential felt by an electron at a small radial distance is:

$$\phi(r_i \to 0) = \frac{Ze}{4\pi\varepsilon_0 r_i} . \tag{3.41}$$

If we assume that the atom has a boundary, with radius R, the net charge of the atom, felt outside this, is $Z - N_e$, with N_e being the number of electrons. This results in the asymptotic potential:

$$\phi(r_i \to \infty) = \frac{(Z - N_e)e}{4\pi\varepsilon_0 r_i} , \tag{3.42}$$

which for a neutral atom is zero.

The expression in (3.40) is the *Thomas-Fermi equation*, except that it is typically presented in a more general form, using a couple of substitutions. We define:

$$x \equiv \frac{4r}{a_0}\left(\frac{2Z}{p\pi^2}\right)^{1/3} , \tag{3.43}$$

and

$$\chi(x) \equiv \frac{4\pi\varepsilon_0}{Ze} r \left[\phi(r) + \frac{E_{\mathrm{tot}}}{e}\right] . \tag{3.44}$$

This leads us to the standard formulation of the *Thomas-Fermi equation*:

$$\frac{\mathrm{d}^2\chi(x)}{\mathrm{d}x^2} = \sqrt{\frac{\chi^3(x)}{x}} . \tag{3.45}$$

A solution to (3.45) gives us the effective electric potential $\phi(r)$, and the potential energy, within the Thomas-Fermi approximation. Equation (3.45) does not have any easy analytical solutions, but it can be integrated numerically. In table 3.2, we present some solutions for a neutral atom.

3.3.2 Results and Extensions

The electric potential that we calculate with the Thomas-Fermi equation corresponds to a time average of the contributions from the nucleus and all electrons. The potential energy going into the Schrödinger equation may then be expressed in a purely radial form, with an effective nuclear charge Z_{eff}, as:

$$V(r) = \frac{e^2 Z_{\mathrm{eff}}(r)}{4\pi\varepsilon_0 r} . \tag{3.46}$$

This assumption is almost the same as the one we will make in connection with the central-field approximation (see chapter 5). The effective charge in (3.46) is the nuclear charge screened by a gradually larger part of the entire electronic cloud.

Table 3.2 Numerical solutions to the Thomas-Fermi equation (3.45) for a neutral atom ($N_e = Z$), see, for example, [18, 19] or [17].

x	$\chi(x)$	x	$\chi(x)$	x	$\chi(x)$	x	$\chi(x)$
0.000	1.000	0.200	0.793	1.2	0.375	3.0	0.157
0.010	0.985	0.300	0.721	1.4	0.333	4.0	0.108
0.020	0.972	0.400	0.660	1.6	0.297	5.0	0.0788
0.030	0.959	0.500	0.607	1.8	0.268	7.5	0.0408
0.040	0.947	0.600	0.562	2.0	0.244	10.0	0.0244
0.050	0.935	0.700	0.521	2.2	0.221	20.0	0.0058
0.060	0.924	0.800	0.485	2.4	0.202	30.0	0.0022
0.080	0.902	0.900	0.453	2.6	0.185	40.0	0.0011
0.100	0.882	1.000	0.425	2.8	0.170	50.0	0.00061

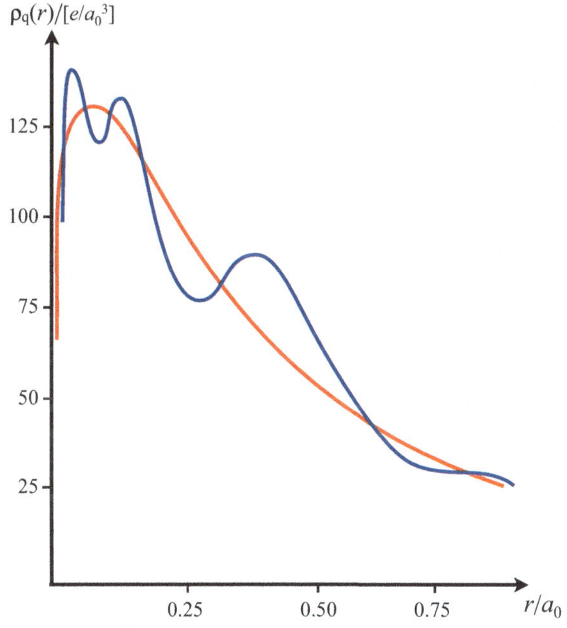

Fig. 3.3 Radial charge density of a neutral Hg atom (red curve), calculated with the Thomas-Fermi model. As a comparison, the blue trace shows a numerical calculation with the Hartree model. [17].

At the immediate vicinity of the nucleus, Z_{eff} equals Z, while for an asymptotically large r it will, for a neutral atom, vanish.

In figure 3.3, we show an example (taken from [17]) of a calculation of the radial charge distribution (derived via the electronic potential energy) for neutral Hg. This is compared to a numerical calculation with the Hartree method (see chapter 14). It shows how the effective nuclear attraction felt by an electron falls off with the radial distance as a consequence of the screening of the remainder of the electronic charge cloud. The agreement is surprisingly good, given the simplicity of the model. One thing that we can note, however, is that the shell structure of the atom is not reproduced.

3.4 Other Approaches — The Need for Approximations in Atomic Structure Theory

Already in chapter 2, it became clear that we will not get very far in a theoretical analysis of atomic structure without approximations. Even with just a second electron added to the atom, the exact Schrödinger equation has solutions that are intrinsically entangled, and the three-body system has no known analytical solutions.

In the present chapter, we have predominantly used perturbation theory and the variational principle in order to calculate wave functions and/or eigenenergies. In doing this, we have gained a lot of useful insights into the nature of atomic structure. At the same time, we have seen that quantitative results diverge considerably from experimental values, except for the lowest energy states of the very smallest multielectron atoms, and even in those cases, the good results come at the expense of great computational complications.

The reasons for the limitations of the above methods are, in the case of the variational principle, that it is a scheme that is good as long as the physical system under study is not too large. Already with rather few electrons, this limit is reached and the number of parameters that need to be varied becomes untenable. Perturbation theory, on the other hand, is a fine method also for very complex systems, but the oversimplification made in section 3.1 was that the *entire* exchange term in the Hamiltonian was taken as the perturbation, and this is too crude. This will be improved upon in chapter 5.

It is also possible to combine perturbation theory and the variational principle in the variation-perturbation method (see [11] or [20]). What this essentially means is that perturbation theory is used, and energy corrections are calculated by taking the matrix elements of the perturbation Hamiltonian with the unperturbed wave functions. When doing this, variational parameters can be added in order to get better perturbed wave functions, or indeed to make it possible to obtain them at all. This has been applied to small multielectron atoms, giving very good results (see, for example, [4]).

There are many more approximation methods, most of which will not be covered in this volume. One can attempt to directly numerically integrate the Schrödinger equation, but it rarely leads to satisfactory results. Very impressive results, however, are obtained with iterative methods such as the Hartree and Hartree–Fock methods. These will be superficially covered in chapter 14, but for more extensive treatments we refer to other works. In section 3.3, we briefly introduced the Thomas-Fermi approach. A relativistic version of the latter is the *Thomas-Fermi-Dirac* model, which gives significantly better results [17]. Yet another approximative method for solving the Schrödinger equation, not covered here, is the *WKB approximation*, named after G. Wentzel, H. A. Kramers and L. Brillouin (see, for example, [10]). This is actually a general method for finding approximative solutions to differential equations.

Further Reading

The theory of atomic spectra, by Condon & Shortley [21]
Quantum theory of atomic structure, by Slater [18]
Quantum mechanics of atomic spectra and atomic structure, by Mizushima [17]
The theory of two-electron atoms: between ground state and complete fragmentation, by Tanner, Richter & Rost [22]
Physics of Atoms and Molecules, by Bransden & Joachain [11]
Springer Handbook of Atomic, Molecular and Optical Physics, by Drake [23]

References

1. D.Z. Kandula, C. Gohle, T.J. Pinkert, W. Ubachs, K.S.E. Eikema, Phys. Rev. Lett. **105**, 063001 (2010)
2. A. Unsöld, Annalen der Physik **387**, 355 (1927)
3. J.C. Slater, Phys. Rev. **32**, 349 (1928)
4. C.W. Scherr, R.E. Knight, Rev. Mod. Phys. **35**, 436 (1963)
5. A. Kramida, Y. Ralchenko, J. Reader, and NIST ASD Team. NIST Atomic Spectra Database (ver. 5.3). [Online]. Available: http://physics.nist.gov/asd (2018). Accessed: 2019-07-14
6. D.C. Morton, Qixue Wu, G.W.F. Drake, Canadian Journal of Physics **84**, 83 (2006)
7. B.A. Bushaw, W. Nörtershäuser, G.W.F. Drake, H.J. Kluge, Phys. Rev. A **75**, 052503 (2007)
8. G.W. Drake, Canadian Journal of Physics **66**, 586 (1988)
9. G.W. Erickson, J. Phys. Chem. Ref. Data **6**, 831 (1977)
10. J.J. Sakurai, J.J. Napolitano, *Modern Quantum Mechanics*, 2nd edn. (Pearson, Harlow, 2010)
11. B.H. Bransden, C.J. Joachain, *Physics of Atoms and Molecules*, 2nd edn. (Prentice Hall, Harlow, England, 2003)
12. G.W.F. Drake, in *Springer Handbook of Atomic, Molecular, and Optical Physics*, ed. by G.W.F. Drake (Springer-Verlag, New York, 2006), p. 199
13. G.W.F. Drake, Zong-Chao Yan, Phys. Rev. A **46**, 2378 (1992)
14. E. Hylleraas, Z. Physik **54**, 347 (1929)
15. Zong-Chao Yan, G.W.F. Drake, Phys. Rev. A **52**, 3711 (1995)
16. D.C. Clary, N.C. Handy, Phys. Rev. A **14**, 1607 (1976)
17. M. Mizushima, *Quantum mechanics of atomic spectra and atomic structure* (W. A. Benjamin, New York, 1970)
18. J.C. Slater, *Quantum theory of atomic structure* (McGraw-Hill, New York, 1960)
19. P. Gombas, *Die Statistische Theorie des Atoms und ihre Anwendungen* (Springer-Verlag, Wien, 1949)
20. H. Kleinert, Physics Letters A **173**, 332 (1993)
21. E.U. Condon, G.H. Shortley, *The theory of atomic spectra* (Cambridge University Press, Cambridge, 1935)
22. G. Tanner, K. Richter, J.M. Rost, Rev. Mod. Phys. **72**, 497 (2000)
23. G.W.F. Drake (ed.), *Springer Handbook of Atomic, Molecular, and Optical Physics* (Springer-Verlag, New York, 2006)

Chapter 4
The Spin–Orbit Interaction

We have in the preceding chapters seen examples of how the electron spin matters for the multielectron atomic state, even when it is not explicitly included in the Hamiltonian. It plays an important role for the symmetry of the wave function, and as a consequence, it can change energy eigenvalues.

To get further with the treatment of atomic structure we have to take into account that the electron spin is associated with a magnetic moment, which is bound to interact with other moments, charges, and fluxes of charges, in the system. More broadly, we have to introduce relativistic effects directly into the atomic Hamiltonian, with the electron spin being one of these.

In the present chapter, we will first return to the one-electron atom and introduce the single electron spin–orbit Hamiltonian — the intrinsic magnetic interaction between an electron's spin and its own orbital angular momentum. We then complement the hydrogenic Hamiltonian of (1.8) with the spin–orbit interaction taken as a perturbation. A more rigorous treatment is to form the relativistic version of the full Schrödinger equation with a central potential. This is known as the *Dirac equation*. The Dirac equation falls outside of the scope of this volume, and we refer the reader instead to the suggested further reading, for example [1] or [2], and to overviews of the Dirac equation in appendix E and of electronic magnetic interactions in appendix F.

The Dirac equation has analytical solutions, but the relativistic contributions are small enough to be treated as perturbations in the first order, with a rather good precision (provided that the nuclear charge, Z, is not too large). The relativistic contributions can, in turn, be split into three parts, all of the same order of magnitude for the case of hydrogen: the spin–orbit interaction, the relativistic correction to the kinetic energy term, and the so-called Darwin term (see for example [1] or [2]).

In the case of hydrogen, all these three energy contributions can be estimated by first-order perturbation theory, using the non-relativistic hydrogen wave functions of (1.23) as the zero-order solutions. For a multielectron system, this becomes much harder, and it may make more practical sense to include some of them in

A. Kastberg, *Structure of Multielectron Atoms*, Springer Series on Atomic, Optical, and Plasma Physics 112, https://doi.org/10.1007/978-3-030-36420-5_4

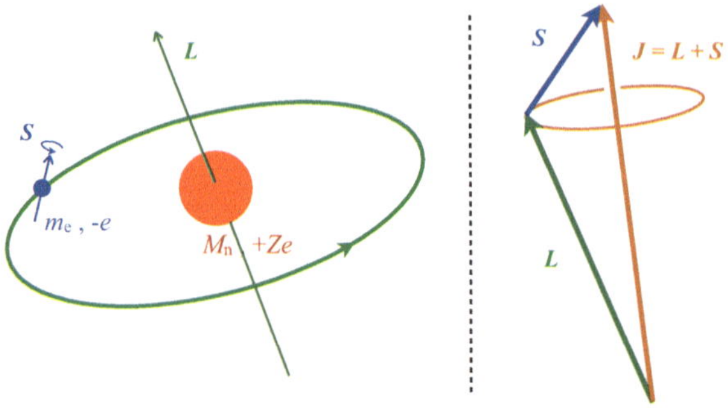

Fig. 4.1 Left: an electron with charge $-e$ and mass m_e in a classical orbit around a nucleus of charge $+Ze$ and mass $M_n \gg m_e$. Right: The corresponding two electronic angular momenta are the orbital angular momentum **L**, and the spin **S**. Due to the spin–orbit coupling, both **L** and **S** will precess. Their sum, the total angular momentum for a single electron, **J**, will be a constant of motion.

an approximative zero-order Hamiltonian, which is purely central. The latter is the essence of the central-field approximation, which will be covered in chapter 5. In particular, such a treatment is often resorted to for the relativistic kinetic energy and the Darwin term — they are included in a central-field approximation Hamiltonian. The spin–orbit term, however, depends strongly on angular momentum, and cannot be a part of an approximately radial potential — in particular for valence electrons. This interaction plays an important role for the structure of multielectron atoms, and it has to be treated separately in some detail.

For one-electron systems, the spin–orbit interaction leads to modifications to the energies found in chapter 1, and some degeneracies are broken. In this chapter, after having introduced the effect for the hydrogenic case, we generalise it to the multi-electron atom. Thenceforth, the spin–orbit interaction will be one of the principal ingredients in chapters to follow. It is also the origin for fine-structure in the *LS*-coupling scheme (see chapters 6 and 7), and it plays an even more important role for *jj*-coupling (see chapters 6 and 8).

4.1 Spin–Orbit Interaction for One-Electron Atoms

We begin by describing the spin–orbit effect with a classical analogy, illustrated in figure 4.1. A light negative charge is in a stable, elliptical orbit around a massive positive charge. The negative charge has an orbital angular momentum **L**, and a spin, **S** — a spin which will give rise to a magnetic moment.

Consider now a coordinate system that moves with the translation of the negative charge. In this, the positive charge will be seen as being in an orbit. The corresponding current will provide an effective magnetic field, which the magnetic moment of the negative charge will couple to. We thus have a coupling between **L** and **S**. The interaction will cause these two vectors to precess around one another, and neither of them will be a constant of motion. The quantity that will be preserved is instead their sum, for which we introduce the notation $\mathbf{J} \equiv \mathbf{L} + \mathbf{S}$. The vector $\mathbf{J}$ obeys the eigenvalue equations (in SI-units):

$$\begin{aligned} \mathbf{J}^2 \Psi_{jm_j}(\mathbf{q}) &= j(j+1)\hbar^2 \Psi_{jm_j}(\mathbf{q}) \\ J_z \Psi_{jm_j}(\mathbf{q}) &= m_j \hbar \Psi_{jm_j}(\mathbf{q}) \ , \end{aligned} \tag{4.1}$$

where $\mathbf{q}$ is used as composite notation for all spatial and spin degrees of freedom.

This interaction between $\mathbf{L}$ and $\mathbf{S}$ is the essence of the *spin–orbit interaction*, and the classical vector coupling image is illustrated in the right part of figure 4.1. In a quantum mechanical terminology, the interaction Hamiltonian does not commute with the individual components of $\mathbf{L}$ and $\mathbf{S}$, for example L_z and S_z. The vector $\mathbf{J}$ is the operator which describes the *total angular momentum* of the electron, and a quantum number basis that will render the Hamiltonian diagonal is: $|n,l,s,j,m_j\rangle$.

A real atom calls for a more sophisticated model than the one with the classical trajectory in figure 4.1, but the reasoning above gives a good qualitative picture of what is going on. For an atom with many electrons, the level of complexity increases greatly. Magnetic interactions can result in magnetic couplings between the different angular momenta, $\mathbf{L}_i$ and $\mathbf{S}_i$, for all electrons. Added to that is the fact the different $\mathbf{L}_i$ will couple through the Coulomb interaction. For a complete description, a Hamiltonian must be formed which somehow includes the couplings between *all* orbital and spin angular momenta.

In appendix F, we formulate a complete electromagnetic interaction Hamiltonian and identify various terms. For a practical deciphering of multielectron atomic structure, however, it is more fruitful to use perturbation theory, and in order to enable this, it is useful to begin with a less stringent formalism. In order to enable an understanding of the spin–orbit effect, which will be sufficient for the first stages of the vector model that will follow in chapters 6, 7 and 8, we will proceed by first looking closer at the magnetic moment associated with the electron spin, and the effective magnetic field caused by the orbital angular momentum.

Note that from now on, we will use indices with the angular momentum vectors ($\mathbf{L}_i$, $\mathbf{S}_i$ and $\mathbf{J}_i$) to indicate the angular momenta for *individual electrons*. This will be adhered to also in the following chapters, whereas we will use the notations without indices ($\mathbf{L}$, $\mathbf{S}$ and $\mathbf{J}$) when we refer to the total electronic angular momenta (the vector sums of the individual contributions).

4.1.1 Magnetic Moments and the Effective Magnetic Field

The magnetic dipole moment of an electron is proportional to its spin. In SI-units, it is:

$$\boldsymbol{\mu}_s = -\frac{g_s \mu_\mathrm{B}}{\hbar} \mathbf{S}_i \ . \tag{4.2}$$

The orientation of the moment is opposite to that of the spin, since the charge is negative. If the magnetic moment had arisen from a classical charge with an orbital angular momentum, the multiplicative factor would have been unity, but here, with a relativistic spin, we have to use the electronic spin g-factor, $g_s \approx 2$ (see appendix E for a relativistic derivation of this).

In an equally qualitative fashion, the effective magnetic field that the electron feels due to its motion in the electric field $\boldsymbol{\mathcal{E}}(\mathbf{r})$, caused by the charged nucleus, is:

$$\mathbf{B}_{\mathrm{eff},i}(\mathbf{r}_i) = -\mu_0 \varepsilon_0 \, \mathbf{v_i} \times \boldsymbol{\mathcal{E}}(\mathbf{r}_i) = -\frac{1}{c^2} \mathbf{v_i} \times \boldsymbol{\mathcal{E}}(\mathbf{r}_i) \ , \tag{4.3}$$

where, $\mathbf{v}$ is the electron orbital velocity. Since the field is entirely central, we have

$$\boldsymbol{\mathcal{E}}(\mathbf{r}_i) = \frac{\mathbf{r}_i \, \mathscr{E}(\mathbf{r}_i)}{|r_i|} \ , \tag{4.4}$$

and with $\mathbf{L}_i \equiv \mathbf{r}_i \times \mathbf{p}_i$, we obtain:

$$\mathbf{B}_{\mathrm{eff},i}(\mathbf{r}_i) = \frac{\mathscr{E}(\mathbf{r}_i)}{m_{\mathrm{e}} \, c^2 \, r_i} \, \mathbf{L}_i \ . \tag{4.5}$$

The potential energy associated with the interaction of a magnetic dipole with a magnetic field is the negative scalar product of the magnetic moment with the field strength. If we write this as a Hamiltonian, we get:

$$H = -\boldsymbol{\mu}_{s,i} \cdot \mathbf{B}_{\mathrm{eff},i}(\mathbf{r}_i) = g_s \frac{\mu_{\mathrm{B}}}{m_{\mathrm{e}} \, c^2 \, \hbar} \frac{\mathscr{E}(\mathbf{r}_i)}{r_i} \, \mathbf{S}_i \cdot \mathbf{L}_i \ . \tag{4.6}$$

This is the spin–orbit Hamiltonian for a hydrogenic atom. There are only two angular momenta present, and thus only one interaction.

This heuristic calculation qualitatively explains the phenomenon, but it gives an interaction energy which is roughly twice as big as it should be. The root of this discrepancy is a purely relativistic effect called Thomas precession. If the Hamiltonian is derived in a proper relativistic fashion, the correct factor appears, and it turns out that rather than dividing (4.6) with 2, it is more correct to replace g_s with $g_s - 1 \approx 1$. While this is demonstrated in appendix E, we will here just correct the equation by dividing by two — or rather, we will replace g_s in (4.6) by unity. This simplification will suffice for the objective at hand (see [1] or [3] for more details).

Since we are for the moment only concerned with a one-electron atom, we can use for $\mathscr{E}(\mathbf{r}_i)$ the field emanating from a classical point charge. This gives us, for the one-electron spin–orbit interaction:

$$H_{\mathrm{SO}-i} = \frac{e \, \mu_{\mathrm{B}}}{4\pi\varepsilon_0 \, m_{\mathrm{e}} \, c^2 \, \hbar} \frac{Z}{r_i^3} \, \mathbf{S}_i \cdot \mathbf{L}_i \ . \tag{4.7}$$

We rewrite this in atomic units, while taking into account that the Bohr magneton is one half in a.u.. We get:

$$H_{SO-i} = \frac{Z}{2c^2 r_i^3} \mathbf{L}_i \cdot \mathbf{S}_i = \frac{\alpha^2 Z}{2 r_i^3} \mathbf{L}_i \cdot \mathbf{S}_i \, , \tag{4.8}$$

where after the second equality, α is the *fine-structure constant* (A.2).

In (4.8) we have reversed the order of $\mathbf{S}_i$ and $\mathbf{L}_i$ in order to conform with a more standard notation. One should note, however, that strictly speaking, $\mathbf{S}_i{\cdot}\mathbf{L}_i$ and $\mathbf{L}_i{\cdot}\mathbf{S}_i$ differ by a phase factor — see appendix C, following (C.96), for a discussion on this. However, since we have yet to define a phase, this can be ignored at this stage.

If the Hamiltonian (4.8) is assumed to be a small (in relative terms) addition to the total energy, its effect can be treated with perturbation theory. The first order spin–orbit energy correction is then the matrix element:

$$E_{SO-i} = \langle H_{SO-i} \rangle = \frac{\alpha^2 Z}{2} \left\langle \frac{1}{r_i^3} \right\rangle \langle \mathbf{L}_i \cdot \mathbf{S}_i \rangle \, . \tag{4.9}$$

The expectation value of r^{-3} can be found in (B.39).

4.1.2 Fine-Structure in Hydrogen

For hydrogen, we can take the wave functions found in (1.23), and solve (4.9) analytically. This is further facilitated by the fact that (4.9) is already factorised in radial and angular parts. The radial factor will be an analytically solvable integral, but to start with we look at the angular contribution, and its expectation value: $\langle \mathbf{L}_i{\cdot}\mathbf{S}_i \rangle$.

The precession of $\mathbf{L}_i$ and $\mathbf{S}_i$ means that neither L_{zi}, nor S_{zi} will commute with the interaction Hamiltonian (see figure 4.1). They will not be constants of motion, or with another terminology, m_l and m_s will not be good quantum numbers. Instead, we have to turn to the total single electron angular momentum, $\mathbf{J}_i \equiv \mathbf{L}_i + \mathbf{S}_i$, in order to quantise the solutions. This quantity will be conserved (in the absence of external fields, ambient effects and dissipation).

The factor $\langle \mathbf{L}_i{\cdot}\mathbf{S}_i \rangle$ an be expressed in terms of $\mathbf{L}_i^2$, $\mathbf{S}_i^2$ and $\mathbf{J}_i^2$ via:

$$\mathbf{J}_i^2 = \mathbf{L}_i^2 + 2\,\mathbf{L}_i \cdot \mathbf{S}_i + \mathbf{S}_i^2 \, . \tag{4.10}$$

For a state which is common eigenfunction to $\mathbf{L}^2$, $\mathbf{S}^2$ and $\mathbf{J}^2$, this leads to:

$$\langle \mathbf{L}_i \cdot \mathbf{S}_i \rangle = \frac{1}{2} \langle \mathbf{J}_i^2 - \mathbf{L}_i^2 - \mathbf{S}_i^2 \rangle = \frac{1}{2} \left[j(j+1) - l(l+1) - s(s+1) \right] \, , \tag{4.11}$$

and to a straightforward confirmation that $\mathbf{J}^2$ and J_z both commute with the total Hamiltonian. The sum, $m_j = m_l + m_s$, is a constant of motion, even though m_l and m_s are not, as could be expected from a conservative interaction.

Note that we have found the angular factor of the energy perturbation as a rather simple function of three quantum numbers. The expectation value is technically an integral, but we did not need to solve this mathematically. What we did do was to change the basis, from $|l,s,m_l,m_s\rangle$ to $|l,s,j,m_j\rangle$, which is made necessary by the interaction between $\mathbf{L}_i$ and $\mathbf{S}_i$. We could then get a quantitative result with algebraic methods. The lack of dependence of the energy correction on m_j in (4.11) is a natural consequence of the spherical symmetry. An external field would break this degeneracy (see chapters 11 and 12). For a detailed account of the coupling of two angular momenta to a third one, and the ensuing change of basis, see appendix C.

Equation (4.11) can be simplified even further, since we know that for a single electron, $s = 1/2$. As a consequence, there are two possibilities for the total angular momentum, namely $j = l \pm 1/2$. This corresponds to the respective cases of $\mathbf{L}_i$ and $\mathbf{S}_i$ being parallel or anti-parallel (see also chapter 6 for a classical discussion on the addition of quantum mechanical angular momenta). We can deduce that for a one-electron atom, the angular factor of the spin–orbit Hamiltonian reduces to (for $l \neq 0$):

$$\langle \mathbf{L}_i \cdot \mathbf{S}_i \rangle = \begin{cases} \frac{1}{2}\, l & \text{for } \mathbf{L}_i \parallel \mathbf{S}_i \\ -\frac{1}{2}\,(l+1) & \text{for } \mathbf{L}_i \perp \mathbf{S}_i \end{cases} . \tag{4.12}$$

A conclusion is that if the radial factor is invariant of the spin, as it should be for hydrogenic wave functions, the spin–orbit interaction will split up an energy level, initially specified only by the quantum number n, into a doublet with a separation that depends on l. This is the *fine-structure* for hydrogen. If the sole electron has $l = 0$, the only alternative is $j = 1/2$, and there is no fine-structure splitting. This is natural, since in that case the spin does not on average have any orbital angular momentum to interact with.

In order to quantify the splitting in absolute terms, we must also study the radial factor. Radial integrals involving positive or negative powers of r and hydrogenic wave functions can be solved exactly, and many such solutions are given in appendix B.3. From there, we find that the expectation value of the radial factor in (4.9) is:

$$\left\langle \frac{1}{r_i^3} \right\rangle = \frac{2Z^3}{n^3\, l\,(2l+1)\,(l+1)} . \tag{4.13}$$

We have kept Z in order to allow for an arbitrary one-electron system. The full expression for the spin–orbit energy becomes:

$$E_{\mathrm{SO}-i} = \frac{\alpha^2 Z^4\,[\,j(j+1) - l(l+1) - s(s+1)\,]}{2\,n^3\, l\,(2l+1)(l+1)} . \tag{4.14}$$

Taking into account the sole possibility for s, as we did in (4.12), (4.14) reduces to:

$$E_{\mathrm{SO}-i} = \frac{\alpha^2 Z^4}{2n^3}\,\frac{1}{2l+1} \times \begin{cases} \dfrac{1}{l+1} & \text{for } \mathbf{L}_i \parallel \mathbf{S}_i \\ -\dfrac{1}{l} & \text{for } \mathbf{L}_i \perp \mathbf{S}_i \end{cases} . \tag{4.15}$$

The splitting of this *doublet* (which only occurs for $l \neq 0$) is the difference between these two energies, and it is:

$$\Delta E_{\mathrm{SO}-i} = \frac{\alpha^2 Z^4}{2n^3 l(l+1)} . \tag{4.16}$$

We can see in (4.15) and (4.16) a pronounced scaling with Z, which is an effect of the increasing importance of relativistic effects for higher nuclear charge. We also see that the spin–orbit interaction breaks the degeneracy in the quantum number l. In order to calculate the entire fine-structure analytically for a one-electron atom, other relativistic effects must also be taken into account. For such a derivation, we refer to other volumes, such as [1, 2], or [3]. For the current treatment, which is mainly meant as a preparation for the analysis of multielectron atoms, we will instead proceed directly by generalising (4.9) to bigger atoms. This will be done in the following section.

4.2 Spin–Orbit Interaction for Multielectron Atoms

For an atom with many electrons, the intrinsic spin–orbit interaction described by (4.9) remains *almost* the same for each electron individually, but even that comes with the caveat that $\mathbf{L}_i$ will not depend just on the electrostatic interaction with the nucleus. With more than one electron, the electrostatic repulsion between the electrons will result in a coupling between the various $\mathbf{L}_i$, and for an outer electron, the inner ones will constitute an effective charge screening. Apart from affecting the spin–orbit interaction, the energy contribution from the mutual Coulomb repulsion will also have an energy contribution in itself, often greater than the spin–orbit interaction. Any attempt to decouple the various effects — by successive perturbation theory — must be done with great care, and is not always possible.

For a full analytical treatment, it is also necessary to take into account smaller interactions, such as interactions between spins, the coupling between the spin of one electron with the orbital angular momentum of another electron, and the magnetic interaction between two different electronic orbital moments (see section 9.1 and appendix F). The one thing that remains a fact (if we neglect the possible spin and structure of the nucleus and external fields) is that the total sum of *all* electron angular momenta remains constant under the plethora of interactions.

We can loosely define the *total electronic angular momentum* for an atom with N electrons as:

$$\mathbf{J} \equiv \sum_{i=1}^{N} \mathbf{L}_i + \sum_{i=1}^{N} \mathbf{S}_i . \tag{4.17}$$

This definition is loose in the sense that for the moment we reserve the possibility that the different terms in (4.17) may have to be added in some specific order. For an exact analytical approach, this would not be an issue, but the complete formulation will not provide us with analytical solutions. If we separate out the different interactions between the angular momenta and arrange them in order of relative energy contributions, we can instead approach the problem with successive layers of perturbation theory. This may give us good approximative values for the energy bestowal of different effects, and also provide us with a good understanding of atomic structure and of atomic spectra. However, it does mean that the order of summations in (4.17) will have significance.

The ways in which we can split up the different angular contributions will be the subject of chapter 6, and thereafter following chapters. Presently, we will simply assume that it may be justified to separate out all individual intrinsic spin–orbit interactions into one total spin–orbit Hamiltonian, regardless of whether the energy contribution of this will turn out to be greater or smaller than other terms.

A complete expression for the electronic spin–orbit effect Hamiltonian is derived in appendix F. For the combined effect of all electrons, it is:

$$H_{\mathrm{SO}} = \frac{\alpha^2}{2}\left\{\sum_{i=1}^{N}\left[\frac{Z\,\mathbf{L}_i\cdot\mathbf{S}_i}{r_i^3} - \sum_{j\neq i}^{N}\frac{1}{r_{ij}^3}\left(\mathbf{L}_{ij}\cdot\mathbf{S}_i + \mathbf{L}_{ji}\cdot\mathbf{S}_j\right)\right]\right\} \quad . \tag{4.18}$$

r_i is the radial coordinate of electron i, and $r_{ij} = |\mathbf{r}_{ij}| \equiv |\mathbf{r}_i - \mathbf{r}_j|$ is the separation between electrons i and j. $\mathbf{L}_{ij}$ is defined as the part of the orbital angular momentum of electron i caused by the Coulomb interaction with electron j, and vice versa for $\mathbf{L}_{ji}$ — see (F.16) for strict definitions. This equation, for which the sum goes over all electrons, contains first of all the spin–orbit effect for each electron, resulting from its angular momentum around the nucleus (the coordinate origin). The following contribution is the deviation from this due to the torque caused by the Coulomb interaction with other electrons.

Interelectronic interactions will to a large extent screen the full nuclear charge also for the spin–orbit effect, by providing torques of opposite signs. Another important property is that inner electronic orbitals that are fully occupied will have overall orbital and spin angular momenta that time averages to zero. This will be demonstrated in section 6.2, and it is a component of the central-field approximation — the subject of chapter 5.

These considerations make it reasonable to formulate an approximate form of the spin–orbit Hamiltonian in terms of an effective nuclear charge, Z_{eff}, as:

$$H_{\mathrm{SO}} = \frac{\alpha^2}{2}\sum_{i=1}^{N_\mathrm{v}}\frac{Z_{\mathrm{eff}-i}}{r_i^3}\,\mathbf{L}_i\cdot\mathbf{S}_i = \sum_{i=1}^{N_\mathrm{v}}\xi(r_i)\,\mathbf{L}_i\cdot\mathbf{S}_i\,. \tag{4.19}$$

The sums in (4.19) must now be taken only over the N_v electrons that are outside of filled orbitals — the valence electrons.

In the final step in (4.19) we have introduced the *radial fine-structure function*, $\xi(r_i)$. This involves important elements of averaging, and we typically do not easily get access to this parameter by analytical means. It can, however, be empirically derived from atomic spectra or numerically computed.

4.3 Spin–Orbit Coupling in Atoms and Other Physical Systems

The magnetic interaction between the orbital angular momentum and the spin of a particle is a general phenomenon. In the present chapter, we have described it in a setting of electrons bound to an atomic nucleus, which is arguably the most obvious way to describe the effect, and it is certainly the one relevant for this book. However, various forms of spin–orbit coupling occur for many other physical systems.

spin–orbit interaction in molecules works in essentially the same way as has been described here. The effect also plays an important role in nuclear physics, where it contributes to the shell structure of the nucleus. In semiconductors, and in solids in general, the spin–orbit interaction presents a perturbation that changes the band structure of the material — something that also brings with it technological applications.

In the case of atoms, we will see in the chapters that follow that spin–orbit coupling plays a crucial role in atomic structure and for atomic spectra. It is the term that, when treated perturbatively, will give rise to atomic fine-structure (in *LS*-coupling, see chapters 6 and 7). The relatively simple form of the Hamiltonian in (4.19) makes it well adapted to perturbation theory. However, in order to take that on, we first need to consider the relative energy contribution of the effect compared to other angular contributions, such as the torque brought about due to the inter-electronic Coulomb repulsion. It is also adamant that we properly formulate zero-order basis functions, even if they are approximations, before the perturbation calculation can commence. This, we will deal with in chapter 5.

Further Reading

Quantum Mechanics of One- and Two-Electron Atoms, by Bethe & Salpeter [1]
Quantum theory of atomic structure, by Slater [3]
Atomic Many-Body Theory, by Lindgren & Morrison [4]
Physics of Atoms and Molecules, by Bransden & Joachain [2]

References

1. H.A. Bethe, E.E. Salpeter, *Quantum Mechanics of One- and Two-Electron Atoms* (Springer-Verlag, Berlin, 1957)
2. B.H. Bransden, C.J. Joachain, *Physics of Atoms and Molecules*, 2nd edn. (Prentice Hall, Harlow, England, 2003)
3. J.C. Slater, *Quantum theory of atomic structure* (McGraw-Hill, New York, 1960)
4. I. Lindgren, J. Morrison, *Atomic Many-Body Theory*, 2nd edn. (Springer Verlag, Berlin, 1986)

Chapter 5
The Central-Field Approximation

In the absence of simple theoretical tools for solving the Schrödinger equation in (2.8), and with the shortcomings and/or complexities of the approximation methods described in chapter 3, we will regroup and see if we, with a modified perturbative analysis, can gain more insight into multielectron atomic structure. We saw in section 3.1 that forming a zero-order Hamiltonian with only the electron–nucleus interactions taken into account brings the advantage of having separable zero-order basis functions. However, this comes at a price. Quantitatively, the approximation is too crude, and for all but the smallest atoms, the variational method approximation becomes forbiddingly complex. In this chapter, and in most of what follows in this volume, we retain the ambition to have separable zero-order functions, emanating from a purely radial potential, but we do this without leaving the *entire* electron–electron interaction H_{ee} as a perturbation.

What this amounts to is to split up the $1/r_{ij}$ term in the Hamiltonian into a radial and an angular part. The former goes into the zero-order Hamiltonian, and the latter, together with the spin–orbit interactions, are treated perturbatively. The division of the electron–electron interaction term cannot be made in a simple analytical way, but nevertheless this approach turns out to give ample understanding of atomic structure and atomic spectra, and it makes possible an analysis of the finer part of atomic structure in terms of coupling of angular momenta. Such couplings can be visualised with a vector model (see chapter 6), without applying much quantum formalism, and the description still holds well, qualitatively and quantitatively, when put on a quantum mechanical footing.

In this so-called *central-field approximation* (CFA), each electron is initially thought of as moving in a purely central potential, and having a wave function akin to hydrogenic electron orbitals. For an atom with many electrons, the zero-order approximation of its state will thus be a product of electron orbitals, whose occupation numbers must adhere to the Pauli principle. Such product states are called *electron configurations*, and with an understanding of ground state electron configurations for different atoms, the periodic system of elements can be constructed.

A. Kastberg, *Structure of Multielectron Atoms*, Springer Series on Atomic, Optical, and Plasma Physics 112, https://doi.org/10.1007/978-3-030-36420-5_5

In this chapter we will focus entirely on the CFA, and the spin–orbit interaction, which was introduced in chapter 4, is yet again be momentarily ignored. We will reintroduce it, within the context of the CFA, in chapter 6.

5.1 The Principle of the Central-Field Approximation

The goal of the central-field approximation is to use a zero-order Hamiltonian that provides separable solutions to the Schrödinger equation. That is, we want to be able to write the total atomic wave function as a product state of single-electron wave functions. When that is done, the parts of the full Hamiltonian that have been neglected will be added as perturbations. In order to do this we have to somehow neutralise the $1/r_{ij}$ term in the Hamiltonian in (2.8). The assumption that will go into the zeroth order approximation is that each electron moves in a central potential consisting of the electrostatic attraction to the nucleus, plus *the radial part of the time-averaged sum of the electrostatic repulsions from all the other electrons.*

In the case of a sole electron in the outermost orbital — an alkali atom — this approximation is rather good. Take Na as an example. Here we have 10 electrons in lower lying orbitals, and a single valence electron lying further out. Approximating the effective potential that this eleventh electron will feel with the charge of the nucleus, $+11e$, minus the screening of 10 electrons, means that the actual potential will be close to the hydrogen Hamiltonian. The slight departure of alkali energies from hydrogenic ones can be quantified by an empirical parameter called the quantum defect, further described in section 5.4.

In most other cases the approximation will be a bit more severe. However, taking the potential felt by one electron as the radial part of the time-averaged sum of all electrostatic interactions, and then treating the ignored part of the interactions as a perturbation, will be an approximation that is sufficient for many purposes. The effective potential, albeit radial, will not have a functional form completely analogous to the hydrogenic wave functions described in chapter 1.

From (2.8), the total Hamiltonian is:

$$H = \sum_{i=1}^{N} \left(-\frac{\nabla_{r_i}^2}{2} - \frac{Z}{r_i} + \sum_{j>i}^{N} \frac{1}{r_{ij}} \right) . \tag{5.1}$$

We now write this as a separable and radial *central-field Hamiltonian* plus a residual term, which represents the part of the Coulomb repulsion that provides a mutual torque on a pair of electrons:

$$H = H_{\mathrm{CF}} + H_{\mathrm{to}} . \tag{5.2}$$

The *Coulomb torque Hamiltonian* H_{to} will then be treated as a perturbation.

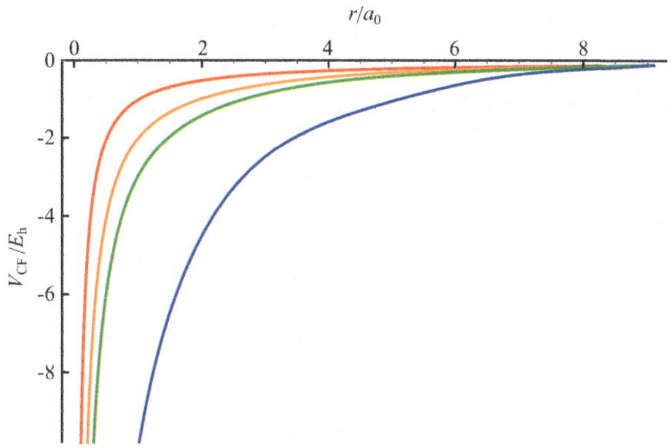

Fig. 5.1 Approximate interpolated central-field potential for different values of Z (red: $Z=1$, orange: $Z=2$, green: $Z=3$, blue: $Z=10$). The effective nuclear charge $Z_{\text{eff}}(r_i)$ — see (5.4) — is a function of r, asymptotically approaching Z for $r \to 0$ and 1 for $r \to \infty$.

The central-field Hamiltonian is:

$$H_{\text{CF}} = \sum_{i=1}^{N} \left[-\frac{\nabla_{r_i}^2}{2} + V_{\text{CF}}(r_i) \right] , \tag{5.3}$$

where

$$V_{\text{CF}}(r_i) = -\frac{Z}{r_i} + V_{\text{scr}}(r_i) = -\frac{Z_{\text{eff}}(r_i)}{r_i} . \tag{5.4}$$

The screening potential, $V_{\text{scr}}(r_i)$, is the radial part of the time average of the electrostatic interaction with all other electrons. This, in turn, means that the Coulomb torque Hamiltonian is:

$$H_{\text{to}} = \sum_{i=1}^{N} \left[\sum_{j>i}^{N} \frac{1}{r_{ij}} - V_{\text{scr}}(r_i) \right] . \tag{5.5}$$

We have here split up the electron–electron Hamiltonian of (3.2) into two parts:

$$H_{\text{ee}} = \left[\sum_{i=1}^{N} V_{\text{scr}}(r_i) \right] + H_{\text{to}} , \tag{5.6}$$

whereafter we have let the first term be a part of the zero-order function, and taken the second part as a perturbation. In the last line of (5.4), we have introduced an *effective nuclear charge*, which is a function of the radial coordinate. This closely corresponds to the Z_{eff} that we used in sections 3.2 and 3.3.

We can make a qualitative estimate of the form of $V_{\text{scr}}(r_i)$ by looking at the asymptotic behaviour of V_{CF}. This is illustrated in figure 5.1. At asymptotically large radial distance, the charge screening of all other electrons means that the valence electron effective potential approaches $-1/r_i$ (for a neutral atom). For a very small

r_i, the potential will instead be near to $-Z/r_i$. For intermediate r_i, an estimate can be attained by numerical interpolation, and this may in turn be a starting point for numerical calculations of wave functions (see chapter 14 and literature such as, for example, [1, 2] or [3]).

Since H_{CF} can be separated into N one-electron Hamiltonians, we can write its corresponding Schrödinger equation as N independent single-electron equations, with purely radial potentials. The solutions to these will in turn be separable into radial and angular parts, with their energies uniquely determined by the radial parameter, and the total central-field wave functions will be non-entangled product states. Thus, we have:

$$H_{CF}\,\psi_{CF}(\mathbf{r}_1,\mathbf{r}_2,\ldots,\mathbf{r}_N) = E_{CF}\,\psi_{CF}(\mathbf{r}_1,\mathbf{r}_2,\ldots,\mathbf{r}_N)\,, \tag{5.7}$$

with

$$\psi_{CF}(\mathbf{r}_1,\mathbf{r}_2,\ldots,\mathbf{r}_N) = \prod_{i=1}^{N} \psi_{n_i l_i m_{li}}(\mathbf{r}_i)\,, \tag{5.8}$$

and

$$E_{CF} = \sum_{i=1}^{N} E_{n_i l_i}\,. \tag{5.9}$$

The single-electron wave functions, $\psi_{n_i l_i m_{li}}(\mathbf{r}_i)$, will have angular parts identical to the spherical harmonics described in section 1.3, in this zero-order approach. For the radial part, it is typically only possible to come up with approximate solutions.

Even though the radial functions will not be the same as those for the hydrogen atom in section 1.2, they will be similar enough to justify using the same notation as for hydrogen orbitals. Therefore, we will label the functions with the symbols for n and l, as in (1.16), and as described in section 1.5. This means that the lowest energy function will be ψ_{1s}, followed by ψ_{2s}, ψ_{2p} and so on. With this notation, and applying the Pauli principle, we are now ready to introduce electron configurations.

5.2 Electron Configurations

We have established that the lowest single-electron energy eigenstate in the CFA is ψ_{1s}. For atoms with many electrons, the Pauli principle inhibits us from having all electrons in the same state. For an electron in the 1s-orbital, or in any $l = 0$ orbital, the only thing that can vary is the orientation of the spin, where we have two options, and thus the maximum number of electrons that can occupy an s-orbital is two.

For a two-electron system, such as helium, the spatial part of the CFA ground state is:

$$\psi^{(CF)}(\mathbf{r}_1,\mathbf{r}_2) = [\psi_{1s}(\mathbf{r}_1)]\,[\psi_{1s}(\mathbf{r}_2)] \equiv \psi_{1s^2}(\mathbf{r}_1,\mathbf{r}_2)\,, \tag{5.10}$$

and the spins of the two electrons must be opposite. Thus, the electron configuration of the helium ground state is $1s^2$. When we proceed to an atom with three electrons, such as neutral lithium, the third electron is prevented by the Pauli principle to go into the 1s-orbital, and the electron configuration yielding the lowest energy is $1s^2\,2s$.

For atoms with increasingly more electrons, the lowest energy states will have electrons in progressively higher energy orbitals; the added electrons will be more loosely bound. This is the so-called *aufbau principle*, with which orbitals are gradually filled up with electrons. The 2s-orbital also has room for two electrons, so the ground state electron configurations for Li and Be (the latter having four electrons) are $1s^2\,2s$ and $1s^2\,2s^2$, respectively. For the next atom, B with 5 electrons, we have the ground state configuration $1s^2\,2s^2\,2p$.

A p-orbital is synonymous with $l=1$, providing three options for the eigenvalue of L_z: $m_l=0,\pm1$. With the two possible spin directions, a p-orbital is able to hold a maximum of 6 electrons. For a general orbital, it will take $2(2l+1)$ electrons to fill it up: 10 for a d-orbital ($l=2$), 14 for an f-orbital ($l=3$) and so on.

The electron configurations are the zero-order wave functions in the central-field approximation. With these as our starting point, we will eventually apply what has been left out of the Coulomb interaction Hamiltonian, plus the spin–orbit interaction and other interactions that may apply, as perturbations. This will help us to decipher substantial parts of the energy level structure, even in the absence of knowledge of the actual functional form of the CFA wave function.

5.2.1 Ground State Configurations

The salient idea with the aufbau principle is that the ground state electronic configuration of any atom can be found by gradually filling up the most tightly bound orbitals. The appropriate order of the orbitals — the energy scaling — can in turn be found by making comparisons with hydrogenic wave functions. This means that orbitals with higher principal quantum numbers n will have their electrons gradually more loosely bound, and the filling up has to begin with the low values of n. In fact, for a multielectron atom the energy scaling with n is much more pronounced than it is for hydrogen. This is because an electron in the 1s-orbital feels the full nuclear charge, $+Ze$, and will be very tightly bound, whereas for electrons in higher orbitals, part of the nuclear charge will be screened by the inner electrons. As a consequence, the scaling is steeper than n^{-2}.

Orbitals with different l are not degenerate for a multielectron atom. In figure 1.3, we showed the radial probability distribution for hydrogenic orbitals. If we compare orbitals with different l, for the same n, we see that for a lower l there is a higher probability that the electron is close to the nucleus, due to anti-nodes of probability for small r. For a multielectron atom, this means that an electron with lower angular momentum will be less screened by inner electrons; it will feel a stronger effective nuclear charge, and thus it will be more tightly bound. This produces a strong dependence on l. Another effect of this is that for a multielectron atom, the mean value of r will typically be larger for higher l, in contrast to the behaviour for hydrogen (see table 1.1).

The general energy order of the lowest orbitals is: 1s, 2s, 2p, 3s, 3p, 4s, 3d, 4p, 5s, 4d, 5p, 6s, 4f, 5d, 6p, 7s and 5f. Some of these are almost degenerate (for example, 4s and 3d), which leads to anomalies in the aufbau principle for some elements. The fact that the 3d state has (slightly) smaller binding energy than 4s is precisely because of the strong dependence on l, described in the preceding paragraph. The same is true for the ordering of many of the following orbitals. With this in mind, we present the ground state configurations of neutral elements in table 5.1.

Table 5.1: Ground electronic configurations for the 118 first elements in the periodic table. The notations in square brackets correspond to filled orbitals, as in the respective element. The table also shows the *LS*-coupling atomic terms of the ground state (see chapters 6 and 7 for a definition of the atomic term), the first ionisation energies and the lowest energy states of the first ionised levels. Some states are described with *jj*-coupling (see chapter 8). A few lanthanide ions have exotic states, with a notation that will be discussed in chapter 9. The empiric data in the table are taken from [4]. For the heaviest elements, the table entries are based on various theoretical predictions [5–8].

Z		Ground state configuration	Atomic term	E_{ion} (eV)	Ionised level
1	H	1s	$^2S_{1/2}$	13.598	
2	He	$1s^2$	$^1S_0$	24.587	$1s\ ^2S_{1/2}$
3	Li	[He] 2s	$^2S_{1/2}$	5.392	$1s^2\ ^1S_0$
4	Be	[He] $2s^2$	$^1S_0$	9.323	$2s\ ^2S_{1/2}$
5	B	[Be] 2p	$^2P_{1/2}$	8.298	$2s^2\ ^1S_0$
6	C	[Be] $2p^2$	$^3P_0$	11.260	$2p\ ^2P_{1/2}$
7	N	[Be] $2p^3$	$^4S_{3/2}$	14.534	$2p^2\ ^3P_0$
8	O	[Be] $2p^4$	$^3P_2$	13.618	$2p^3\ ^4S_{3/2}$
9	F	[Be] $2p^5$	$^2P_{3/2}$	17.423	$2p^4\ ^3P_2$
10	Ne	[Be] $2p^6$	$^1S_0$	21.565	$2p^5\ ^2P_{3/2}$
11	Na	[Ne] 3s	$^2S_{1/2}$	5.139	$2p^6\ ^1S_0$
12	Mg	[Ne] $3s^2$	$^1S_0$	7.646	$3s\ ^2S_{1/2}$
13	Al	[Mg] 3p	$^2P_{1/2}$	5.986	$3s^2\ ^1S_0$
14	Si	[Mg] $3p^2$	$^3P_0$	8.152	$3p\ ^2P_{1/2}$
15	P	[Mg] $3p^3$	$^4S_{3/2}$	10.487	$3p^2\ ^3P_0$
16	S	[Mg] $3p^4$	$^3P_2$	10.360	$3p^3\ ^4S_{3/2}$
17	Cl	[Mg] $3p^5$	$^2P_{3/2}$	12.968	$3p^4\ ^3P_2$
18	Ar	[Mg] $3p^6$	$^1S_0$	15.760	$3p^5\ ^2P_{3/2}$
19	K	[Ar] 4s	$^2S_{1/2}$	4.341	$3p^6\ ^1S_0$
20	Ca	[Ar] $4s^2$	$^1S_0$	6.113	$4s\ ^2S_{1/2}$

Continued on next page

continued from previous page

Z		Ground state configuration	Atomic term	E_{ion} (eV)	Ionised level
21	Sc	[Ar] $3d4s^2$	$^2D_{3/2}$	6.561	$3d4s\ ^3D_1$
22	Ti	[Ar] $3d^24s^2$	$^3F_2$	6.828	$3d^24s\ ^4F_{3/2}$
23	V	[Ar] $3d^34s^2$	$^4F_{3/2}$	6.746	$3d^4\ ^5D_0$
24	Cr	[Ar] $3d^54s$	$^7S_3$	6.767	$3d^5\ ^6S_{5/2}$
25	Mn	[Ar] $3d^54s^2$	$^6S_{5/2}$	7.434	$3d^54s\ ^7S_3$
26	Fe	[Ar] $3d^64s^2$	$^5D_4$	7.902	$3d^64s\ ^6D_{9/2}$
27	Co	[Ar] $3d^74s^2$	$^4F_{9/2}$	7.881	$3d^8\ ^3F_4$
28	Ni	[Ar] $3d^84s^2$	$^3F_4$	7.640	$3d^9\ ^2D_{5/2}$
29	Cu	[Ar] $3d^{10}4s$	$^2S_{1/2}$	7.726	$3d^{10}\ ^1S_0$
30	Zn	[Ar] $3d^{10}4s^2$	$^1S_0$	9.394	$3d^{10}4s\ ^2S_{1/2}$
31	Ga	[Zn] $4p$	$^2P_{1/2}$	5.999	$4s^2\ ^1S_0$
32	Ge	[Zn] $4p^2$	$^3P_0$	7.899	$4p\ ^2P_{1/2}$
33	As	[Zn] $4p^3$	$^4S_{3/2}$	9.789	$4p^2\ ^3P_0$
34	Se	[Zn] $4p^4$	$^3P_2$	9.752	$4p^3\ ^4S_{3/2}$
35	Br	[Zn] $4p^5$	$^2P_{3/2}$	11.814	$4p^4\ ^3P_2$
36	Kr	[Zn] $4p^6$	$^1S_0$	14.000	$4p^5\ ^2P_{3/2}$
37	Rb	[Kr] $5s$	$^2S_{1/2}$	4.177	$4p^6\ ^1S_0$
38	Sr	[Kr] $5s^2$	$^1S_0$	5.695	$5s\ ^2S_{1/2}$
39	Y	[Kr] $4d5s^2$	$^2D_{3/2}$	6.217	$5s^2\ ^1S_0$
40	Zr	[Kr] $4d^25s^2$	$^3F_2$	6.634	$4d^25s\ ^4F_{3/2}$
41	Nb	[Kr] $4d^45s$	$^6D_{1/2}$	6.759	$4d^4\ ^5D_0$
42	Mo	[Kr] $4d^55s$	$^7S_3$	7.092	$4d^5\ ^6S_{5/2}$
43	Tc	[Kr] $4d^55s^2$	$^6S_{5/2}$	7.119	$4d^55s\ ^7S_3$
44	Ru	[Kr] $4d^75s$	$^5F_5$	7.361	$4d^7\ ^4F_{9/2}$
45	Rh	[Kr] $4d^85s$	$^4F_{9/2}$	7.459	$4d^8\ ^3F_4$
46	Pd	[Kr] $4d^{10}$	$^1S_0$	8.337	$4d^9\ ^2D_{5/2}$
47	Ag	[Kr] $4d^{10}5s$	$^2S_{1/2}$	7.576	$4d^{10}\ ^1S_0$
48	Cd	[Kr] $4d^{10}5s^2$	$^1S_0$	8.994	$4d^{10}5s\ ^2S_{1/2}$
49	In	[Cd] $5p$	$^2P_{1/2}$	5.786	$5s^2\ ^1S_0$
50	Sn	[Cd] $5p^2$	$^3P_0$	7.344	$5p\ ^2P_{1/2}$
51	Sb	[Cd] $5p^3$	$^4S_{3/2}$	8.608	$5p^2\ ^3P_0$
52	Te	[Cd] $5p^4$	$^3P_2$	9.010	$5p^3\ ^4S_{3/2}$

Continued on next page

continued from previous page

Z		Ground state configuration	Atomic term	E_{ion} (eV)	Ionised level
53	I	[Cd] $5p^5$	$^2P_{3/2}$	10.451	$5p^4\ ^3P_2$
54	Xe	[Cd] $5p^6$	$^1S_0$	12.130	$5p^5\ ^2P_{3/2}$
55	Cs	[Xe] 6s	$^2S_{1/2}$	3.894	$5p^6\ ^1S_0$
56	Ba	[Xe] $6s^2$	$^1S_0$	5.212	$6s\ ^2S_{1/2}$
57	La	[Xe] $5d6s^2$	$^2D_{3/2}$	5.577	$5d^2\ ^3F_2$
58	Ce	[Xe] $4f5d6s^2$	$^1G_4$	5.539	$4f5d^2\ ^4H_{7/2}$
59	Pr	[Xe] $4f^36s^2$	$^4I_{9/2}$	5.470	$4f^3(^4I)\,6s\,[9/2,1/2]_4$
60	Nd	[Xe] $4f^46s^2$	$^5I_4$	5.525	$4f^46s\ ^6I_{7/2}$
61	Pm	[Xe] $4f^56s^2$	$^6H_{5/2}$	5.577	$4f^56s\ ^7H_2$
62	Sm	[Xe] $4f^66s^2$	$^7F_0$	5.644	$4f^66s\ ^8F_{1/2}$
63	Eu	[Xe] $4f^76s^2$	$^8S_{7/2}$	5.670	$4f^76s\ ^9S_4$
64	Gd	[Xe] $4f^75d6s^2$	$^9D_2$	6.150	$4f^75d6s\ ^{10}D_{5/2}$
65	Tb	[Xe] $4f^96s^2$	$^6H_{15/2}$	5.864	$4f^9(^6H)\,6s\,[15/2,1/2]_8$
66	Dy	[Xe] $4f^{10}6s^2$	$^5I_8$	5.939	$4f^{10}(^5I)\,6s\,[8,1/2]_{17/2}$
67	Ho	[Xe] $4f^{11}6s^2$	$^4I_{15/2}$	6.021	$4f^{11}(^4I)\,6s\,[15/2,1/2]_8$
68	Er	[Xe] $4f^{12}6s^2$	$^3H_6$	6.108	$4f^{12}(^3H)\,6s\,[6,1/2]_{13/2}$
69	Tm	[Xe] $4f^{13}6s^2$	$^2F_{7/2}$	6.184	$4f^{13}(^2F)\,6s\,[7/2,1/2]_4$
70	Yb	[Xe] $4f^{14}6s^2$	$^1S_0$	6.254	$4f^{14}6s\ ^2S_{1/2}$
71	Lu	[Xe] $4f^{14}5d6s^2$	$^2D_{3/2}$	5.426	$6s^2\ ^1S_0$
72	Hf	[Xe] $4f^{14}5d^26s^2$	$^3F_2$	6.825	$5d6s^2\ ^2D_{3/2}$
73	Ta	[Xe] $4f^{14}5d^36s^2$	$^4F_{3/2}$	7.550	$5d^36s\ ^5F_1$
74	W	[Xe] $4f^{14}5d^46s^2$	$^5D_0$	7.864	$5d^46s\ ^6D_{1/2}$
75	Re	[Xe] $4f^{14}5d^56s^2$	$^6S_{5/2}$	7.834	$5d^56s\ ^7S_3$
76	Os	[Xe] $4f^{14}5d^66s^2$	$^5D_4$	8.438	$5d^66s\ ^6D_{9/2}$
77	Ir	[Xe] $4f^{14}5d^76s^2$	$^4F_{9/2}$	8.967	$5d^76s\ ^5F_5$
78	Pt	[Xe] $4f^{14}5d^96s$	$^3D_3$	8.959	$5d^9\ ^2D_{5/2}$
79	Au	[Xe] $4f^{14}5d^{10}6s$	$^2S_{1/2}$	9.226	$5d^{10}\ ^1S_0$
80	Hg	[Xe] $4f^{14}5d^{10}6s^2$	$^1S_0$	10.438	$5d^{10}6s\ ^2S_{1/2}$
81	Tl	[Hg] 6p	$^2P_{1/2}$	6.108	$6s^2\ ^1S_0$
82	Pb	[Hg] $6p^2$	$[6p^2_{1/2}]_0$	7.417	$6p\ ^2P_{1/2}$
83	Bi	[Hg] $6p^3$	$^4S_{3/2}$	7.286	$6p^2\ ^3P_0$
84	Po	[Hg] $6p^4$	$^3P_2$	8.414	$6p^3\ ^4S_{3/2}$

Continued on next page

continued from previous page

Z		Ground state configuration	Atomic term	E_{ion} (eV)	Ionised level
85	At	[Hg] $6p^5$	$^2P_{3/2}$	9.318	$6p^4\ ^3P_2$
86	Rn	[Hg] $6p^6$	$^1S_0$	10.749	$6p^5\ ^2P_{3/2}$
87	Fr	[Rn] $7s$	$^2S_{1/2}$	4.073	$6p^6\ ^1S_0$
88	Ra	[Rn] $7s^2$	$^1S_0$	5.278	$7s\ ^2S_{1/2}$
89	Ac	[Rn] $6d7s^2$	$^2D_{3/2}$	5.380	$6p^67s^2\ ^1S_0$
90	Th	[Rn] $6d^27s^2$	$^3F_2$	6.307	$6d7s^2\ ^2D_{3/2}$
91	Pa	[Rn] $5f^26d7s^2$	$^4K_{11/2}$	5.89	$5f^27s^2\ ^3H_4$
92	U	[Rn] $5f^36d7s^2$	$^5L_6$	6.194	$5f^47s^2\ ^4I_{9/2}$
93	Np	[Rn] $5f^46d7s^2$	$^6L_{11/2}$	6.266	$5f^46d7s\ ^7L_5$
94	Pu	[Rn] $5f^67s^2$	$^7F_0$	6.026	$5f^67s\ ^8F_{1/2}$
95	Am	[Rn] $5f^77s^2$	$^8S_{7/2}$	5.974	$5f^77s\ ^9S_4$
96	Cm	[Rn] $5f^76d7s^2$	$^9D_2$	5.991	$5f^77s^2\ ^8S_{7/2}$
97	Bk	[Rn] $5f^97s^2$	$^6H_{15/2}$	6.198	$5f^97s\ ^7H_8$
98	Cf	[Rn] $5f^{10}7s^2$	$^5I_8$	6.282	$5f^{10}7s\ ^6I_{17/2}$
99	Es	[Rn] $5f^{11}7s^2$	$^4I_{15/2}$	6.368	$5f^{11}7s\ ^5I_8$
100	Fm	[Rn] $5f^{12}7s^2$	$^3H_6$	6.50	$5f^{12}7s\ ^4H_{13/2}$
101	Md	[Rn] $5f^{13}7s^2$	$^2F_{7/2}$	6.58	$5f^{13}7s\ ^3F_4$
102	No	[Rn] $5f^{14}7s^2$	$^1S_0$	6.66	$5f^{14}7s\ ^2S_{1/2}$
103	Lr	[Rn] $5f^{14}7s^27p$	$^2P_{1/2}$	4.96	$7s^2\ ^1S_0$
104	Rf	[Rn] $5f^{14}6d^27s^2$	$^3F_2$	6.02	$6d7s^2\ ^2D_{3/2}$
105	Db	[Rn] $5f^{14}6d^37s^2$	$^4F_{3/2}$	7.0	$6d^27s^2\ ^3F_2$
106	Sg	[Rn] $5f^{14}6d^47s^2$	$^5D_0$	8.2	$6d^37s^2\ ^4F_{3/2}$
107	Bh	[Rn] $5f^{14}6d^57s^2$	$^6S_{5/2}$	8.0	$6d^47s^2$, $(J=0)$
108	Hs	[Rn] $5f^{14}6d^67s^2$	$^5D_4$	8.5	$6d^57s^2$, $(J=5/2)$
109	Mt	[Rn] $5f^{14}6d^77s^2$	$^4F_{9/2}$	9.9	$6d^67s^2$, $(J=4)$
110	Ds	[Rn] $5f^{14}6d^87s^2$	–	11.2	$6d^77s^2$
111	Rg	[Rn] $5f^{14}6d^97s^2$	–	12.2	$6d^87s^2$
112	Cn	[Rn] $5f^{14}6d^{10}7s^2$	$(J=0)$	13.1	$6d^97s^2$
113	Nh	[Cn] $7p$	$(J=1/2)$	7.3	–
114	Fl	[Cn] $7p^2$	$(J=0)$	8.6	–
115	Mc	[Cn] $7p^3$	–	5.6	–
116	Lv	[Cn] $7p^4$	–	6.6	–
117	Ts	[Cn] $7p^5$	–	7.6	–
118	Og	[Cn] $7p^6$	$^1S_0$	9.4	–

5.3 The Periodic System

The building up of the electron orbitals provides a practical way to arrange the elements, first introduced by Mendeleev [9, 10]. In the periodic system (shown in figure 5.2), starting from the top and going from left to right, we gradually fill up new electron orbitals, with progressively lower binding energies. Every time we are about to open up a new s-orbital, we begin on a new line. The horizontal positioning of the elements makes atoms with similar qualities appear in vertical columns. Each such column corresponds to a certain number of electrons in the outermost orbital, with a few minor exceptions. The periodic chart has proved to be of enormous utility for enhancing the understanding of the structure and the properties of matter, both in chemistry and in physics.

In the following, we shall take a closer look at the periodic chart, from an essentially qualitative point of view. This will help us to identify classes of atoms with similar structural properties, and we will be able to see general trends. For this analysis, we will scrutinise the table both across and down.

5.3.1 Horizontal View of the Periodic System — The Gradual Evolution of Electronic Configurations

We have established that each hydrogen-like nl-orbital can have a maximum electron occupation of $2(2l+1)$, and that in terms of energy, we build up orbitals from the bottom. This means that we start with the lowest orbital energy, which is the same as the tightest binding. We begin a new row in the periodic system immediately before we start to fill up a new s-orbital, with the rationale being that that will coincide with a relatively big decrease in binding energy.

We have also seen that when we are not dealing with orbitals in one-electron systems, but rather with valence electrons outside closed orbitals in the CFA, the degeneracy in l is broken. A low l means more penetration inside the charge screening closed orbitals, and therefore stronger binding. For example, an ns-orbital will have lower energy than an n'd-orbital, when $n'=n-1$.

To be more concrete, when we have filled up the 1s, 2s, 2p, 3s and 3p orbitals, we have to first open the 4s orbital for elements 19 and 20 — K and Ca. For the following ten elements, the 4s and 3d electrons will be almost degenerate, and without a careful analysis, it is not obvious which of the total 12 available spots in the 4s- and 3d-orbitals that will be occupied in the ground states for the atoms from Sc to Zn. A corresponding situation is seen for atoms with valence electrons in 5s and 4d: Y to Cd.

When we have filled up the 6s-orbital with elements 55 and 56, Cs and Ba, the situation is such that the 4f and 5d orbitals are very close in energy, and those are just a little higher than 6s. This can be seen in the ground state configuration of the next elements in line (see table 5.1), La: $5d6s^2$, Ce: $4f5d6s^2$ and Pr: $4f^3 6s^2$.

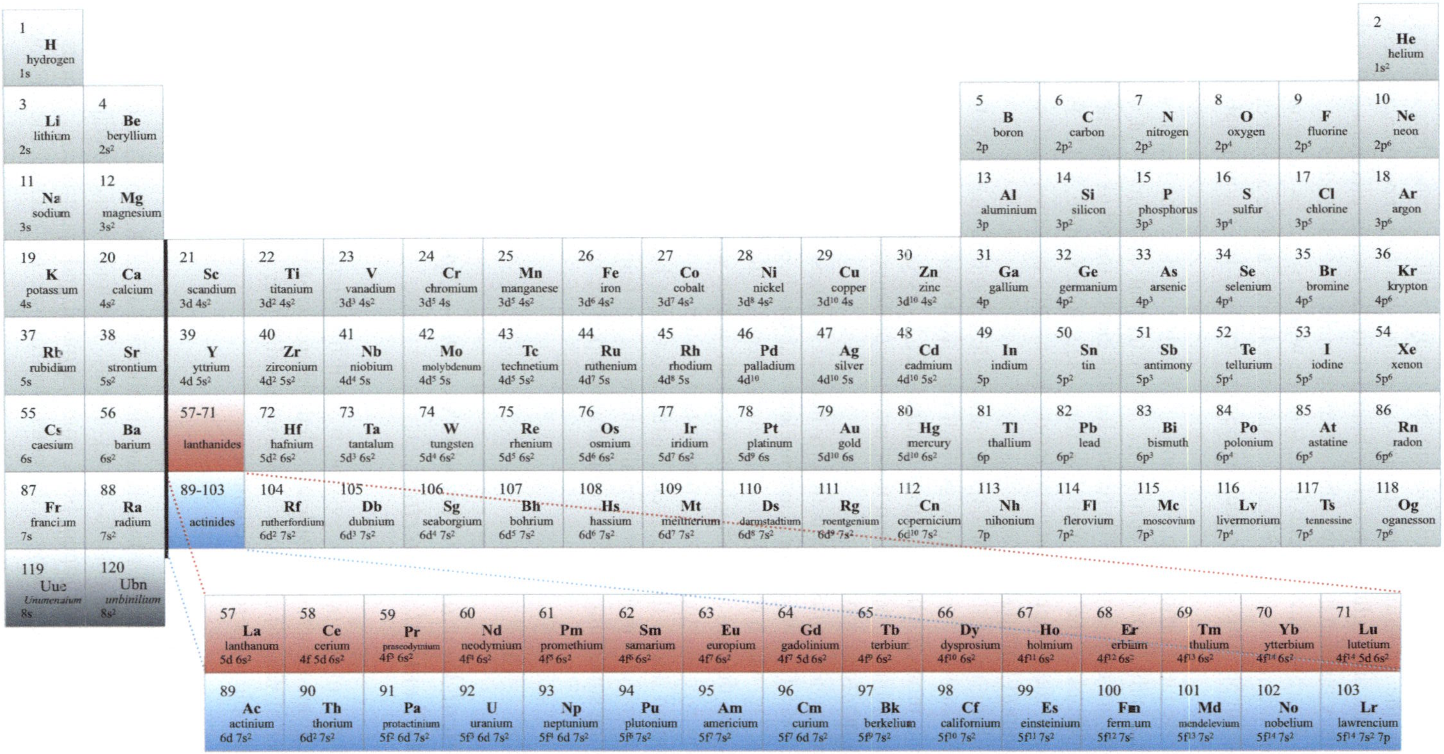

Fig. 5.2 The periodic system of elements. Note that at the time that this book is written (2019), the two heaviest elements in the figure — numbers 119 and 120 — have not yet been synthesised. For many of the other heaviest atoms, nuclei have been produced and observed, but the electronic structure shown in the figure are theoretical predictions.

After that, the 4f-orbital gradually fills up, but always with 5d being close in energy. The 4f-series, the lanthanides, are special in the sense that they have optically active 4f-electrons that are slightly less tightly bound than the two 6s-electrons, but at the same time, they have an average radial distance smaller than the latter.

With the 4f orbital having been filled (ending with Yb), the 6s electrons will for the following atoms have higher energies than the 4f ones and be almost degenerate with the ones in 5d. This can be seen in the next series of transition metals from Hf to Hg.

In the subsequent orbital, 6p, we run out of stable nuclei. The heaviest element that still has a stable isotope is number 82, Pb. After that, all isotopes of all the following atoms are radioactive. The reason for this can be traced to nuclear structure and the properties of the strong nuclear force (see, for example, [11]). For a very heavy nucleus, the attractive potential inside the nucleus becomes shallow and will no longer support bound states.

For the filling up of the orbitals, the pattern described for the preceding atoms continues in essentially the same way, with the caveat that relativistic effects gradually become more important. Atom number 92, U, is generally regarded as the heaviest primordial element, albeit trace amounts of heavier nuclei may be formed naturally by neutron capture.

For even heavier atoms, production by nuclear synthesis is the only way to observe them. As of 2019, the heaviest nucleus that has been synthesised is that of element 118, Og, which in the form of a neutral atom should have a filled 7p orbital. This means that the next atom should be an alkali, with a sole 8s electron. It should be noted, however, that these latter ground state configurations are predictions, made less than certain due to the large relativistic effects that become apparent for very heavy atoms, and also that experimentally there is a big leap from synthesising a nucleus to having something akin to a neutral atom open for spectroscopic or chemical studies.

5.3.2 *Vertical View of the Periodic System — Spectroscopic and Chemical Properties*

In each column of the periodic system, we will find elements that have the same type of electronic structure outside all filled electronic orbitals. They will have the same number of valence electrons, which means that they will have many chemical and spectroscopic properties in common.

5.3.2.1 The Alkalis

In the first column, we have hydrogen and the alkalis (Li, Na, K, Rb, Cs and Fr), with a single s-valence electron outside closed orbitals. The lone valence electron makes these atoms similar to hydrogen, both from chemical and spectroscopic points of view. It also means that as long as the inner electrons are not excited, the CFA will be a good approximation.

Another effect of the single valance electron is that the emission and absorption spectra will be relatively simple, with strong and distinct spectral lines. In terms of chemistry, alkalis are very reactive — the valence electron can easily be dissociated to leave behind a cation. The alkalis will be treated with some more detail in section 5.4, where we will introduce the concept of the *quantum defect*.

5.3.2.2 Alkaline Earths

Atoms in column two are the alkaline earths: Be, Mg, Ca, Sr, Ba and Ra. They have filled s-orbitals, and the two electrons will in the lowest energy states be paired as in the He examples in chapters 2 and 3. The ground and excited electron configurations will split up in spin triplets and spin singlets. In terms of chemistry, both s-electrons will be relatively loosely bound, which means that these elements form cations of charge $+2$.

5.3.2.3 Transition Metals

In the centre of the periodic chart, we have the transition metals, with their outermost orbitals being 3d, 4d or 5d, often close to degeneracy with s-orbitals of higher n. The presence of many valence electrons has as consequence that in the solid state, such electrons become delocalised over the entire bulk, which is indeed what gives these elements their metallic properties, such as high electric conductivity. The presence of many unpaired d-electrons also makes many transition metals paramagnetic.

5.3.2.4 p-Block Elements

The group of elements in the six right-most columns includes atoms with a great variety of properties. In an approximately diagonal band, starting from B and then going down and to the right — and including Si, Ge, As, Sb, Te and At — we have metalloids. The definition of a metalloid is loose, and far from consistent in the literature. What it essentially means is that these atoms are not quite metals, and not quite non-metals either; they have a mixture of the typical properties of those two groups. Note that a metalloid is not the same thing as a semiconductor, since the latter specifically refers to a certain electronic band structure in the bulk material.

The metalloids form a dividing line between metals and non-metals. The formers are below and to the left in the periodic table, and the p-elements among those (Al, Ga, In, Sn, Tl, Pb, Bi and Po) have a lot in common with the transition metals. The atoms above and to the right of the metalloids are C, N, P, O, S, Se, and the elements in the two right-most columns, which will be described separately.

5.3.2.5 Halogens

The halogens are atoms in the penultimate column in the periodic table. They have one valence electron too little to make up a filled outermost orbital. This means that they have a high electron affinity and they will easily form anions. The halogens are F, Cl, Br, I. The final one, At, is by some authors seen as a halogen and by others as a metalloid. This is largely a semantic question.

As we will see in the following chapters, atoms that have one p-electron missing in order to have a filled orbital will be spectroscopically similar to those that have one single p-electron.

5.3.2.6 Rare Gases

To the far right in the chart, we have the noble gases: He, Ne, Ar, Kr, Xe and Rn. Their filled orbitals make them essentially inert. This means that they remain gaseous until very low temperatures, and they rarely form compounds. Structurally, they have high ionisation potentials and substantial energy gaps between their respective ground states and the lowest excited states.

5.3.2.7 Rare Earths and Actinides

In the descriptions above, we have left out the 4f- and 5f-elements. The 4f-atoms are known as lanthanides or rare earths. They have chemical properties similar to the transition metals, and they were covered in the horizontal view of the periodic chart. The actinides, the 5f-elements, are also similar in terms of chemical properties, but they are all radioactive, and have low (or non-existent) natural abundance.

5.3.3 *Physical Trends in the Periodic System*

We have introduced the periodic system, and we have looked at it both horizontally and vertically. We have already seen hints to the effect that the way in which the elements are organised in the periodic chart means that a number of physical properties vary from element to element in a way that is commensurate with the presentation of the table.

5.3.3.1 Atomic Radius

Horizontally, atomic radii decrease from left to right; vertically, they increase from top to bottom

The first issue to observe about the atomic radius is that it is a concept that is not well defined. The radial wave function is a continuous function, and so is its probability distribution. Thus defining a 'size' of an atom does not make any sense unless that concept is better specified.

In a solid, the radius of the atom can be taken as the half distance between two neighbouring atoms. In a molecule, it could be connected to the range of a covalent bond. For free atoms, it is rather more intuitive to specify the maxima of the probability distributions and charge distributions. However, this could refer to only the outermost electron, or to some kind of weighted average. Such different definitions, and their relevance, are partially discussed in, for example, [12, 13] and [14]. Either way, all these measures scale in more or less the same way, and here we are only interested in showing general trends, without any rigour.

The picture in figure 1.3 shows the probability radius increasing with principal quantum number. This is not quite reflected in the progression of the atomic radius across the periodic table, since when we are moving to the right, and down, in the periodic table, we primarily change the nuclear charge. This means that if we move from left to right within one orbital — say 3p — we keep adding electrons into the same orbital, with approximately the same amount of screening, but since the nuclear charge gradually increases the electrons get progressively more strongly bound. For the example with the 3p-atoms, the smallest atom is Ar with the approximate charge radius $1.3\,a_0$. Then follows in order Cl, S, P, Si and Al, with the latter being the biggest one in the series having radius $2.6\,a_0$.

As we move straight down in the table, however, we get essentially the same valence electron structure, but with larger principal quantum numbers. We then get progressively larger atoms, but the increase is rather slight. To take the alkalis as an example, Cs has a radius of $5.1\,a_0$ (available data for Fr are scarce), and for Li it is $3.1\,a_0$ (the values in this paragraph and the preceding one are taken from [14], and refer to calculations of the radii of the maximum charge density for the outermost electron).

5.3.3.2 Ionisation Energy

Horizontally, ionisation energies increase from left to right; vertically, they decrease from top to bottom

The first ionisation energy is the amount of work needed in order to remove the least bound electron from an atom and to leave a cation behind. The scaling of the ionisation energies is approximately opposite to that for the atomic radii, and the physical arguments are essentially the same, since a more strongly bound electron tends to be closer to the nucleus. A difference is that for the ionisation energy, this is particularly pronounced when moving from left to right within one period (one orbital). This is due to the added stability that comes with having a filled, or almost filled, outermost orbital. Thus, alkalis are most reactive and have low ionisation energies. For Cs, it is about 3.9 eV. On the other extreme, noble gases are very stable and are difficult to ionise. For He, the first ionisation energy is 24.6 eV and for Ne, it is 21.6 eV [4].

5.3.3.3 Electron Affinity/Negativity

Horizontally, electron affinities increase from left to right; vertically, they decrease from top to bottom

Electron affinity is the amount of energy released when a neutral atom attracts an electron and becomes an anion. Electron negativity is a very similar concept — it is the 'willingness' of an atom to attract an electron, but it is a more qualitative concept than electron affinity.

As can be expected, electron affinity scales in the same general manner as the ionisation energy. An exception is the noble gases. These have filled electron orbitals, and do not readily become anions. Other exceptions are the lanthanides and the actinides, for whom the chemistry is more complicated.

5.3.3.4 Metallic Character

Horizontally, the metallic character decreases from left to right; vertically, it increases from top to bottom

The metallic character is another qualitative concept, and it refers to how easily an atom gives up an electron to the bulk, which means that the latter will contribute to the electric conductivity. Loosely, one can say that low ionisation energy means more distinct metallic character, and the qualitative arguments given above still hold. The difference is that when we speak about metallic character, we are talking about solid materials, whereas atomic ionisation energies typically refer to atoms in gas phase.

5.4 The Alkalis — Quantum Defect

The alkali atoms have a sole valence electron outside filled orbitals. From the CFA point of view, this makes these atoms similar to hydrogen atoms. When the valence electron of a neutral alkali is at large radial distances from the nucleus, the total electrostatic potential it feels will approach $-1/r_i$, regardless of the nuclear charge. For such a large r_i, the solutions to the Schrödinger equation should indeed be almost identical to those of hydrogen.

The deviation from hydrogenic behaviour for wave functions of an alkali atom will mainly appear at small r_i, where the effective attractive potential will be stronger than the hydrogenic one, due to the less efficient screening. This will also depend on the orbital angular momentum. For a larger l, the centrifugal barrier will prevent the valence electron from penetrating the nucleus, and the atomic energy eigenstates will be closer to those for a true one-electron atom. On the other hand, this behaviour will break the degeneracy in l, particularly for low n. This is illustrated in figure 5.3, for the atoms lithium, sodium and potassium. The energy deviation from a hydrogen atom can be qualitatively understood as arising from a polarisation of the core of screening electrons, caused by the interaction with the outer electron.

Also apparent from figure 5.3 is that the ionisation energies of Li, Na and K are of the same orders of magnitude, and only decrease slightly with increasing Z. This is further shown in table 5.2. The binding energies of the respective lone ground state valence electrons are around 4–5 eV for all atoms in the sequence. For Fr, the binding energy is slightly higher than it is for Cs. This departure from the otherwise monotonous decrease with Z happens because relativistic effects associated with the heavy nucleus start to become important for Fr. It is predicted that the next alkali — element 119 — will have a higher binding energy still.

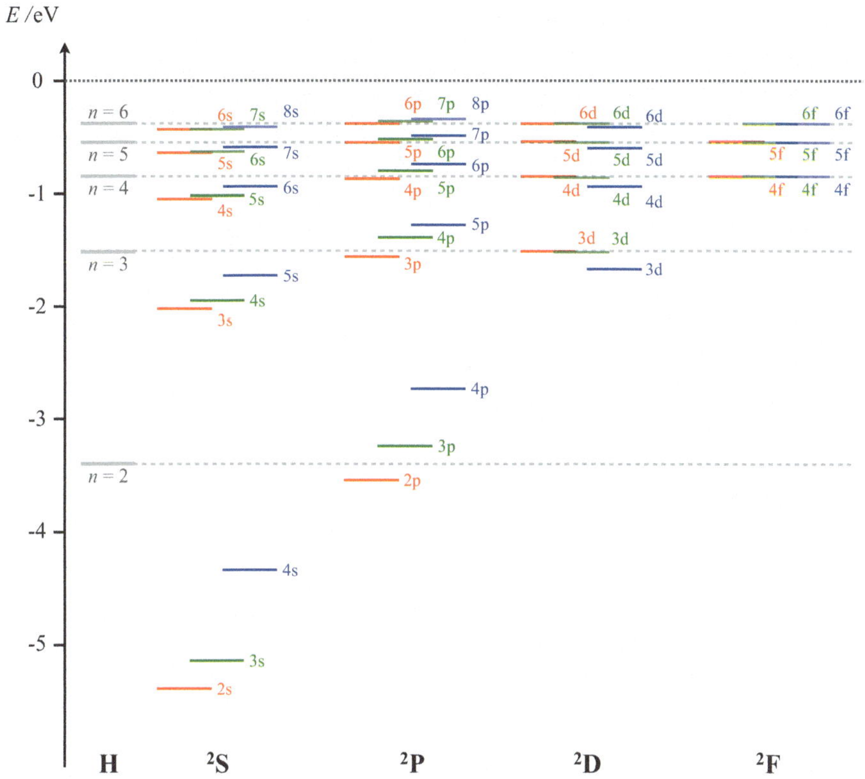

Fig. 5.3 Experimentally reported energies (in eV) of the lowest energy states, with $l \leq 3$, for Li (red), Na (green) and K (blue) [4]. To the left, in grey, are the energy levels for H. In this graph, the respective ionisation limits have been chosen as a common zero point for the energy scale. The states 2s, 3s and 4s are the electronic ground states for Li, Na and K, and the energies of these correspond to the first ionisation energy for the respective atoms. For low energy states, the valence electron is more strongly bound for the shown alkalis than the hydrogenic version of the same level. This is due to significant penetration of the wave function into the screening charge cloud of the filled orbital electrons. The horizontal scale is organised after the value of l of the valence electron (and the corresponding atomic term — see chapter 6).

The electrons in the closed orbitals of an alkali atom have binding energies of 20 eV or more. The large difference between this and the energies of the valence electrons shown in table 5.2 means that optical spectra of alkalis are totally dominated by the lone electron, which in its ground state is in an ns-orbital. This makes the analogy with the hydrogen atom even more suggestive, and it has led to an empirical model including the so-called *quantum defect*.

Table 5.2 Measured ionisation energies for alkali atoms from Li to Fr [4]. Also included in the table is a calculated estimate for element 119 [7]. The initial trend is the expected non-relativistic one. The increasing values for Fr and element 119 originate from relativistic effects.

Element	Ionisation energy
Li	5.39 eV
Na	5.14 eV
K	4.34 eV
Rb	4.18 eV
Cs	3.89 eV
Fr	4.07 eV
119	4.53 eV

5.4.1 The Quantum Defect

Some specific chemical properties of alkalis have been known since antiquity (the etymological origin of the word *alkali* is an Arabic word for calcined ashes). The spectroscopic similarities to hydrogen were well noted already in the 19th century. These comparisons — done before quantum mechanics was developed, and before much of atomic structure was understood — led to the concept of the quantum defect. To a large extent, this theory is phenomenological. It works, in the sense that it predicts many optical spectra and ionisation energies of alkali atoms well, but it does not give a very deep insight into atomic structure, at least not in its initial avatar.

The similarities between the hydrogen energy structure and that of the alkalis, shown in figure 5.3, suggest that a relatively slight modification of the one-electron formula in (1.24) might suffice in order to reproduce the energies of the alkali. The successful route to do this turns out to be that the principal quantum number in the denominator is empirically corrected with an amount that depends on l, as in:

$$E_{nl} = -\frac{1}{2n_{\text{eff}}^2} = -\frac{1}{2(n-\delta_{nl})^2} \,. \tag{5.11}$$

The *effective principal quantum number* is:

$$n_{\text{eff}} \equiv n - \delta_{nl} \,. \tag{5.12}$$

The *quantum defect*, δ_{nl}, is a function of the quantum numbers n and l, and is specific for each alkali atom. It should be noted that in (5.11) we assume a neutral atom, where the screened charge of the nucleus is $+e$. For an ion with a single valence electron — thus isoelectronic with an alkali atom — the numerator in (5.11) must be adjusted.

The quantum defect for a neutral atom is always positive since the net screening of the nuclear charge can never be more than $(Z-1)e$. The dependence of δ_{nl} on l is more pronounced than that on n, and for many purposes, it is enough to specify the quantum defect (for a specific element) only as a function of the orbital angular momentum. However, a more precise value can be obtained by an iterative formula, given by Ritz [15–17]:

$$n_{\text{eff}} = n - \alpha(l) - \frac{\beta(l)}{(n-\delta_{nl})^2} \ . \tag{5.13}$$

In table 5.3, we show some experimentally deduced values for the quantum defect. It can be seen that the trends in the dependencies on n and l are those alluded to above, and we also see that the quantum defect increases with Z.

For positively charged ions with a single valence electron, inner orbital electrons will be more tightly bound, which leads to the valence electron penetrating the core to a lesser extent. Therefore, the quantum defect decreases with increasing Z in an isoelectronic series. An example of this is given in figure 5.4, for Na and a series of anions with the same ground state electronic configuration: $1s^2 2s^2 2p^6 3s$.

Besides the alkali atoms, the quantum defect picture also works well for atoms where one electron is highly excited — the so-called Rydberg atoms (see chapter 13). The high excitation means that such an electron will have most of its probability density located well outside those of all other electrons, and thus the inner electrons will have the effect of a very nearly spherical screening of the nuclear charge. The key characteristic is that the total Coulomb potential felt by the con-

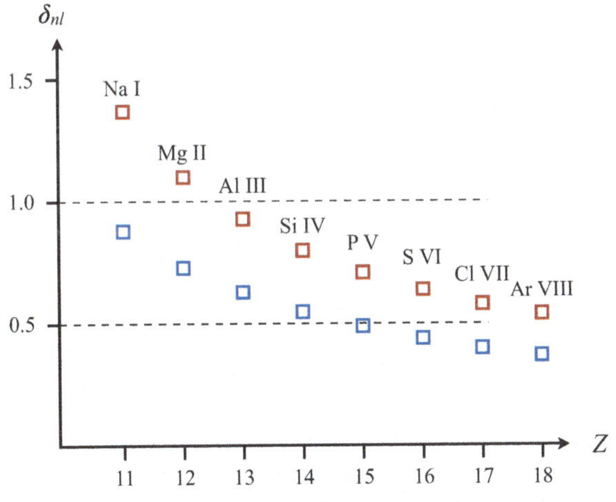

Fig. 5.4 Experimentally determined quantum defects for the isoelectronic series of Na, as function of Z, calculated from data in [4]. Red squares show the values for the 3s ground state, and the blue ones those of the 3p excited state.

Table 5.3 Values for quantum defects ($\delta_{nl} = n_{\text{eff}} - n$) for alkalis from Li to Cs, and for $l \leq 3$. The data have been compiled using [4] and [15]. For some of the table entries, the three decimal places shown are more than the level at which the quantum defect approximation is relevant, but this precision has been retained in the table in order to illustrate the trends.

Li					
	$n = 2$	3	4	5	6
$l = 0$	0.411	0.404	0.402	0.402	0.401
1	0.040	0.045	0.046	0.047	0.048
2		0.002	0.002	0.003	0.003
3			0.000	0.00	

Na					
	$n = 3$	4	5	6	7
$l = 0$	1.373	1.358	1.353	1.352	1.352
1	0.884	0.868	0.863	0.860	0.859
2	0.011	0.013	0.014	0.015	0.015
3		0.002	0.002	0.002	0.003

K					
	$n = 3$	4	5	6	7
$l = 0$		2.230	2.199	2.191	2.187
1		1.768	1.738	1.728	1.723
2	0.147	0.204	0.232	0.247	0.255
3		0.008	0.009	0.009	0.010

Rb					
	$n = 4$	5	6	7	8
$l = 0$		3.196	3.156	3.144	3.140
1		2.721	2.684	2.672	2.667
2	1.234	1.295	1.318	1.328	1.334
3	0.012	0.014	0.015	0.017	0.017

Cs					
	$n = 4$	5	6	7	8
$l = 0$			4.131	4.081	4.067
1			3.671	3.627	3.612
2		2.453	2.473	2.476	2.476
3	0.023	0.027	0.029	0.031	0.032

sidered electron is close to being proportional to $-1/r$. This is what gives us the possibility to analyse the atom in terms of the quantum defect.

During the last decades, there have been theoretical works that cast the quantum defect theory in a less phenomenological form (see, for example, [18] and comments to that article, [19] and [17]). This involves defining an effective potential that has solid physical justifications, and from which analytical solutions to the Schrödinger equation can be found, and also elaborations about the iterative expression (5.13). For further studies of these aspects, we refer to the general literature, and to the works cited above.

5.5 The Allure of the Central Potential

The CFA is a key to the understanding of atomic structure for all atoms except the very lightest ones (and even for these it helps). It bridges the gap between the exact treatment of the hydrogen atom, the in principle exact variational analysis for helium and couplings between angular momenta for bigger atoms. Thereby the CFA gives the latter concept a platform from which it can disentangle a majority of observed phenomena in atomic spectra and atomic structure.

The ability to express electronic interactions as an effective potential, albeit approximately, presents us with the possibility to classify atoms according to the periodic system. By providing us with a zero-order picture in which electrons are assigned to specific *orbitals*, the CFA also becomes an enabling theory for molecular physics and physical chemistry, as well as for solid-state physics.

In the reminder of this book, we will rely heavily on the CFA. Backed up by this, we will use perturbation theory to clarify correlation effects and magnetic interactions, and their influence on atomic structure. This will begin with a first look, in the following chapter, at how to couple the profusion of all electronic angular momenta in a multielectron atom.

Further Reading

The theory of atomic spectra, by Condon & Shortley [20]
Quantum theory of atomic structure, by Slater [16]
Atomic spectra, by Kuhn [15]
Quantum mechanics of atomic spectra and atomic structure, by Mizushima [21]
Atomic Many-Body Theory, by Lindgren & Morrison [2]
Physics of Atoms and Molecules, by Bransden & Joachain [22]
Springer Handbook of Atomic, Molecular, and Optical Physics, by Drake [3]

References

1. R.D. Cowan, *The theory of atomic structure and spectra* (University of California press, Berkeley, 1981)
2. I. Lindgren, J. Morrison, *Atomic Many-Body Theory*, 2nd edn. (Springer Verlag, Berlin, 1986)
3. G.W.F. Drake (ed.), *Springer Handbook of Atomic, Molecular, and Optical Physics* (Springer-Verlag, New York, 2006)
4. A. Kramida, Y. Ralchenko, J. Reader, and NIST ASD Team. NIST Atomic Spectra Database (ver. 5.3). [Online]. Available: http://physics.nist.gov/asd (2018). Accessed: 2019-07-14
5. B.G.C. Lackenby, V.A. Dzuba, V.V. Flambaum, Phys. Rev. A **99**, 042509 (2019)
6. V.A. Dzuba, Phys. Rev. A **93**, 032519 (2016)
7. D. Hoffman, D. Lee, V. Pershina, in *The Chemistry of the Actinide and Transactinide Elements*, ed. by L. Morss, N. Edelstein, J. Fuger (Springer, Dordrecht, 2008), p. 1652
8. B.G.C. Lackenby, V.A. Dzuba, V.V. Flambaum, Phys. Rev. A **98**, 042512 (2018)

9. J.W. van Spronsen, *The periodic system of chemical elements. A history of the first hundred years*. (Elsevier, Amsterdam, 1969)
10. W.B. Jensen, *Mendeleev and the periodic law* (Dover publications, Mineola, 2002)
11. E.M. Henley, A. Garcia, *Subatomic Physics*, 3rd edn. (World Scientific, 2007)
12. P. Indelicato, J.P. Santos, S. Boucard, J.P. Desclaux, Eur. Phys. J. D **45**, 155 (2007)
13. Tanmoy Chakraborty, Kamarujjaman Gazi, Dulal C. Ghosh, Molecular Physics **108**, 2081 (2010)
14. M. Guerra, P. Amaro, J.P. Santos, P. Indelicato, Atomic Data and Nuclear Data Tables **117–118**, 439 (2017)
15. H.G. Kuhn, *Atomic spectra* (Longmans, London, 1969)
16. J.C. Slater, *Quantum theory of atomic structure* (McGraw-Hill, New York, 1960)
17. G.W.F. Drake, in *Springer Handbook of Atomic, Molecular, and Optical Physics*, ed. by G.W.F. Drake (Springer-Verlag, New York, 2006), p. 199
18. V.A. Kostelecký, M.M. Nieto, Phys. Rev. A **32**, 3243 (1985)
19. I. Martin, International Journal of Quantum Chemistry **74**, 479 (1999)
20. E.U. Condon, G.H. Shortley, *The theory of atomic spectra* (Cambridge University Press, Cambridge, 1935)
21. M. Mizushima, *Quantum mechanics of atomic spectra and atomic structure* (W. A. Benjamin, New York, 1970)
22. B.H. Bransden, C.J. Joachain, *Physics of Atoms and Molecules*, 2nd edn. (Prentice Hall, Harlow, England, 2003)

Chapter 6
Coupling of Angular Momenta – The Vector Model

In the previous chapter, we introduced the central-field approximation (CFA) as a way to obtain separable basis functions for the energy eigenstates of multielectron atoms. These product states are solutions to a Schrödinger equation with a purely radial potential, and they are annotated with an electron orbital nomenclature (see sections 1.5 and 5.2.1). What we need to do now is to introduce all that was left out from the full Hamiltonian in that analysis, by applying perturbation theory, and while using the electron configurations as zero-order functions.

When we initially introduced the CFA in chapter 5 and established electron configurations with the aufbau principle, we mainly ignored two important contributions to the total Hamiltonian, namely all spin–orbit interactions and what is left of the sum of all the electron–electron repulsion terms, when we have separated out the mean of the radial part. The issue now at hand is to add these contributions. Eventually, we will then be able to further add other interactions, for example with external fields and higher electromagnetic moments of the nucleus.

What we have to deal with presently is the interactions between the orbital and spin angular momenta of *all* electrons. A model for doing this was developed by early spectroscopists, before quantum mechanics and before the salient features of atomic structure were known. It is a phenomenological model based on the addition of vectors, describing the different angular momenta, with constraints concerning the discrete nature of the vector sums. The latter rules were initially entirely based on empirical observations of spectra.

When studying atomic spectra, it was noted that groups of energy levels (derived from spectral lines) appeared together and that some specific types of atoms always had such groupings with particular multiplicities. For example, alkali atoms had states appearing in doublets, alkaline earths were shown to have singlets and triplets and nitrogen had its energy levels arranged in quartets and doublets. Working backwards, it was hypothesised that this has to do with interacting angular momenta. The ensuing model was developed before quantum mechanics, but it turns out that when the addition of angular momenta was subsequently put on a quantum

A. Kastberg, *Structure of Multielectron Atoms*, Springer Series on Atomic, Optical, and Plasma Physics 112, https://doi.org/10.1007/978-3-030-36420-5_6

mechanical footing, this vector model still gives excellent qualitative results, and it greatly facilitates a close study of basic atomic structure and atomic spectroscopy.

In the following, we will introduce the *vector model*, and quickly proceed to the application of quantum mechanical perturbation analyses. When doing this, we have to isolate some special situations, where some interactions dominate over others. The extremes of these special cases are the so-called *LS*-coupling and *jj*-coupling schemes. They are introduced in sections 6.3 and 6.4, and then treated in more detail in chapters 7 and 8. Albeit the above schemes are limiting cases, they will help us to understand a majority of atomic structure, and via interpolation schemes we can get a good grasp also of various intermediate situations.

6.1 The Concept of the Vector Model

The central-field approximation is one key to the justifications to the vector model. In the pre-quantum era, and after the introduction of the atomic models by Rutherford and Bohr, the formulation was rather along the lines that only a few of the negative charges in an atom contribute to the finer features of atomic structures. With an atomic model based on discrete orbitals, we could also say that only electrons in the outermost occupied orbitals give angular contributions to the structure, whereas the inner ones give a spherically symmetric contribution to the overall energy of the atom. This hypothesis was for early spectroscopists based just on observations. Today, we can give better rational arguments for it, and we will have a detailed look at the spherical symmetry of fully occupied orbitals in section 6.2.

Central potentials give comparatively simple solutions to the Schrödinger equation. The observation that atomic spectra can still be very complicated then naturally leads to the hypothesis that these complications have to do with valence electrons. Fully occupied orbitals combine to a spherical force field, but only valence electrons will exert net torques on each other. From this, we can gather that the number of valence electrons must have an important bearing on the energy level structure, and thus it is understandable why atoms found in the same column in the periodic system have qualitatively similar spectra.

The angular effects must, from a Newtonian point of view, be possible to describe as interactions between different angular momenta. These have the effect that the interacting momenta will all vary with time, whereas their overall sum will remain constant (for the moment we rule out dissipation). We have

$$\frac{\mathrm{d}\mathbf{L_i}}{\mathrm{d}t} = \boldsymbol{\tau} \ , \tag{6.1}$$

where $\mathbf{L_i}$ is the orbital angular momentum of one electron, and $\boldsymbol{\tau}$ is the torque. For two electrons, the Coulomb repulsion between them has the effect that neither $\mathbf{L}_1$ nor $\mathbf{L}_2$ will be constant in time. They will both precess about their vector sum, $\mathbf{L} = \mathbf{L}_1 + \mathbf{L}_2$, as illustrated in figure 6.1

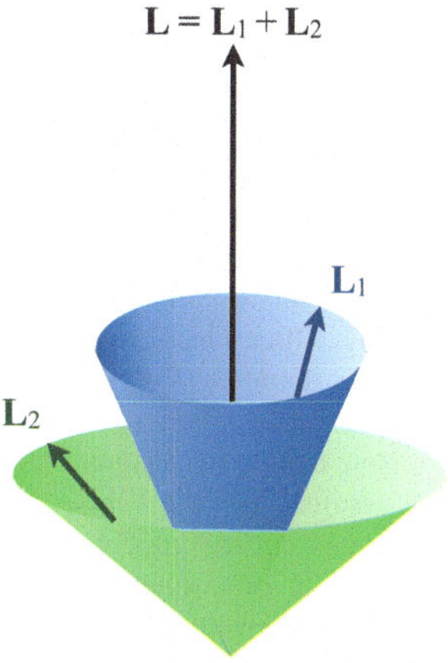

Fig. 6.1 Schematic illustration of the orbital angular momenta of two electrons, $\mathbf{L}_1$ and $\mathbf{L}_2$, which are coupled trough the electrostatic interaction. As a result, they will both precess, whereas their sum, $\mathbf{L}$, remains constant in time.

In quantum terminology, $\mathbf{L}_1^2$, $\mathbf{L}_2^2$, $\mathbf{L}^2$ and L_z commute with the Hamiltonian describing the interaction, whereas L_{1z} and L_{2z} do not. Analogously, for an electron spin interacting with the orbital angular momentum, via the spin–orbit effect, $\mathbf{L}_i$ and $\mathbf{S}_i$ will precess around their sum $\mathbf{J}_i$, as described in chapter 4.

Sums of angular momenta clearly play important roles, and thus we have to know with what restrictions these additions can be made. To start with, a sum of angular momenta is in itself an angular momentum. A general quantum mechanical angular momentum, $\mathbf{J}$, must be quantised such that the eigenvalues of $\mathbf{J}^2$ are $J(J+1)$, with J being either zero, a positive integer, or a positive half-integer. The eigenvalues of J_z are M_J, with possible values between $-J$ and J, separated by integer numbers (see appendix C).

With this, we can illustrate the addition rules with a few examples. More rigorous justifications for these rules are given in appendix C (see also [1]). Consider a number of valence electrons, with spins $\mathbf{S}_i$, interacting to form a total spin $\mathbf{S}=\sum_i \mathbf{S}_i$. For every individual electron, $s_i = 1/2$, and we assume that the orbital angular momenta (and/or the principal quantum numbers) are such that we do not have to bother about the Pauli principle. We write the quantum numbers associated with $\mathbf{S}^2$ and S_z as S and M_S, respectively (see section 1.5). The possible resulting values of S are illustrated in figure 6.2.

For an atom with two valence electrons, the possible results for the total spin are $S=0$ or $S=1$ — the two spins are either parallel or anti-parallel. This corresponds to the singlets and triplets described in section 2.2. With a third electron, the additional spin-vector must be added to the vector sum of the first two, either increasing or decreasing the total spin, and the possible values become $S=1/2$ or $S=3/2$. That gives spectral lines appearing in doublets and quartets, and we see that there is indeed a logical link between the multiplicity, and the quantity $2S+1$, as stated in earlier chapters (see section 1.5). It should be noted that some of the addition paths in figure 6.2 may in some instances be inhibited by the Pauli principle (depending on other quantum numbers).

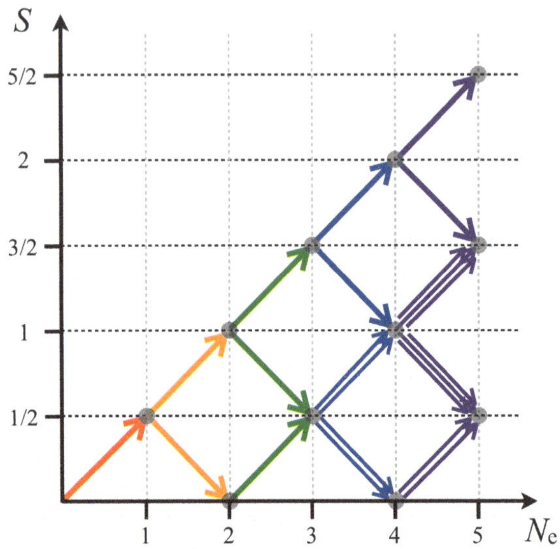

Fig. 6.2 Illustration of the vector addition of electron spins. On the x-axis is the number of electrons, and on the y-axis the total spin (the vector sum). With two electrons, only singlets ($S = 0$) and triplets ($S = 1$) are possible. Three electrons produces doublets ($S = 1/2$) and quartets ($S = 3/2$), and so on. Note that in this diagram, it is assumed that the Pauli principle is not an impediment (some of the other quantum numbers are assumed to be non-identical). Note also that for three or more electrons, some values of S can be reached by degenerate paths.

Addition of orbital angular momenta works in the same way, except that the different l_i cannot be half-integers. Take an example with one d-electron ($l_1 = 2$) and one p-electron ($l_2 = 1$). The possible values of the quantum number for the total angular momentum L are 3, 2 and 1. If there is a third electron, its angular momentum is added in the same way to the different sums of the first two. At this stage, the order of the summation does not matter.

In the above paragraphs, the different $\mathbf{L}_i$ and $\mathbf{S}_i$ couple, respectively, to $\mathbf{L}_i$ and $\mathbf{S}$. Remaining to take into account is the spin–orbit interaction. Here we are faced with a choice. We can first couple all the individual $\mathbf{L}_i$ and $\mathbf{S}_i$ to a number of different $\mathbf{J}_i = \mathbf{L}_i + \mathbf{S}_i$, whereafter all the individual $\mathbf{J}_i$ are summed into a grand total angular momentum, $\mathbf{J} = \sum_i \mathbf{J}_i$. Alternatively, we first form $\mathbf{L} = \sum_i \mathbf{L}_i$ and $\mathbf{S} = \sum_i \mathbf{S}_i$, and then we get to the same $\mathbf{J} = \mathbf{L} + \mathbf{S}$, having taken a different route. This is illustrated in figure 6.3 for an example with two electrons.

What path we take to $\mathbf{J}$ is actually crucial. Even when we exclude angular effects from inner, closed, orbitals we will have two angular momenta per valence electron, and all these will interact with each other. We are on the route to calculate valence electron energy contributions with perturbation theory, using CFA closed orbital configurations as zero-order states. The perturbation Hamiltonian is:

$$H_{\text{valence}} = H_{\text{ee}} + H_{\text{SO}} , \tag{6.2}$$

where the subscript ‘ee’ signifies the electron–electron repulsion. The two included terms are those from (5.6) and (4.19), and the sums therein are exclusively over the valence electrons. The most rigorous way to apply this is to take the two terms as *one* single perturbation, but that route is rarely practical. It typically leads to huge secular equations, and to problems that are mathematically too cumbersome to yield useful results.

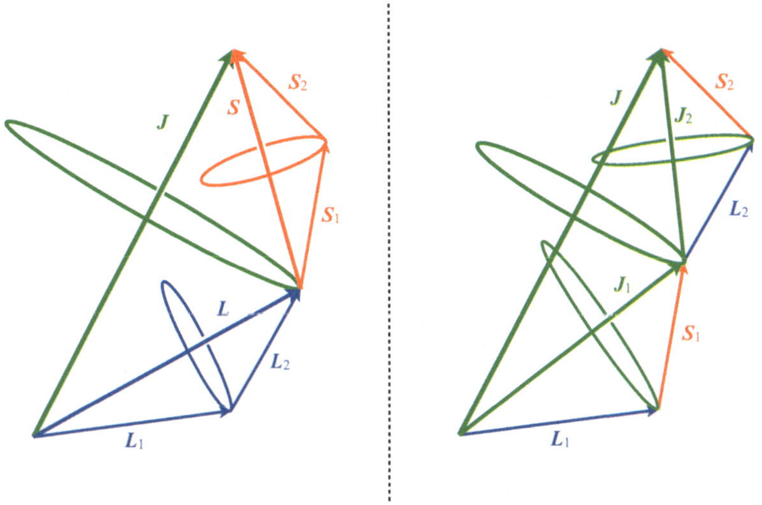

Fig. 6.3 Two different ways to couple the four electronic angular momenta of a two-electron atom ($\mathbf{L}_1$, $\mathbf{S}_1$, $\mathbf{L}_2$ and $\mathbf{S}_2$) in order to form the total electronic angular momentum $\mathbf{J}$. In the example in the left panel, the strongest angular interaction is that *between* the electrons. The individual $\mathbf{L}_i$ and $\mathbf{S}_i$ first couple to form $\mathbf{L}$ and $\mathbf{S}$, and then the spin–orbit interaction gives us $\mathbf{J}$. This is the *LS*-coupling approximation. In the right panel, the spin–orbit interaction is the most pronounced one. In this case — the *jj*-coupling one — the two individual $\mathbf{J}_i$ are first formed, and these then couple due to the torque involved in the electron–electron interaction, and this results in $\mathbf{J}$.

A better way forward is to instead treat the two terms in (6.2) one by one. For a perturbative calculation, that will only work if the largest contribution is treated first, and the second is taken as a subsequent perturbation. This is actually what is illustrated in figure 6.3. The two parts of the figure represent the two Hamiltonians H_{ee} and H_{SO} added as perturbations in different orders. Both pictograms represent approximations, and they constitute two limiting cases.

Accordingly, we need to find the relative importance of H_{ee} and H_{SO}, and the answer to that is not universal. For light atoms, and for a considerable part of the periodic system, the electrostatic interaction between valence electrons is much more important than the spin–orbit interaction. For such atoms, we first take into account H_{ee}, and the result is states referred to as *atomic terms*. For these, the first stage on the path to $\mathbf{J}$ has been to form $\mathbf{L}$ and $\mathbf{S}$ from the angular momenta of the individual electrons. This is referred to as *LS-coupling*, and it will be treated in section 6.3 and in chapter 7. In the final step (ignoring external fields and nuclear effects), the term H_{SO} is applied as a perturbation to the atomic terms. This reveals the *fine-structure*, which is characterised by the quantum numbers J and M_J. The good representation of the states in *LS*-coupling (the one that diagonalises the interaction Hamiltonians) is:

$$|\gamma, LSJM_J\rangle \;, \tag{6.3}$$

where γ is used as shorthand for the electron configuration.

The fine-structure Hamiltonian H_{SO} scales as Z^4 — see (4.14) — whereas H_{ee} has a linear scaling with Z (see table D.2). As a consequence, the approximation that the spin–orbit term is small compared to the electron–electron interaction may not hold for very heavy atoms. For such elements, we will instead approach the other limit, where we should inverse the order in which the two Hamiltonians are applied. That is what is illustrated in the right part of figure 6.3. This is called *jj-coupling* since we begin by forming the individual $\mathbf{J}_i$, and then we sum all these to form $\mathbf{J}$. The quantum numbers L and S will then not commute with the total Hamiltonian, and the diagonal representation must rather be:

$$|\gamma, j_1 \ldots j_N J M_J\rangle . \tag{6.4}$$

For three electrons, a more precise terminology for the scheme would be *jjj*-coupling and so on, but there is a value also in keeping a notation brief and simple. We will treat *jj*-coupling in section 6.4 and in chapter 8.

Obviously, both *LS*-coupling and *jj*-coupling are approximations. However, with these descriptions as a basis, we are able to describe also many intermediate scenarios. We will look closer at such atomic states in chapter 9.

6.2 Closed Orbitals

Before we begin to detail the *LS*- and *jj*-coupling schemes, where only valence electrons are involved, we will take a closer look at the electrons in inner, filled orbitals. We show that the contribution to atomic structure from electrons in closed orbitals is spherically symmetric, at the relevant order of approximation. It provides an energy contribution common to all states within an electronic configuration, but does not cause any splitting of levels within a configuration. Thus, restricting the summations inherent in the perturbation Hamiltonian (6.2) to valence electrons is justified.

A closed orbital is an electron orbital, designated by the quantum number combination nl, which has the maximum possible number of electrons in it, and the fact that orbitals can be thought of as 'filled' at all is a consequence of the Pauli principle. In the CFA, we designate each electron with a certain set of unique quantum numbers. From a stringent point of view, this is incorrect. In the real multielectron atom, all electrons are entangled, and each set of quantum numbers represent *occurrences*. Assigning specific quantum number labels to each electron is strictly speaking unphysical. Nevertheless, thinking of electron configurations in this way will work fine for figuring out most of the salient features of the atomic structure.

The Pauli principle tells us that no two electrons in an ensemble may have the same set of quantum numbers, since that would make the wave function exchange symmetric. This means that for a given set of n, l, and m_l, there can be only two electrons — with opposite spins. For each *orbital*, nl, we can allow $2(2l+1)$ electrons. Thus, an s-orbital is closed, or filled, when it has two electrons; a p-orbital takes six electrons, a d-orbital ten and an f-orbital fourteen.

This way of filling up the available space (the *aufbau principle* — see chapter 5) means that in a closed orbital, for every electron that has a positive value of m_l there must be another that is identical except for the opposite sign of m_l. Likewise, for every spin-up electron, there must be a spin-down one. As a consequence, all m_l and all m_s for all electrons in all closed orbitals will add up to zero. The projection quantum numbers are the only ones that can contribute to an orientation, and thus closed orbitals must be spherically symmetric.

For the spin–orbit Hamiltonian (4.19), where all vectors $\mathbf{L}_i$ and $\mathbf{S}_i$ are included in the summation, we can therefore ignore all electrons in closed orbitals — their contribution to the overall spin–orbit coupling will be zero. When formulating the perturbative Hamiltonian (4.19), it will thus suffice to include only valence electrons.

It is intuitively reasonable to assume that the angular contribution to the electron–electron interaction from closed orbitals will also be zero, based precisely on the observation that these filled orbitals are spherically symmetric. However, mathematically this is less obvious, and we will therefore, in the following section, put this argument on a more solid mathematical footing.

6.2.1 Electron–Electron Repulsion Involving Closed Orbitals

The complete electron–electron repulsion Hamiltonian (3.2), for an N-electron atom, can be split up into three terms. We consider an N-electron atom, and we assume that the first N_c electrons are in closed orbitals. The valence electrons are then the ones numbered between N_c+1 and N. The three terms, corresponding, respectively, to interactions between pairs of closed orbital electrons, to terms with one closed orbital electron and one valence electron, and finally to pairs of valence electrons, are:

$$
\begin{aligned}
H_{ee}^{(cc)} &= \sum_{i=1}^{N_c} \sum_{j>i}^{N_c} \frac{1}{r_{ij}} \\
H_{ee}^{(cv)} &= \sum_{i=1}^{N_c} \sum_{j=N_c+1}^{N} \frac{1}{r_{ij}} \\
H_{ee}^{(vv)} &= \sum_{i=N_c+1}^{N} \sum_{j>i}^{N} \frac{1}{r_{ij}} \ .
\end{aligned}
\tag{6.5}
$$

To calculate the respective energy contributions, we need to multiply both sides of the operators with Slater determinant wave functions corresponding to the relevant ensemble of electrons, and then integrate over all electron coordinates. This will leave us with sums of two-electron Coulomb and exchange integrals, of the type introduced in section 3.1.2 and in (3.10), and described in more detail in appendix D (see also sections 7.2.3 and 8.2.2). An important difference from chapter 3 is that the wave functions are not exactly hydrogenic, but rather ones emanating from the CFA.

The different terms in (6.5) will contain sums of six different types of two-electron integrals:

$$
\begin{aligned}
&J^{(\mathrm{cc})}_{n_il_i:n_jl_j} \;, \quad K^{(\mathrm{cc})}_{n_il_i:n_jl_j} \\
&J^{(\mathrm{cv})}_{n_il_i:n_jl_j} \;, \quad K^{(\mathrm{cv})}_{n_il_i:n_jl_j} \\
&J^{(\mathrm{vv})}_{n_il_i:n_jl_j} \;, \quad K^{(\mathrm{vv})}_{n_il_i:n_jl_j} \;.
\end{aligned}
\tag{6.6}
$$

The J's and K's in (6.6) are defined in (3.10). The goal of the present section is to show that only the pure valence electron integrals — that is $J^{(\mathrm{vv})}_{n_il_i:n_jl_j}$ and $K^{(\mathrm{vv})}_{n_il_i:n_jl_j}$ — contribute to the complexity of the atomic structure, and that these terms are the only ones that on average have an angular dependence. To this end, we will attempt to prove the relative disinterest of the other four types of terms, or at least their spherical symmetry.

Without loss of generality, this can be reduced to a study between *one* valence electron (indexed 'v', and with the quantum numbers n_v, l_v, $m_{l\mathrm{v}}$, and $m_{s\mathrm{v}}$) and *all* the electrons in *one* closed orbital. The latter are $2(2l_\mathrm{c}+1)$ in number, and they share the quantum numbers n_c and l_c. All possible values of $m_{l\mathrm{c}}$ will be present, from $m_{l\mathrm{c}} = -l_\mathrm{c}$ to $m_{l\mathrm{c}} = l_\mathrm{c}$, and for each of those there is one spin-up and one spin-down electron.

Every contribution of this type will contain sums of J- and K-integrals, each having $2(2l_\mathrm{c}+1)$ terms. For brevity we introduce the notations $J^{(\mathrm{cv})}$ and $K^{(\mathrm{cv})}$ (without subscripts) for these sums. This reduction of the problem will suffice for our study, since the same analysis can then be applied to for every valence electron and every closed orbital. The number of sums $J^{(\mathrm{cv})}$ and $K^{(\mathrm{cv})}$ will, for each closed orbital, equal the number of valence electrons. Towards the end of the section, we will argue that the same formalism can be used to average out also the contributions $J^{(\mathrm{cc})}_{n_il_i:n_jl_j}$ and $K^{(\mathrm{cc})}_{n_il_i:n_jl_j}$.

The Coulomb integral sums, with one valence electron interacting with all electrons in one filled orbital will be of the form:

$$
J^{(\mathrm{cv})} = \sum_{m_{l\mathrm{c}},m_{s\mathrm{c}}} \iint \left|\Psi_{n_\mathrm{c}l_\mathrm{c}m_{l\mathrm{c}}m_{s\mathrm{c}}}(\mathbf{q}_1)\right|^2 \left|\Psi_{n_\mathrm{v}l_\mathrm{v}m_{l\mathrm{v}}m_{s\mathrm{v}}}(\mathbf{q}_2)\right|^2 \frac{1}{r_{12}}\,\mathrm{d}\mathbf{q}_1\mathrm{d}\mathbf{q}_2 \,. \tag{6.7}
$$

The double sum in (6.7) goes from $-l_c$ to l_c in $m_{l\mathrm{c}}$, and from $-1/2$ to $1/2$ in $m_{s\mathrm{c}}$. The wave functions $\Psi_{nlm_lm_s}(\mathbf{q}_i)$ are products of the spatial parts, $\psi_{nlm_l}(\mathbf{r}_i)$, and the spin wave functions, $\zeta(m_{si})$. This is allowed at this level of approximation, since we are leaning on the CFA for the zero-order wave functions, and the small spin–orbit coupling is for the moment ignored.

The integrals over $\mathbf{r}_i$ are taken over all space for both electrons in a pair, and the inclusion of the spin contributes with a discrete sum. The latter will simply result in a factor of two. We thus rewrite (6.7) as:

$$J^{(\mathrm{cv})} = 2 \iint \sum_{m_{lc}=-l_c}^{l_c} \left|\psi_{n_c l_c m_{lc}}(\mathbf{r}_1)\right|^2 \left|\psi_{n_v l_v m_{lv}}(\mathbf{r}_2)\right|^2 \frac{1}{r_{12}} \, \mathrm{d}\mathbf{r}_1 \mathrm{d}\mathbf{r}_2 \,. \tag{6.8}$$

This sum can be factorised in radial and angular parts since:

$$\sum_{m_{lc}=-l}^{l} \left|\psi_{n_c l_c m_{lc}}(\mathbf{r}_1)\right|^2 = |R_{n_c l_c}(r_1)|^2 \sum_{m_{lc}=-l}^{l} \left|Y_{l_c m_{lc}}(\theta_1, \varphi_1)\right|^2 . \tag{6.9}$$

The summations in (6.8) and (6.9) run over $(2l_c+1)$ electrons with the same spin.

We now look at the summation of spherical harmonics in (6.9). Since the sum is over all possible values of m_l associated with a particular l, we can directly apply the addition theorem for spherical harmonics (C.53). Moreover, all the factors $|R_{n_c l_c}(r_i)|^2$ in the sum will be the same, since we are only considering one orbital (all n_c and l_c are the same). With all this taken into account, 6.9 simplifies to:

$$\sum_{m_{lc}=-l}^{l} \left|\psi_{n_c l_c m_{lc}}(\mathbf{r}_1)\right|^2 = \frac{2l_c+1}{4\pi} |R_{n_c l_c}(r_1)|^2 . \tag{6.10}$$

The next step is to express the factor $1/r_{12}$ as a sum of polynomials. To do this, we again take help from appendix C.3. According to (C.55) and (C.56), the operator can be written as:

$$\begin{aligned}
\frac{1}{r_{12}} &= \sum_{k=0}^{\infty} \sum_{q=-k}^{k} \frac{(k-|q|)!}{(k+|q|)!} \frac{(r_<)^k}{(r_>)^{k+1}} P_k^{(|q|)}(\cos\theta_1)\, P_k^{(|q|)}(\cos\theta_2)\, \mathrm{e}^{\mathrm{i}q(\varphi_1-\varphi_2)} \\
&= \sum_{k=0}^{\infty} \sum_{q=-k}^{k} \frac{4\pi}{2k+1} \frac{(r_<)^k}{(r_>)^{k+1}} Y_{kq}^*(\theta_1, \varphi_1)\, Y_{kq}(\theta_2, \varphi_2) \;,
\end{aligned} \tag{6.11}$$

with the $P_k^{(|q|)}(\cos\theta_i)$ being associated Legendre polynomials. In the fractions of radial powers, $r_<$ signifies the smaller of the radial parameters r_1 and r_2, and $r_>$ the larger.

Putting everything together, we get:

$$\begin{aligned}
J^{(\mathrm{cv})} = 2 \iint & \frac{(2l_c+1)}{4\pi} |R_{n_c l_c}(r_1)|^2 \, |R_{n_v l_v}(r_2)|^2 \, |Y_{l_v m_v}(\theta_2, \varphi_2)|^2 \\
& \times \sum_{k=0}^{\infty} \sum_{q=-k}^{k} \frac{4\pi}{2k+1} \frac{(r_<)^k}{(r_>)^{k+1}} Y_{kq}^*(\theta_1, \varphi_1)\, Y_{kq}(\theta_2, \varphi_2)\, \mathrm{d}\mathbf{r}_1 \mathrm{d}\mathbf{r}_2 \;,
\end{aligned} \tag{6.12}$$

or if we split up the integration:

$$J^{(\mathrm{cv})} = 2(2l_\mathrm{c}+1)\sum_{k=0}^{\infty}\sum_{q=-k}^{k}\frac{1}{2k+1}\int_0^\infty\int_0^\infty |R_{n_\mathrm{c}l_\mathrm{c}}(r_1)|^2\,|R_{n_\mathrm{v}l_\mathrm{v}}(r_2)|^2\,\frac{(r_<)^k}{(r_>)^{k+1}}\,r_1^2r_2^2\,\mathrm{d}r_1\mathrm{d}r_2$$
$$\times\int_0^\pi\int_0^{2\pi} Y_{kq}^*(\theta_1,\varphi_1)\,\sin\theta_1\,\mathrm{d}\theta_1\mathrm{d}\varphi_1$$
$$\times\int_0^\pi\int_0^{2\pi} |Y_{l_\mathrm{v}m_\mathrm{v}}(\theta_2,\varphi_2)|^2\,Y_{kq}(\theta_2,\varphi_2)\,\sin\theta_2\,\mathrm{d}\theta_2\mathrm{d}\varphi_2\;. \tag{6.13}$$

In the penultimate line of (6.13), we have an integral over all angular space of a single spherical harmonic Y_{kq}^*. This will return zero except for $k=q=0$, and for the latter values the integral equals $2\sqrt{\pi}$. Thus, only a single term in the double sum will be retained. In this, the product of the angular integrals in the last line will be $Y_{00}=1/\sqrt{4\pi}$ multiplied by the absolute value of a normalised spherical harmonic. This leaves us with the following final expression:

$$\begin{aligned} J^{(\mathrm{cv})} &= 2(2l_\mathrm{c}+1)\int_0^\infty\int_0^\infty |R_{n_\mathrm{c}l_\mathrm{c}}(r_1)|^2\,|R_{n_\mathrm{v}l_\mathrm{v}}(r_2)|^2\,\frac{1}{r_>}\,r_1^2r_2^2\,\mathrm{d}r_1\mathrm{d}r_2 \\ &= 2(2l_\mathrm{c}+1)\,F^{(0)}_{n_\mathrm{c}l_\mathrm{c}:n_\mathrm{v}l_\mathrm{v}}\;. \end{aligned} \tag{6.14}$$

In the last step, we have used a compact notation for the radial integral known as a *Slater F-integral* [2]. Such integrals will be properly introduced and treated in chapter 7 and in appendix D — see (7.25) and (D.22). However, the one important conclusion from (6.14), in the present context, is that the integral is independent of $m_{l\mathrm{v}}$, and thereby of the orientation of the valence electron. There will be an energy contribution from the interaction between the valence electron and the closed orbitals, but this will be the same for all electrons in the open orbitals. This is our desired result, which shows that integrals of the type $J^{(\mathrm{cv})}_{n_il_i:n_jl_j}$ can safely be omitted in the upcoming study of the energy splitting of states belonging to a single electronic configuration.

The next type of term is the corresponding one with exchange integrals involving closed orbitals electrons and one valence electron, that is $K^{(\mathrm{cv})}$. In the analysis of Li in section 3.1.3, we showed that exchange integrals are zero for pairs of electrons with opposite spins. This will be further elaborated on in section 7.1. For the moment, we make use of this intelligence in order to avoid having to sum over the spins. The expression then becomes:

$$K^{(\mathrm{cv})} = \sum_{m_{l\mathrm{c}}=-l_\mathrm{c}}^{l_\mathrm{c}}\iint \psi^*_{n_\mathrm{c}l_\mathrm{c}m_{l\mathrm{c}}}(\mathbf{r}_1)\,\psi^*_{n_\mathrm{v}l_\mathrm{v}m_{l\mathrm{v}}}(\mathbf{r}_2)\,\frac{1}{r_{12}}\,\psi_{n_\mathrm{v}l_\mathrm{v}m_{l\mathrm{v}}}(\mathbf{r}_1)\,\psi_{n_\mathrm{c}l_\mathrm{c}m_{l\mathrm{c}}}(\mathbf{r}_2)\,\mathrm{d}\mathbf{r}_1\mathrm{d}\mathbf{r}_2\,. \tag{6.15}$$

The wave functions are now just the spatial ones. The sum over m_{lc} still goes from $-l_c$ to l_c, but it is understood that the only electrons included in the sum are those that have $m_{sc} = m_{sv}$.

There are no differential operators in the integrand in (6.15), and thus we may take the factors in any order we want. We can then begin by taking the sum over the product of the closed orbital electrons, in a similar manner as was done for (6.9), but this time with two different electron coordinates. The summation only runs over the angular functions, and the sum of the products of spherical harmonics for two different coordinates is exactly what is given by the addition theorem for spherical harmonics (C.52). This leads to:

$$\sum_{m_{lc}=-l_c}^{l_c} \psi^*_{n_c l_c m_{lc}}(\mathbf{r}_1)\, \psi_{n_c l_c m_{lc}}(\mathbf{r}_2) = \frac{2l_c+1}{4\pi}\, R_{n_c l_c}(r_1)\, R_{n_c l_c}(r_2)\; P_{l_c}(\cos\omega)\,. \qquad (6.16)$$

The parameter ω in the Legendre polynomial P_{l_c} is here defined as the angle between the position vectors $\mathbf{r}_1$ and $\mathbf{r}_2$ (see further appendix C.3).

The factor $1/r_{12}$ will again be expressed in angular coordinates, but now we write also this with Legendre polynomials (C.55):

$$\frac{1}{r_{12}} = \sum_{k=0}^{\infty} \frac{r_<^k}{r_>^{k+1}}\, P_k(\cos\omega)\,. \qquad (6.17)$$

Multiplication of (6.17) with (6.16) yields:

$$\sum_{m_{lc}=-l_c}^{l_c} \psi^*_{n_c l_c m_{lc}}(\mathbf{r}_1)\, \psi_{n_c l_c m_{lc}}(\mathbf{r}_2)\, \frac{1}{r_{12}}$$
$$= \frac{2l_c+1}{4\pi} \sum_{k=0}^{\infty} R_{n_c l_c}(r_1)\, R_{n_c l_c}(r_2)\, \frac{r_<^k}{r_>^{k+1}}\, P_{l_c}(\cos\omega)\, P_k(\cos\omega)\,. \qquad (6.18)$$

Included in (6.18) is a product of two Legendre polynomials with the same argument. Such a product can be expanded as (see, for example, [2, 3] or [1]):

$$P_{l_c}(\cos\gamma)\, P_k(\cos\gamma) = \sum_{\lambda=0}^{\infty} A(k,l_c,\lambda)\, P_l(\cos\gamma)$$
$$= \sum_{\lambda=0}^{\infty} A(k,l_c,\lambda) \left[\sum_{\mu=-\lambda}^{\lambda} \frac{(\lambda-|\mu|)!}{(\lambda+|\mu|)!}\, P_\lambda^{(|\mu|)}(\cos\theta_1)\, P_\lambda^{(|\mu|)}(\cos\theta_2)\, \mathrm{e}^{\mathrm{i}\mu(\varphi_1-\varphi_2)} \right], \qquad (6.19)$$

where $A(k,l_c,\lambda)$ is a coefficient that depends on the orbital angular momentum quantum numbers, and where we now use associated Legendre Polynomials. The coefficient can be estimated in terms of radial integrals, but for the moment, we do not need to know its numerical values. The objective here is only to annul the directional dependence for the valence electron.

In the last step in (6.19) we have used the addition theorem yet again, and as a result we now have functions separable in r, θ and φ coordinates for the two electrons. Substituting (6.19) in (6.18), and then that result into (6.15), leads us to the following long expression:

$$
\begin{aligned}
K^{(\mathrm{cv})} = & \sum_{k=0}^{\infty} \sum_{\lambda=0}^{\infty} \sum_{\mu=-\lambda}^{\lambda} A(k, l_\mathrm{c}, \lambda) \frac{(2l_\mathrm{v}+1)(l_\mathrm{v}-|m_{l\mathrm{v}}|)!\,(2l_\mathrm{c}+1)(\lambda-|\mu|)!}{(l_\mathrm{v}+|m_{l\mathrm{v}}|)!\,(\lambda+|\mu|)!} \\
& \times \int_0^\infty \int_0^\infty R^*_{n_\mathrm{c}l_\mathrm{c}}(r_1)\, R^*_{n_\mathrm{c}l_\mathrm{c}}(r_2)\, R_{n_\mathrm{v}l_\mathrm{v}}(r_1)\, R_{n_\mathrm{v}l_\mathrm{v}}(r_2)\, \frac{r_<^k}{r_>^{k+1}}\, r_1^2 r_2^2\, \mathrm{d}r_1 \mathrm{d}r_2 \\
& \times \int_0^\pi P_{l_\mathrm{v}}^{(|m_{l\mathrm{v}}|)}(\cos\theta_1)\; P_\lambda^{(|\mu|)}(\cos\theta_1)\, \frac{\sin\theta_1}{2}\, \mathrm{d}\theta_1 \\
& \times \int_0^\pi P_{l_\mathrm{v}}^{(|m_{l\mathrm{v}}|)}(\cos\theta_2)\; P_\lambda^{(|\mu|)}(\cos\theta_2)\, \frac{\sin\theta_2}{2}\, \mathrm{d}\theta_2 \\
& \times \int_0^{2\pi} \frac{\exp[\mathrm{i}(\mu-m_{l_\mathrm{v}})\varphi_1]}{2\pi}\, \mathrm{d}\varphi_1 \int_0^{2\pi} \frac{\exp[-\mathrm{i}(\mu-m_{l_\mathrm{v}})\varphi_j]}{2\pi}\, \mathrm{d}\varphi_2 \; .
\end{aligned} \tag{6.20}
$$

This can be simplified considerably. Using the orthogonality of spherical harmonics, we see that the integrals over the zenith angles θ_1 and θ_2 return zero unless $\lambda = l_\mathrm{o}$, in which case each of them equals:

$$
\frac{(l_\mathrm{v}+|m_{l\mathrm{v}}|)!}{(2l_\mathrm{v}+1)(l_\mathrm{v}-|m_{l\mathrm{v}}|)!} \; . \tag{6.21}
$$

This will exactly cancel the prefactors containing $m_{l\mathrm{v}}$ in (6.20), and it also means that we can remove the summation over λ. The sum (over μ) of the integrals of the azimuthal angles will return exactly unity, since the integrals will be zero for $\mu \neq m_{l_\mathrm{v}}$. We can then rewrite (6.20) as:

$$
\begin{aligned}
K^{(\mathrm{cv})} = & \sum_{k=0}^{\infty} A(k, l_\mathrm{c}, l_\mathrm{v}) \frac{(2l_\mathrm{c}+1)}{(2l_\mathrm{v}+1} \\
& \times \int_0^\infty \int_0^\infty R^*_{n_\mathrm{c}l_\mathrm{c}}(r_1)\, R^*_{n_\mathrm{v}l_\mathrm{v}}(r_2)\, R_{n_\mathrm{c}l_\mathrm{c}}(r_2)\, R_{n_\mathrm{v}l_\mathrm{v}}(r_1)\, \frac{r_<^k}{r_>^{k+1}}\, r_1^2 r_2^2\, \mathrm{d}r_1 \mathrm{d}r_2 \\
= & \sum_{k=0}^{\infty} A(k, l_\mathrm{c}, l_\mathrm{v}) \frac{2l_\mathrm{c}+1}{2l_\mathrm{v}+1}\, G^{(k)}_{n_\mathrm{c}l_\mathrm{c}:n_\mathrm{v}l_\mathrm{v}} \; .
\end{aligned} \tag{6.22}
$$

This time, the radial integral is expressed as $G^{(k)}_{n_\mathrm{c}l_\mathrm{c}:n_\mathrm{v}l_\mathrm{v}}$ which is a *Slater G-integral*, defined in chapter 7 and in appendix D — see (7.26) and (D.23).

In (6.22), we see that $K^{(\mathrm{cv})}$ is independent of m_{l_v}, as was the case for $J^{(\mathrm{cv})}$ in (6.14). With this, we have confirmed that for any valence electron, the interaction with inner (closed) orbital electrons is spherically symmetric. It will add a constant multiplicative factor to the energies, but it will not contribute to the angular complexity.

The analysis that we have done for the interaction between a valence electron and entire closed orbitals, containing the integrals $J^{(cv)}_{n_il_i:n_jl_j}$ and $K^{(cv)}_{n_il_i:n_jl_j}$ can be made in the same way for the interaction between a closed orbital electron and all the other closed orbitals. The integrals $J^{(cc)}_{n_il_i:n_jl_j}$ and $K^{(cc)}_{n_il_i:n_jl_j}$ can be computed in an analogue way to that which led to (6.14) and (6.22). Also within a filled orbital, the presence of the whole range of m_{lc} and m_{sc} will assure that a spherically symmetric contribution. Thus the objective with this section has been achieved. We have justified that we, in the initial treatments of the angular effects of *LS*- and *jj*-couplings, need only to consider valence electrons.

One thing that should be noted, however, is that the preceding discussion only considered diagonal matrix elements. As shown in section 2.3, for a valence orbital involving two or more electrons, Slater determinant wave functions should be used, which means that also non-diagonal matrix elements can be non-zero. Using the same techniques as those used in this section, it can be shown that non-diagonal elements for interactions between valence electrons and closed orbitals will cancel for configurations where only one unfilled orbital is involved [2]. For states including interactions between configurations (which will be briefly covered in section 9.5), such contributions have to be considered.

6.3 *LS*-coupling

In the *LS*-coupling approximation, we assume that the interaction *between* valence electrons is so much stronger than the spin–orbit interaction, intrinsic for each electron, that the latter can be ignored in the first perturbation analysis. Once we have specified the electronic configuration, the following step is thus to form the quantum numbers L and S, and we will assume that the corresponding operators, $\mathbf{L}^2$ and $\mathbf{S}^2$, commute with the part of the Hamiltonian used at this stage. More generally, we will take this as a definition of *LS*-coupling schemes; that is, ones for which the Hamiltonian is diagonal in L and S.

We have shown in section 6.2 that we can ignore electrons in closed orbitals. The process of adding up the different $\mathbf{L}_i$ and $\mathbf{S}_i$ for the valence electrons, in order to form L and S, should be performed as was explained in section 6.1, and as illustrated in figure 6.2 for S.

Since the individual quantum numbers l_i are always zero or positive integers, this will be true also for L. The standard notation for L is analogous to that of the orbital angular momentum of individual electrons (see section 1.5), but with capital letters. A total angular momentum of $L = 0$ will be annotated 'S', $L = 1$ 'P', $L = 2$ 'D' and so on. This letter symbol will be the base of the *LS-coupling term*.

With the same reasoning, the total spin, S, must be half-integer for an odd number of electrons, and integer or zero for an even number. In an LS-coupling term, the value of S is indicated by writing the numerical value of $2S+1$ as a superscript to the left of the letter symbol for L. The quantity $2S+1$ is called the *multiplicity* of the term.

6.3.1 Atomic Terms

The atomic terms in LS-coupling are specified by the quantum numbers L and S. The standardised notation (see also section 1.5) is:

$$^{2S+1}L \;, \tag{6.23}$$

where a letter symbol is used for the value of L, as described in the previous section. The numbers L and S represent orbital and spin angular momenta that are vector sums of all valence electrons, each one described by four quantum numbers. Thus, it appears inadequate to use just two quantum numbers to fully describe the atomic state, and therefore we will now look more closely at the plurality of quantum numbers (for the complete atom) needed to describe an atomic state.

For an individual electron, we need the four quantum numbers n_i, l_i, m_{li} and m_{si} to describe a state, which corresponds to four degrees of freedom. For an N-electron atom, we can therefore assume that we have $4N$ degrees of freedom, and accordingly a need for 4N quantum numbers. Suppose that the total number of atoms in all closed orbitals is N_c and that the number of valence electrons — electrons in some open orbital — is N_v, with $N_\mathrm{c}+N_\mathrm{v}=N$. Within the CFA, we describe the closed orbitals with their corresponding part of the electron configuration, which will give us N_c pairs of n_i and l_i. In light of the Pauli principle — which is the effect making these orbitals 'closed' in the first place — this will fix all four relevant quantum numbers for all these electrons. We have no flexibility in choosing our N_c combinations of m_{li} and m_{si}. With the closed orbital electrons thus taken care of, this leaves us with another $4N_\mathrm{v}$ quantum numbers to be specified in order to fully describe a state.

The values of n_i and l_i for the N_v remaining electrons are given by the electron configuration in the CFA, providing us with $2N_\mathrm{v}$ quantum numbers. For the still outstanding $2N_\mathrm{v}$ numbers, things now depend on how many valence electrons we have. We typically specify the state (in this case synonymous with the atomic term) with the quantum numbers for the *total* orbital angular momentum and the *total* spin, plus either their projections or their vector sum — that is either as the kets $|LSM_LM_S\rangle$ or as $|LSLM_J\rangle$. If there is just one valence electron, only two degrees of freedom remains after the closed orbital electrons have been accounted for, and thus it appears as if the kets above give us too many quantum numbers. This is not actually a problem since, in a case of a single electron, we will always have $L=l$ and $S=s=1/2$, and thus there is no actual redundancy.

For two valence electrons, the sets $|LSJM_J\rangle$ or $|LSM_LM_S\rangle$ will do perfectly to provide a full description. If we have three or more valence electrons, however, the plain *LS*-coupling designations above are incomplete. With many valence electrons, a *parent term* (and sometimes even a *grandparent* one and so on) can be formed from the coupling of the first few electrons, to which the coupling of the last ones result in the final *LS*-coupling term. If the three or more valence electrons are in the same orbital (having identical n_i and l_i), this parentage is typically redundant, since the limitations posed by the Pauli principle will assure that the numbers of degrees of freedom and quantum numbers are commensurate after all.

To conclude, in the *LS*-coupling approximation we specify an atomic term for an N-electron atom as:

$$(n_1 l_1 \dots n_{N_c} l_{N_c}) n_{v1} l_{v1} \dots n_N l_N \;\; {}^{2S+1}L \;, \tag{6.24}$$

where the part of the electron configuration inside the parenthesis represents the closed orbitals electrons. This is more often than not omitted in the representation. If a parent term is needed, the representation will be (with the closed orbital configuration omitted for clarity):

$$n_{v1} l_{v1} \dots n_{N-1} l_{N-1} \, ({}^{2S_p+1}L_p) \; n_N \;\; {}^{2S+1}L \,. \tag{6.25}$$

For an atom with many valence electrons (for example, a d-orbital with four or more electrons), more than one parent term may be needed.

The atomic terms in (6.24) and (6.25) leave two quantum numbers unspecified, either M_L and M_S, or J and M_J. This degeneracy will be lifted when more effects are added as further perturbations, such as, for example, the spin–orbit interaction or interactions with external fields.

6.3.2 Two Non-Equivalent Electrons

To give concrete examples, we start by showing the possible atomic terms for an atom with two valence electrons that are *not* in the same orbital. That is, we have either one or both of the conditions $n_1 \neq n_2$ and $l_1 \neq l_2$ fulfilled (we now ignore the electrons of the inner orbital and index the valence electrons from one to N). This is the simplest case since we do not have to be careful about obeying the Pauli principle. We can never make these two electrons equivalent, regardless of how we specify m_{l1}, m_{l2}, m_{s1} and m_{s2}. We will give two specific examples — the configurations 2p3p and 3d4p.

For the 2p3p-configuration (or any n_1pn_2p with $n_1 \neq n_2$), we have $l_1 = l_2 = 1$ and as always $s_1 = s_2 = 1/2$. The values for m_{l1} and m_{l2} can be $+1$, 0 or -1, and the spins can be spin-up or spin-down. With all possible permutations of the four projection quantum numbers, this should give us 36 different states in total, degenerate in energy or not. When l_1 and l_2 couple due to the angular part of the electrostatic repulsion, the possible values for L are 2, 1 or 0. The two-electron spins, in turn, can be either parallel or anti-parallel, giving the possible values of 1 or 0 for S. This gives us the following possible terms for the configuration 2p3p:

$$^1\mathrm{D}\,,\,^3\mathrm{D}\,,\,^1\mathrm{P}\,,\,^3\mathrm{P}\,,\,^1\mathrm{S}\,,\,^3\mathrm{S}\ . \tag{6.26}$$

Suppose we choose the representation $|LSJM_J\rangle$ for the complete state; then the number of possible combinations of a term plus its corresponding possible values of J and M_J should be 36. The accounting of this is illustrated in table 6.1, where we have used the fact that the possible values for J are between $L+S$ and $|L-S|$, and that M_J ranges from J to $-J$. Had we instead chosen the representation $|LSM_LM_S\rangle$,

Table 6.1 Possible *LS*-coupling terms for a 2p3p-configuration, and possible values for J and M_J. The total number of states is 36.

Term	J	M_J	number of states
^{1}D	2	$-2\ldots2$	5
	3	$-3\ldots3$	7
^{3}D	2	$-2\ldots2$	5
	1	$-1\ldots1$	3
^{1}P	1	$-1\ldots1$	3
	2	$-2\ldots2$	5
^{3}P	1	$-1\ldots1$	3
	0	0	1
^{1}S	0	0	1
^{3}S	1	$-1\ldots1$	3
Total number of states:			36

we would have had to instead combine the terms with the possible values of M_L and M_S, and we could have confirmed that also this gives 36 states.

As a second example, we take a 3d4p-configuration. This time S is still 0 or 1, as always for an atom with two valence electrons. The total orbital angular momentum L can vary between 3 and 1. This means that the possible terms are:

$$^1\mathrm{F}\,,\,^3\mathrm{F}\,,\,^1\mathrm{D}\,,\,^3\mathrm{D}\,,\,^1\mathrm{P}\,,\,^3\mathrm{P}\ . \tag{6.27}$$

If we count the number of possible states, in either of the representations $|LSJM_J\rangle$ or $|LSM_LM_S\rangle$, we get 60.

6.3.3 Two Equivalent Electrons

The examples in the preceding section were comparatively simple since the precepts were such that violations of the need for exchange antisymmetry had already been avoided. However, it is common that some of the valence electrons occupy the same orbital. In particular, this will almost always be the case for atomic ground state configurations.

To exemplify atoms with equivalent valence electrons, and to show how the allowed termsindexatomic term!LS@*LS*-coupling can be identified, we take the configuration $2p^2$ (the ground state configuration for atom number 6, C). Had it not been for the Pauli principle, the possible terms would have been the same as for 2p3p, shown in (6.26). However, this time some states $|LSJM_J\rangle$ are incompatible with exchange antisymmetric wave functions. We must exclude atomic terms that are only possible for electrons with identical sets of n_i, l_i, m_{li} and m_{si}. We must also exclude some combinations of m_{l1}, m_{s1}, m_{l2} and m_{s2} that are identical with another combination for an exchange of two-electron coordinates.

A practical way to illustrate allowed and forbidden terms is to use a table representation of quantum numbers, introduced by Condon & Shortley [4]. Such a table is organised horizontally and vertically by the total angular momentum projection numbers $M_L = m_{l1} + m_{l2}$ and $M_S = m_{s1} + m_{s2}$. The entries in the table are combinations of *spin-orbitals*; essentially pairs of the individual m_{li} and m_{si} values.

To demonstrate how this works, we first show an example for the already treated configuration 2p3p in table 6.2, where the electrons were *not* equivalent. Each entry in this table is a pair of electrons, with the numerical values showing m_{li} and the plus/minus superscripts indicates m_{si}. We see that M_L can vary between 2 and −2,

Table 6.2 Diagram with the possible combinations of individual spin-orbitals (combinations of m_{li} and m_{si}), tabulated against M_L and M_S. This particular table is valid for any n_ipn_jp-configuration, where $n_i \neq n_j$. The total number of possible states is 36. This is in agreement with table 6.1.

2p3p		M_S		
		1	0	−1
	2	$(1^+\ 1^+)$	$(1^+\ 1^-)(1^-\ 1^+)$	$(1^-\ 1^-)$
	1	$(1^+\ 0^+)$ $(0^+\ 1^+)$	$(1^+\ 0^-)(1^-\ 0^+)$ $(0^+\ 1^-)(0^-\ 1^+)$	$(1^-\ 0^-)$ $(0^-\ 1^-)$
M_L	0	$(1^+\ {-1}^+)$ $(0^+\ 0^+)$ $({-1}^+\ 1^+)$	$(1^+\ {-1}^-)(1^-\ {-1}^+)$ $(0^+\ 0^-)(0^-\ 0^+)$ $({-1}^+\ 1^-)({-1}^-\ 1^+)$	$(1^-\ {-1}^-)$ $(0^-\ 0^-)$ $({-1}^-\ 1^-)$
	−1	$(0^+\ {-1}^+)$ $({-1}^+\ 0^+)$	$(0^+\ {-1}^-)(0^-\ {-1}^+)$ $({-1}^+\ 0^-)({-1}^-\ 0^+)$	$(0^-\ {-1}^-)$ $({-1}^-\ 0^-)$
	−2	$({-1}^+\ {-1}^+)$	$({-1}^+\ {-1}^-)({-1}^-\ {-1}^+)$	$({-1}^-\ {-1}^-)$

which means that the maximum value of L is two, and that $S=1$ is the maximum total spin, since M_S is between 1 and -1. We also see that we have 36 states in total, which is fully consistent with what we showed in section 6.3.2.

For the 2p3p-configuration, a diagram such as table 6.2 is superfluous. However, for equivalent electrons it will help. Consider now the same type of table for the $2p^2$ configuration. Initially, we can copy table 6.2, but then a number of terms have to be deleted. We have to remove pairs of identical electrons, such as $(1^+\,1^+)$. Furthermore, with equivalent electrons we have some pairs that are invariant at an exchange of electrons labels, such as $(1^+\,1^-)$ and $(1^-\,1^+)$. Of these, one has to go, since these two spin-orbitals cannot represent two different states (this argument is given a more mathematical form in section 7.4.1). After having done all the necessary deletions, we end up with table 6.3.

Table 6.3 Diagram for identification of *LS*-coupling terms not in conflict with the Pauli principle, for the $2p^2$-configuration (or any np^2-configuration). The entries are individual spin-orbitals, tabulated against M_L and M_S. The total number of allowed states is 15.

$2p^2$		M_S		
		1	0	−1
	2		$(1^+\ 1^-)$	
	1	$(1^+\ 0^+)$	$(1^+\ 0^-)$ $(1^-\ 0^+)$	$(1^-\ 0^-)$
M_L	0	$(1^+\ -1^+)$	$(1^+\ -1^-)$ $(0^+\ 0^-)$ $(1^-\ -1^+)$	$(1^-\ -1^-)$
	−1	$(0^+\ -1^+)$	$(0^+\ -1^-)$ $(0^-\ -1^+)$	$(0^-\ -1^-)$
	−2		$(-1^+\ -1^-)$	

We see that the 36 states possible in the configuration 2p3p have been reduced to just 15 for two *equivalent* p-electrons. A consequence of this is that some terms listed in (6.26), for 2p3p, are not feasible. For example, ^{3}D is not possible, since we cannot have $M_L=2$ and $M_S=1$ at the same time without violating the Pauli principle. Since there remain states with $M_S=1$ and $M_S=-1$, we need to have at least one triplet, and since we have excluded ^{3}D, but still have spin-orbitals for $M_L=1$ and $M_L=-1$, this triplet must be a ^{3}P-term. If we contemplate the possible values of J and M_J for ^{3}P, we find that it corresponds to 9 states or nine entries in table 6.4. We can then remove nine electron pairs from table 6.3, with M_S and M_L values both between 1 and -1.

For the remaining spin-orbital pairs in the table, we see that we must have a term ^{1}D, in order to account for the values $M_L=2$ and $M_L=-2$. This term will correspond to five states in the table, for $M_S = 0$ and M_L between 2 and -2. The only entry still remaining in table 6.3 is now a lone electron pair for $M_L = 0$ and $M_S = 0$. This can only correspond to ^{1}S. We conclude that the possible *LS*-coupling terms for the 2p^2-configuration are:

$$^1\mathrm{D}\,,\,^3\mathrm{P}\,,\,^1\mathrm{S}\ , \tag{6.28}$$

and in table 6.4 we show that this corresponds to a total number of states of 15, as predicted. The table also shows the ranges of J and M_J.

Table 6.4 Possible *LS*-coupling states for a 2p^2-configuration, with values for J and M_J. The total number of states is 15, with other terms present in table 6.1 (for a 2p3p configuration) being forbidden due to the Pauli principle.

Term	J	M_J	number of states
^{1}D	2	$-2\ldots 2$	5
	2	$-2\ldots 2$	5
^{3}P	1	$-1\ldots 1$	3
	0	0	1
^{1}S	0	0	1
Total number of states:			15

6.3.4 *More Than Two Valence Electrons*

Also with three or more valence electrons, it will make a big difference if all or some of the electrons are equivalent. Again, the best way to show how *LS*-coupling terms can be identified is through examples. Suppose that we have three identical p-orbital electrons, such as the ground state configuration in N, 2p^3. Making a diagram such as tables 6.2 or 6.3 may at a first glance appear unattractive, due to the large number of combinations of spin-orbitals, and the ensuing very large table. If the electrons had been inequivalent (such as the exotic doubly excited configuration 2p3p4p), we would have had 256 combinations of the different m_{li} and m_{si}, and thus a very cumbersome table (albeit a superfluous one).

However, with identical electrons, the vast majority of these trios of spin-orbitals must be excluded. Moreover, the symmetry of a diagram like table 6.3 means that it suffices to construct one quadrant of the table, while still including the $M_L = 0$ row and the possible $M_S = 0$ column. For the 2p^3 example, the diagram will then turn out as shown in table 6.5, that is greatly reduced. From this table, we can identify the possible *LS*-coupling terms as:

$$^2\mathrm{D}\,,\,^2\mathrm{P}\,,\,^4\mathrm{S}\,. \tag{6.29}$$

Table 6.5 Diagram for identification of *LS*-coupling terms not in conflict with the Pauli principle, for the $2p^3$ configuration (or any np^3 configuration). The entries are individual spin-orbitals, tabulated against M_L and M_S. The diagram is only partially displayed. The complete version has a mirrored version to the right for the values $M_S = -1/2$ and $-3/2$, and a lower part for the M_L values -1, -2 and -3, which is a mirrored copy of the corresponding positive values. The total number of possible states for this configuration is thus 20.

$2p^3$		M_S	
		$3/2$	$1/2$
M_L	3		
	2		$(1^+\ 1^-\ 0^+)$
	1		$(1^+\ 1^-\ -1^+)$ $(1^+\ 0^+\ 0^-)$
	0	$(1^+\ 0^+\ -1^+)$	$(1^+\ 0^+\ -1^-)$ $(1^+\ 0^-\ -1^+)$ $(1^-\ 0^+\ -1^+)$

If we count the number of possible values of J and M_J for these terms, and add them up, we get ten for ^2D, six for ^2P, and four for ^4S; thus a total of 20. This agrees with the number of spin-orbitals found from table 6.5.

If we instead consider an excited configuration of N, such as $2p^23s$, only a subset of all valence electrons are identical. In this situation, we have to first couple two of the electrons, and then add the third, at every step taking into account the Pauli principle as necessary. The final result is the same no matter which electron pair one starts with, but the most logical procedure, and the standardised one, is to begin with the most tightly bound electrons, and then successively add the more loosely bound.

In the particular case of $2p^23s$, we form a *parent term* from the two equivalent 2p-electrons, which is the same problem as the one we solved in (6.28). To these three parent terms, we then add the 3s-electron, with $l_3 = 0$ and $s_3 = 1/2$ and all possible projections of m_{l3} (in this case zero) and m_{s3}. The result is the following four possible *LS*-coupling terms for the $2p^23s$-configuration:

$$\begin{aligned} &2p^2(^1D)\,3s\ ^2D \\ &2p^2(^3P)\,3s\ ^4P \\ &2p^2(^3P)\,3s\ ^2P \\ &2p^2(^1S)\,3s\ ^2S\,. \end{aligned} \tag{6.30}$$

We have already shown that a $2p^2$-configuration results in 15 possible states, so it is reasonable to assume that the addition of the 3s-electron (with $l_3 = 0$ and having either spin-up or spin-down) should result in that number being doubled. If we look at the possible terms in (6.30), we see that the possibilities for J and M_J give us ten different states for ^{2}D, twelve for ^{4}P, six for ^{2}P and two for ^{2}S; thus a total of 30, as predicted.

6.3.5 Fine-Structure

We end this introduction to the *LS*-coupling scheme by adding the spin–orbit Hamiltonian of (4.19) as the next level of perturbation. To recapitulate, we have first applied the CFA in order to get a separable Schrödinger equation and an electron configuration zero-order state. We have then applied the electron–electron repulsion for the valence electrons, as a perturbation. That has split and/or shifted the configuration states into *LS*-coupling terms, with a degeneracy in energy of $(2L+1)(2S+1)$. We will now remove a part of that degeneracy by applying the spin–orbit interaction as yet another perturbation.

We begin by slightly reformulating and simplifying (4.19):

$$H_{\mathrm{SO}} = \sum_{i=1}^{N} \xi(r_i)\,\mathbf{L}_i \cdot \mathbf{S}_i = \sum_{i=1}^{N} \xi(r_i)\left(\mathbf{L} \cdot \mathbf{S} - \sum_{j \neq i}^{N} \mathbf{L}_i \cdot \mathbf{S}_j\right) \approx \left[\sum_{i=1}^{N} \xi(r_i)\right] \mathbf{L} \cdot \mathbf{S}\,. \quad (6.31)$$

The summation only runs over the valence electrons. For the vectors without indices, the definitions are $\mathbf{L}=\sum_i \mathbf{L}_i$ and $\mathbf{S}=\sum_i \mathbf{S}_i$. The approximation in the last step is justified by the acknowledgement that the coupling between the orbital angular momentum of one electron and the spin of another is a small effect. A complementary point of view is that a spherical average of electron–electron interactions to the spin–orbit coupling is incorporated in the radial fine-structure function $\xi(r_i)$. The so-called *spin-other-orbit interaction* can be treated more accurately, and outlines of this are provided in section 9.1 and appendix F.

The Hamiltonian given in the last line of (6.31) commutes with $\mathbf{J}$. This means that it will be diagonal in J and M_J, and this can also be seen as the essence of the *LS*-coupling approximation. Non-diagonal terms in J would signal a departure from *LS*-coupling, and would make the last equality in (6.31) invalid. It should, however, be noted that by its very nature, the spin–orbit Hamiltonian does not commute with the individual components of $\mathbf{L}$ and $\mathbf{S}$. M_L and M_S will not provide good quantum numbers, but within the approximation that $\mathbf{L^2}$ and $\mathbf{S}^2$ do commute with (6.31), the diagonal representation can be taken as:

$$|\gamma LSJM_J\rangle\,. \quad (6.32)$$

The energy correction caused by the perturbation (6.31) is:

$$E^{(\mathrm{SO})}(\gamma\ ^{2S+1}L_J) = \left\langle \sum_{i}^{N} \xi(r_i) \right\rangle \langle \mathbf{L} \cdot \mathbf{S} \rangle\,. \quad (6.33)$$

The first factor in (6.33) can only be calculated analytically if we know an explicit expression for the wave function, which we typically do not. It can, however, be derived empirically from spectroscopic data, and it can also be estimated by approximative and/or numerical methods (see chapters 7 and 14).

We define the *fine-structure factor* for a certain term as:

$$A(\gamma LS) \equiv \left\langle \gamma LS \middle| \sum_i^N \xi(r_i) \middle| \gamma LS \right\rangle . \tag{6.34}$$

Here, γ is as usual short for the electron configuration. We see that the fine-structure factor depends on L and S, making it specific for each atomic term, but we also see that it is independent of J.

The second factor of (6.33) can be calculated as was done in (4.11). That is, we have:

$$E^{(\mathrm{SO})}(\gamma\, {}^{2S+1}L_J) = \frac{A(\gamma LS)}{2}\left[J(J+1) - L(L+1) - S(S+1)\right] . \tag{6.35}$$

This means that every LS-coupling term will be split up in $2S+1$ *fine-structure levels* (or $2L+1$ if $S > L$). This will be annotated by adding the quantum number J to the LS-coupling term, as in:

$$^{2S+1}L_J . \tag{6.36}$$

The $(2J+1)$-fold degeneracy in M_J will remain.

In figures 6.4 and 6.5, we illustrate in partial energy level diagrams how this can play out. These examples are for the $2p^2$ and $2p^3$ configurations, respectively, for which the terms were found in (6.28) and (6.29). The energy values given are experimentally obtained ones for neutral C ($2p^2$) and N ($2p^3$) [6].

For the doublet $^2D_{5/2}$ and $^2D_{3/2}$ in figure 6.5, we see that the fine-structure factor can be negative (this is called *inverted order* — see section 7.3.1 for a discussion on this). The small splittings in the multiplets, compared to the energy difference between the terms, suggest that LS-coupling are good approximations for both these examples. We will defer a discussion on the energy order of the terms to chapter 7.

For every term with a multiplicity of two or larger, the fine-structure factor $A(\gamma LS)$ can be deduced from experimental data and (6.35). In figure 6.4, we see that the interval between $^3P_2$ and $^3P_1$ is close to twice as great as that between $^3P_1$ and $^3P_0$. This is roughly consistent with (6.35).

We can take a closer look at the energy interval between fine-structure levels in LS-coupling. From (6.35), we can calculate such intervals as:

$$E^{(\mathrm{SO})}(\gamma\, {}^{2S+1}L_J) - E^{(\mathrm{SO})}(\gamma\, {}^{2S+1}L_{J-1}) = A(\gamma LS)\, J . \tag{6.37}$$

This is known as *Landé's interval rule*, which shows that the interval between two adjacent levels is directly proportional to the value of J for the one of them with the highest J. This rule is very useful in empirical analyses of spectra, and it can be used

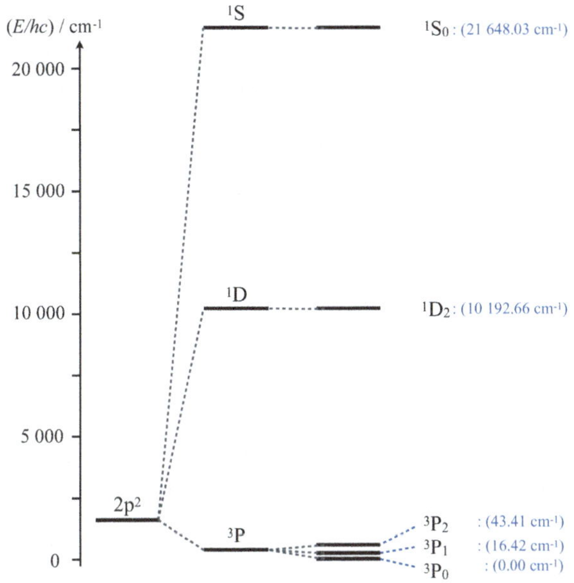

Fig. 6.4 Energy level diagram for the ground state configuration $2p^2$ in C. The diagram illustrates the stepwise perturbation theory approach in *LS*-coupling. An electron configuration within the CFA has first been chosen as a zero-order state. Thereafter, the Hamiltonian H_{ee} has been applied as the principal perturbation, followed by the spin–orbit Hamiltonian, H_{SO}. The energy of the ground state $2p^2\,{}^3P_0$ is taken as zero energy, and other energies (in units of wave numbers) are experimentally obtained ones [5, 6].

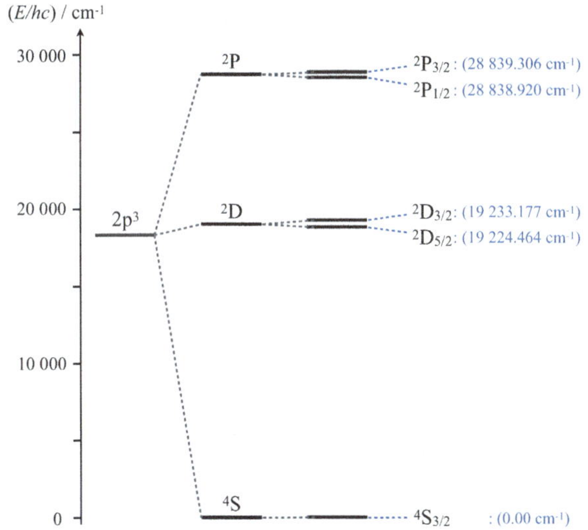

Fig. 6.5 Energy level diagram for the ground state configuration $2p^3$ in N, analogous to figure 6.4. The energy of the ground state $2p^3\,{}^4S_{3/2}$ is taken as zero, and other energies (in units of wave numbers) are experimentally obtained ones [5, 6].

as a measure of the validity of the LS-coupling approximation. If the multiplet is at least a triplet, the following ratio may be tested against experimental data:

$$\frac{E_{J+1} - E_J}{E_J - E_{J-1}} = \frac{J+1}{J} \, . \tag{6.38}$$

For example, for a ^{3}P-term, this ratio should be 2 for a situation well described by LS-coupling. For the case of C in figure 6.4, the ratio is not quite 2, but just over 1.6. A brief discussion on this discrepancy follows in section 6.5.

6.4 *jj*-coupling

The LS-coupling described in the preceding chapter is a limiting case, and at the other extreme end of this scale, we have jj-coupling. This occurs when the energy contribution of $H_{\rm SO}$ (for the valence electrons) in (6.2) is so large that $H_{\rm ee}$ can at first be neglected. From the simple variational analysis in (3.24) in chapter 3, and from table D.2, we can note that the electron–electron interaction Hamiltonian scales linearly with the atomic number. From chapter 4 and (D.2), we learn that the spin–orbit Hamiltonian, in contrast, is proportional to the fourth power of Z. The conclusion is that the LS-coupling approximation will get progressively worse for heavier atoms, while we at the same time approach the jj-coupling regime.

In terms of the vector model, a jj-coupling situation means that we first have to couple the $\mathbf{L}_i$ and $\mathbf{S}_i$ of each individual valence electron (we can still take the average effect of all closed orbitals as spherically symmetric). That is, we form the quantum numbers j_i and m_{ji} from all the l_i, m_{li}, s_i and m_{si}, following the rules for quantum mechanical vector addition, outlined in section 6.1. For two valence electrons, this is what is illustrated in the right panel on figure 6.3. The quantum numbers m_{li} and m_{si} will no longer be 'good' since their corresponding angular momentum projections are not constants of motion under the interaction.

The Hamiltonian used for forming the terms in jj-coupling is:

$$H_{\rm SO} = \sum_{i=1}^{N_{\rm v}} \xi(r_i) \, \mathbf{L}_i \cdot \mathbf{S}_i \, , \tag{6.39}$$

summed over the valence electrons. The CFA is still applied in the same way. The ensemble of closed orbital electrons, together with the radial single-electron effects for the valence electrons (see chapters 7 and 8 for details) form the zero-order states that are to be perturbed. Those states will be degenerate in m_{li} and m_{si}, and after forming Slater determinant superpositions of combinations of these quantum numbers, the spin–orbit Hamiltonian can be applied as a perturbation.

An important difference between (6.39) and the Hamiltonian used in LS-coupling is the absence of cross-couplings between the electrons. Solutions such as those represented by the ket in (6.4) are at this stage not entangled. Equation (6.39) is a sum of single-electron Hamiltonians, and we can designate the product state solutions as:

$$(n_1 l_1 j_1, \ldots, n_{N_v} l_{N_v} j_{N_v}) . \tag{6.40}$$

In this sequence we only include electrons outside closed orbitals. Sets of quantum numbers such as in (6.40) are the jj-coupling terms. Note that we have ignored the degeneracy in m_{ji} in the description.

The fine-structure in jj-coupling is what we get when the remaining Hamiltonian, H_{ee}, is applied as a subsequent perturbation to (6.39). In the vector model image, all the $\mathbf{J}_i$ now combine to a $\mathbf{J}$, as illustrated in figure 6.3. Just as for *LS*-coupling, we will get fine-structure levels described by quantum numbers J, with a $(2J+1)$-fold degeneracy in M_J.

We have seen that *LS*-coupling is a good approximation for light atoms. In contrast, jj-coupling rarely appears in pure form. Even for very heavy atoms, we often see a mixed case. Some of these will be presented in chapter 9. One set of elements that display almost pure jj-coupling are some highly charged heavy ions. In these, the strong binding energy experienced by the valence electrons will diminish the relative importance of the Coulomb exchange term, whereas relativistic effects (such as the spin–orbit interaction) will be accentuated.

6.4.1 *jj*-coupling Terms

The terms in jj-coupling are made up of the ensemble of all the j_i among the valence electrons. Even though the electrons do not interact at this level of approximation, there will still be some combinations of quantum numbers that are prohibited by the Pauli principle, and the procedure to follow in order to identify these will resemble that used for *LS*-coupling.

The notation for jj-coupling has not become as standardised as that for *LS*-coupling. The one we will use in this book is that where the electrons in open orbitals are divided into groups with common values of n_i, l_i and j_i. The electron configuration of closed orbitals is annotated in the same way as for *LS*-coupling. Two examples of our notation for the terms are the following:

$$\begin{aligned} &[6\mathrm{p}^2_{1/2}] \\ &[5\mathrm{d}^2_{5/2}\, 5\mathrm{d}_{3/2}] . \end{aligned} \tag{6.41}$$

In the first line, we have a term with two valence electrons, with $n_1 = n_2 = 6$, $l_1 = l_2 = 1$, and $j_1 = j_2 = 1/2$. Note that this combination does not violate the Pauli principle. It can (and must) correspond to two different values of m_{ji}: $1/2$ and $-1/2$. These projection quantum numbers could for example correspond, respectively, to spin–orbit coupled orbitals with $m_{l1} = 1$ and $m_{s1} = -1/2$, and $m_{l2} = -1$ and $m_{s2} = 1/2$. In the second example of (6.41), we have three 5d-electrons. Two of them have $j_1 = j_2 = 5/2$, and the third has $j_3 = 3/2$.

For non-equivalent electrons, identifying the possible jj-coupling terms is an easy task. As an example, we take the configuration 6p7p. The possible terms will be:

$$[6p_{3/2}\,7p_{3/2}]\,,\ [6p_{3/2}\,7p_{1/2}]\,,\ [6p_{1/2}\,7p_{3/2}]\,,\ [6p_{1/2}\,7p_{1/2}]\,. \tag{6.42}$$

For the sake of comparison with the LS-coupled 2p3p-configuration shown in table 6.1, we can tabulate the terms in (6.42), with the respective possible J-values, and keep a record of the possible number of states. This is done in table 6.6. We

Table 6.6 Possible jj-coupling states, with values of J, for a 6p7p-configuration. The total number of states is 36.

Term	J	M_J	number of states
$[6p_{3/2}\,7p_{3/2}]$	3	$-3\ldots3$	7
	2	$-2\ldots2$	5
	1	$-1\ldots1$	3
	0	0	1
$[6p_{3/2}\,7p_{1/2}]$	2	$-2\ldots2$	5
	1	$-1\ldots1$	3
$[6p_{1/2}\,7p_{3/2}]$	2	$-2\ldots2$	5
	1	$-1\ldots1$	3
$[6p_{1/2}\,7p_{1/2}]$	1	$-1\ldots1$	3
	0	0	1
Total number of states:			36

find that the number of states is consistent with what it was when we coupled two non-equivalent p-electrons with LS-coupling (see table 6.1).

6.4.2 Equivalent Electrons

An atom with two equivalent electrons, for which jj-coupling plays an important role, is the ground configuration of Pb, $6p^2$. Even in this case, the jj-coupling is not quite pure, but we will use this as an example nevertheless. We know from table 6.6 that had the electrons been non-equivalent, we would have had M_J-values between 3 and -3, and we would have had 36 states altogether. In order to figure out which states that are not forbidden by the Pauli principle, we can tabulate the terms in a manner similar to that used in section 6.3.3. Such a diagram is shown in

Table 6.7 Diagram for identification of *jj*-coupling terms, not in conflict with the Pauli principle, for the $6p^2$-configuration. The entries in the table are *jj*-orbitals, that is combinations of m_{j1} and m_{j2}. They are tabulated against M_J and pairs of j_1 and j_2. Possible values of J for the different terms are given at the foot of the table. The total number of allowed terms is 15.

$6p^2$		j_1 j_2		
		3/2 3/2	3/2 1/2	1/2 1/2
	3			
	2	(3/2 1/2)	(3/2 1/2)	
	1	(3/2 −1/2)	(3/2 −1/2) (1/2 1/2)	
M_J	0	(3/2 −3/2) (1/2 −1/2)	(1/2 −1/2) (−1/2 1/2)	(1/2 −1/2)
	−1	(1/2 −3/2)	(1/2 −3/2) (−1/2 −1/2)	
	−2	(−1/2 −3/2)	(−1/2 −3/2)	
	−3			
possible J		2 , 0	2 , 1	0

table 6.7, where the entries in the table are pairs of the quantum numbers m_{j1} och m_{j2} (*jj-orbitals*). In total we get 15 states that are not prevented by the Pauli principle. This is the same number of states as those possible for the equivalent *LS*-coupling case, exemplified by the $2p^2$-configuration in table 6.3.

A conclusion from table 6.7 is that the allowed *jj*-terms for the $6p^2$-configuration are:

$$
\begin{aligned}
&[6p^2_{3/2}]_2 \\
&[6p^2_{3/2}]_0 \\
&[6p_{3/2}\,6p_{1/2}]_2 \\
&[6p_{3/2}\,6p_{1/2}]_1 \\
&[6p^2_{1/2}]_0 \;.
\end{aligned}
\tag{6.43}
$$

In (6.43), we have also added the values of J (the *jj*-coupling fine-structure) as a subscript to the terms. In chapter 8 we will give a more complete account of which configurations that can lead to which *jj*-coupling terms.

6.4.3 Fine-Structure in jj-coupling

In order to make the jj-coupling description complete, we must add the remaining term in (6.2). The final component to be added as a perturbation is the electron–electron interaction Hamiltonian. H_{ee} commutes with $\mathbf{J}^2$ and J_z, which means that a good diagonal representation is:

$$|n_1 l_1 j_1 \ldots n_N l_n j_N \, JM_J\rangle. \tag{6.44}$$

This time, the electrostatic mutual repulsion couples the different $\mathbf{J}_i$, and this has the effect that the individual m_{ji} will no longer be good quantum numbers.

To form the quantum number J, one has to follow the standard rules for addition of quantum mechanical angular momenta, outlined in section 6.1. As an example, we can anew take a jj-coupled p^2-configuration, as illustrated in table 6.7. If we had had two non-identical p-electrons, both the individual j_i could have been either $3/2$ or $1/2$. That would have given three different combinations. For $j_1 = j_2 = 3/2$, possible values of J are 3, 2, 1 or 0. For $j_1 = 3/2$, $j_2 = 1/2$ (or inversely), the values are instead 2 and 1. For the last combination, $j_1 = j_2 = 1/2$, we have to have either $J = 1$ or $J = 0$. When the electrons are instead identical, we saw in the analysis accompanying table 6.7 that some of these combinations will be excluded, and we end up with the terms shown in (6.43).

In figure 6.6, we show the lowest energy states in Pb, which has the ground state configuration 6p^2. It should be noted that the jj-coupling is not pure in the ground state of Pb. In an LS-coupling scheme (see figure 6.4), the three lowest states would belong to the term ^{3}P, and the two upper ones would have been ^{1}D and ^{1}S. In many treaties of the lead atom, these levels are actually annotated with the LS-coupling notation, simply because that kind of labelling is more succinct, more well known, and generally applied in a more consistent fashion. If the jj-coupling had been purer in Pb, the separation between the jj-coupling terms in figure 6.6 would have been more pronounced, and the fine-structure splitting within the terms would have been smaller.

In chapter 8, we will look into this coupling scheme in more detail. We will study more complex electronic configurations — with more than two electrons — and we add quantitative analyses.

6.5 *LS*- and *jj*-coupling

The vector model of the atom predates the wave function formalism in quantum mechanics. Notwithstanding, the model remains powerful and also greatly facilitates a more rigorous quantum mechanical analysis. Getting acquainted with the vector model, and with the different vector coupling schemes, remains a key to the understanding of the structure of multielectron atoms.

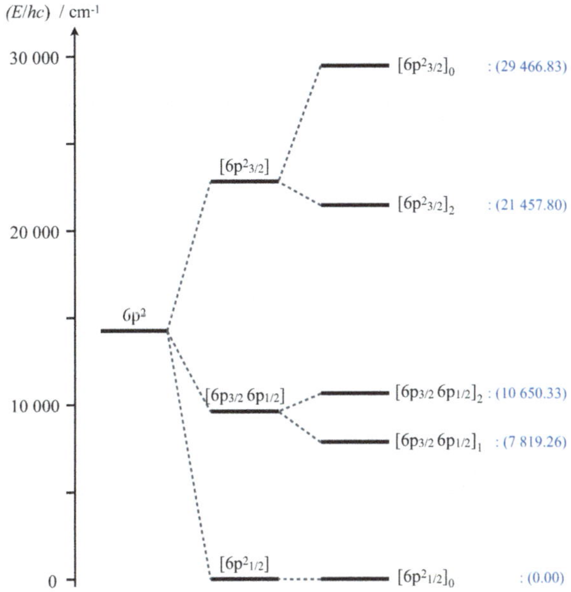

Fig. 6.6 Energy level diagram for the ground state configuration $6p^2$ in Pb. This should be compared to the *LS*-coupling case, shown in figure 6.4 for a light atom. In the present figure, the energy of the ground state $[6p^2_{1/2}]_0$ is taken as zero, and other energies (in units of wave numbers) are experimentally reported ones [6].

The two schemes that have been treated in this chapter are *LS*-coupling and *jj*-coupling. As has been explained, these are really limiting situations that will never be more than approximations, but with a good knowledge of these, one can get quite far with deciphering atomic structure also for intermediate cases.

The division of angular momentum couplings into separate perturbations, made in the two main vector coupling schemes covered in this chapter (and in the two following ones) can be seen as a competition between two effects. The spin–orbit interaction is small compared to the electron repulsion between electrons for small atoms. However, it scales as the fourth power of Z, and its relative importance increases as we advance along the periodic chart.

Different p^2 configurations were analysed in sections 6.3 and 6.4, for the two different principal coupling schemes. As a further comparison, we show in figure 6.7 a diagram of the five levels that belong to the ground states of the elements Si, Ge, Sn and Pb; with the respective configurations $3p^2$, $4p^2$, $5p^2$, and $6p^2$. At the light atom end of the horizontal scale, Si has a typical *LS*-coupling signature, with one triplet and two clearly distinct singlets. As we move towards the heavy Pb, it is clear that the spin–orbit interaction becomes more and more important, and we end up in a situation relatively well described by *jj*-coupling. For Sn, it is evident that a description intermediate between *LS*-coupling and *jj*-coupling should be more appropriate than any of those approximations.

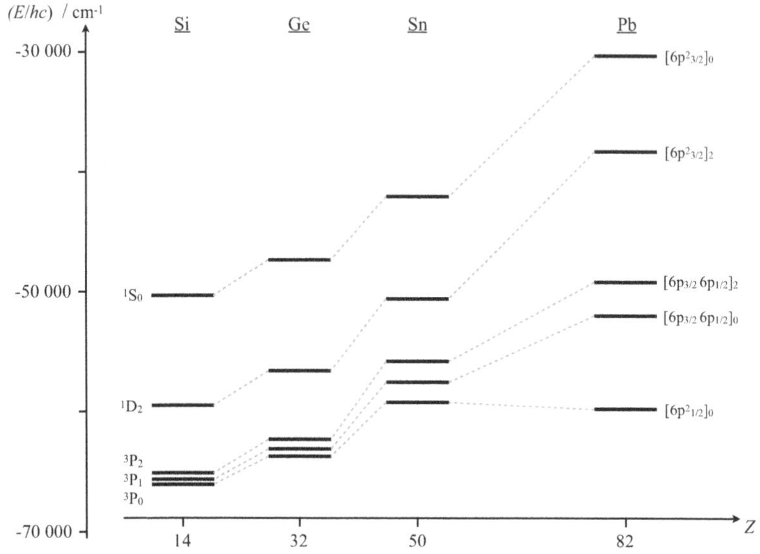

Fig. 6.7 Experimentally obtained energies [6] for levels belonging to the ground state configurations of the atoms Si, Ge, Sn and Pb (the respective configurations are $3p^2$, $4p^2$, $5p^2$ and $6p^2$). On the horizontal axis is the atomic number Z. As a common zero energy, the respective first ionisation limits have been taken. The splitting in the triplet state of Si has been exaggerated in the figure, in order to make it at all visible. All other energies are to scale. For low Z, the five energies are ordered well in keeping with *LS*-coupling — one triplet and two distinct singlets — and the levels are labelled with *LS*-coupling notation. At the other end of the scale, the ordering is more akin to *jj*-coupling, with two pairs and one distinct level, and we use *jj*-coupling designations. In between, we can perceive the gradual transition from *LS*-coupling to *jj*-coupling.

This progression from *LS*-coupling to *jj*-coupling in the p^2 configuration is probably the vertical column in the periodic chart where this progression is the clearest. For most others, the middle part of the table will see more configuration mixing, and the high Z part will not have as marked *jj*-coupling signatures as is the case for Pb. The $2p^2$ ground state element C (left out of figure 6.7) is also well described by *LS*-coupling (see figure 6.4). However, we have not included this in the comparative chart in figure 6.7, because the lack of a d-orbital for C makes that comparison less relevant. Moreover, for C the energies are significantly affected by a spin–spin interaction between the valence electrons (this effect is more prominent for light atoms — see chapter 9), a consequence of which will be a departure from the Landé rule (6.38).

We can also compare the progression illustrated in figure 6.7 by analysing the validity of the Landé interval rule (6.38) for the elements and states in question. In figure 6.8, we show the ratio of the two energy intervals that correspond to the three lowest levels shown for all elements in figure 6.7 — the levels that would be part of a triplet in a case of good *LS*-coupling. For pure *LS*-coupling, this ratio should be exactly two. As can be seen, it is very close to the ideal *LS*-coupling value for Si, but then, with increasing Z, a tentative *LS*-coupling assumption gets progressively worse.

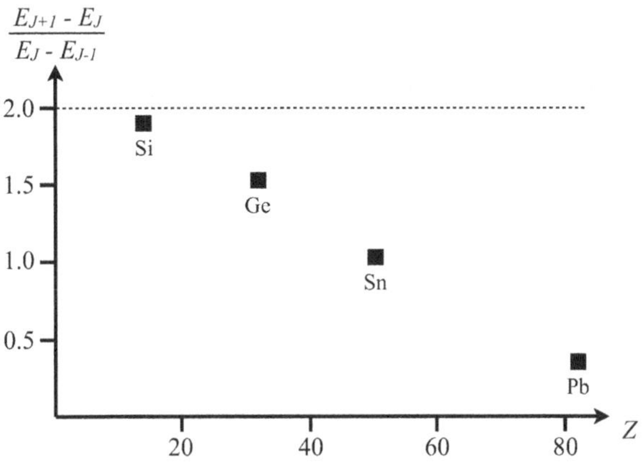

Fig. 6.8 Illustration of the gradual departure from the Landé interval rule for the p^2 elements Si, Ge, Sn and Pb, with increasing Z. The ideal *LS*-coupling value of 2, for the quantity $(E_{J+1}-E_J)/(E_J-E_{J-1})$, is shown as a dotted line. What can be concluded from the graph is that for the lightest element of the four, *LS*-coupling is a very good approximation, whereas for the heaviest, it is very poor. For the intermediate cases, Ge and Sn, the situation is indeed intermediate.

In the upcoming two chapters, we will look into *LS*-coupling and *jj*-coupling in more detail. After that, we will in chapter 9 follow up with an analysis of atomic states for which neither of the two limiting schemes gives a satisfactory description.

Further Reading

The theory of atomic spectra, by Condon & Shortley [4]
Quantum theory of atomic structure, by Slater [2]
Atomic spectra, by Kuhn [7]
Atomic Many-Body Theory, by Lindgren & Morrison [1]
Atomic Physics, by Foot [8]

References

1. I. Lindgren, J. Morrison, *Atomic Many-Body Theory*, 2nd edn. (Springer Verlag, Berlin, 1986)
2. J.C. Slater, *Quantum theory of atomic structure* (McGraw-Hill, New York, 1960)
3. G.B. Arfken, H.J. Weber, F.E. Harris, *Mathematical Methods for Physicists: A Comprehensive Guide*, 7th edn. (Academic Press, Amsterdam, 2012)
4. E.U. Condon, G.H. Shortley, *The theory of atomic spectra* (Cambridge University Press, Cambridge, 1935)
5. C.E. Moore, *Tables of Spectra of Hydrogen, Carbon, Nitrogen, and Oxygen Atoms and Ions* (CRC Press, Boca Raton, 1993)

6. A. Kramida, Y. Ralchenko, J. Reader, and NIST ASD Team. NIST Atomic Spectra Database (ver. 5.3). [Online]. Available: http://physics.nist.gov/asd (2018). Accessed: 2019-07-14
7. H.G. Kuhn, *Atomic spectra* (Longmans, London, 1969)
8. C. Foot, *Atomic Physics* (Oxford University Press, Oxford, 2005)

Chapter 7
LS-Coupling

The *LS*-coupling scheme was introduced in chapter 6 as a special case for the treatment of angular effects on atomic wave functions, within the central-field approximation. Each individual electron has both orbital angular momentum and spin, and when we have N electrons, we have in total $2N$ different angular momenta, which all interact with each other via electromagnetic couplings. This means that for $N > 1$, the full wave function of the system always has a plethora of angular interactions between different electrons, and modelling all of these in one go is prohibitively complex.

When we try to disentangle this complication by first applying the CFA, and by thereafter treating the interactions between angular momenta as perturbations, in decreasing order of magnitude, we get different limiting cases. *LS*-coupling is the one where we assume that we can first apply solely the angular contribution of the electrostatic interaction *between* the electrons as a perturbation, and only after that the spin–orbit interactions.

In this chapter, we will expand the discussion on *LS*-coupling from chapter 6 in several ways. We extend the initial treatment, introduced in section 6.3, to more complex situations; put the theory on a more rigorous theoretical footing, and go into more detail. Initially, we focus on the so-called *LS*-coupling atomic terms, while temporarily ignoring spin–orbit couplings. We will explore the possible terms emanating from different configurations, and make estimates of the term energies. After that, add the fine-structure to the treatment, and eventually, we explore the functional form of the *LS*-coupling states.

In the course of this exertion, we will have to hone the notation used in chapter 6 for Hamiltonians and energies; the complexity and the discrimination between *LS*-coupling and *jj*-coupling require some rigour. This is explained incrementally in the chapter, and the notation will be summarised in figure 7.2 and table 7.12 in section 7.5.

A. Kastberg, *Structure of Multielectron Atoms*, Springer Series on Atomic, Optical, and Plasma Physics 112, https://doi.org/10.1007/978-3-030-36420-5_7

7.1 Atomic Terms

The *LS*-coupling atomic terms are what we get when we apply the first perturbation in the *LS*-coupling scheme to the electronic configuration; that is, the electron–electron Coulomb Hamiltonian H_{ee} summed over all valence electrons. The electron configuration provides the zero-order states, which are separable in individual electron coordinates. It should be re-stressed, however, that even though an electron configuration is an assignment of N hydrogen-like orbitals to N electrons, it is in a strict sense not correct to say that each individual electron *is* in one specific orbital. The cross terms have the effect that the true eigenstate will always be an entangled one.

The electron–electron Hamiltonian (5.6) commutes with all the total angular momentum operators of the atom: $\mathbf{L}^2$, L_z, $\mathbf{S}^2$, S_z, $\mathbf{J}^2$ and J_z. When we take into account the spin-orbit interaction (4.19), the total Hamiltonian will no longer commute with L_z and S_z. Presently, however, with H_{SO} neglected, there is no interaction between the individual orbital angular momenta, $\mathbf{L}_i$, and the individual spins, $\mathbf{S}_i$, of individual electrons. Hence, the corresponding operators for the full atom, $\mathbf{L} = \sum_{i=1}^{N} \mathbf{L}_i$ and $\mathbf{S} = \sum_{i=1}^{N} \mathbf{S}_i$, do not couple, and H_{ee} — the perturbation to which we temporarily restrict the analysis — will not have any non-diagonal elements in M_L or M_S.

7.1.1 Allowed LS-coupling Terms

We recall the notation for *LS*-coupling atomic terms, introduced in section 1.5, and used in section 6.3. The perturbation H_{ee} breaks the degeneracy in quantum numbers L and S, and these two numbers will be used to specify a *term*. L is for historic reasons annotated with a letter — S, P, D, F, G, ... for $L = 0, 1, 2, 3, 4, \ldots$. To indicate S, one writes the *multiplicity* — $2S+1$ — as a superscript to the left of the letter indicating L.

Added to that, atomic data tables often indicate the term parity. An odd term gets a lowercase ‘o’ as a superscript to the *L*-letter. For example, the ground state term of C is the even term $2p^2\,{}^3P$, whereas one of the lower excited states is the odd term $2p3s\,{}^3P^o$. The parity of terms matter when considering radiative transitions or cross-couplings between states, but will mostly be omitted from the notation in this volume.

Using diagrams such as tables 6.2, 6.3 and 6.5, we can deduce the possible *LS*-coupling terms for every conceivable electron. If all electrons are non-equivalent, the task is easy. When they are equivalent, and we have to cater for the Pauli principle, it is more challenging. The case with equivalent electrons is often more relevant, since many ground state configurations will have exclusively equivalent electrons.

We saw in tables 6.3 and 6.5 that with more valence electrons, it gets more complicated to find the allowed terms. This problem is, however, alleviated due to a symmetry consideration. Consider N_v equivalent valence electrons in an orbital that can accommodate maximum N_o electrons. The possible terms for an atom with $N_o - N_v$ valence electrons will be exactly the same as for one with N_v. For example, a p^2-configuration will have the same possible *LS*-coupling terms as p^4, a d^7-configuration will yield an identical group of terms as d^3, and so on.

This could be proven by constructing charts like tables 6.3 and 6.5, but instead of entering the m_{li} and m_{si} values of the occupied spin-orbitals, one could do that for the unoccupied ones. The only difference in the table would be that positive values of m_{li} and m_{si} would correspond to negative values of M_L and M_S, but that does not change the conclusion.

In table 7.1, we provide a table with allowed *LS*-coupling terms for all s- p- and d-configurations of equivalent electrons. Separately, in table 7.2, we show the same thing for f-configurations. It suffices to include the valence electrons in this analysis,

Table 7.1 Allowed *LS*-coupling terms for the electronic configurations s^w, p^w and d^w (w is the number of equivalent electrons). The subscripts in brackets for a few terms symbolise that these occur more than once for the configuration in question. The subscript 'o' indicates odd parity. This is possible only for an odd number of electrons, with an odd value of l.

Electron configuration	Allowed terms				
s		^2S			
s^2	^1S				
p , p^5		$^2P^o$			
p^2 , p^4	^1S , ^1D		^3P		
p^3		$^2P^o$, $^2D^o$		$^4S^o$	
p^6	^1S				
d , d^9		^2D			
d^2 , d^8	^1S , ^1D ^1G		^3P , ^3F		
d^3 , d^7		^2P , $^2D_{(2)}$ ^2F , ^2G ^2H		^4P , ^4F	
d^4 , d^6	$^1S_{(2)}$, $^1D_{(2)}$ ^1F , $^1G_{(2)}$ ^1I		$^3P_{(2)}$, ^3D $^3F_{(2)}$, ^3G ^3H		^5D
d^5		^2S , ^2P $^2D_{(3)}$, $^2F_{(2)}$ $^2G_{(2)}$, ^2H ^2I		^4P , ^4D ^4F , ^4G	^6S
d^{10}	^1S				

Table 7.2 Allowed *LS*-coupling terms for the electronic configurations f^w (where w is the number of equivalent electrons).

Electron config.	Allowed terms						
f , f^{13}		$^2F^o$					
f^2 , f^{12}	^1S , ^1D ^1G , ^1I		^3P , ^3F ^3H				
f^3 , f^{11}		$^2P^o$, $^2D^o_{(2)}$ $^2F^o_{(2)}$, $^2G^o_{(2)}$ $^2H^o_{(2)}$, $^2I^o$ $^2K^o$, $^2L^o$		$^4S^o$, $^4D^o$ $^4F^o$, $^4G^o$ $^4I^o$			
f^4 , f^{10}	$^1S_{(2)}$, $^1D_{(4)}$ ^1F , $^1G_{(4)}$ $^1H_{(2)}$, $^1I_{(3)}$ ^1K , $^1L_{(2)}$ ^1N		$^3P_{(3)}$, $^3D_{(2)}$ $^3F_{(4)}$, $^3G_{(3)}$ $^3H_{(4)}$, $^3I_{(2)}$ $^3K_{(2)}$, ^3L ^3M		^5S , ^5D ^5F , ^5G ^5I		
f^5 , f^9		$^2P^o_{(4)}$, $^2D^o_{(5)}$ $^2F^o_{(7)}$, $^2G^o_{(6)}$ $^2H^o_{(7)}$, $^2I^o_{(5)}$ $^2K^o_{(5)}$, $^2L^o_{(3)}$ $^2M^o_{(2)}$, $^2N^o$ $^2O^o$		$^4S^o$, $^4P^o_{(2)}$ $^4D^o_{(3)}$, $^4F^o_{(4)}$ $^4G^o_{(4)}$, $^4H^o_{(3)}$ $^4I^o_{(3)}$, $^4K^o_{(2)}$ $^4L^o$, $^4M^o$		$^6P^o$, $^6F^o$ $^6H^o$	
f^6 , f^8	$^1S_{(4)}$, ^1P $^1D_{(6)}$, $^1F_{(4)}$ $^1G_{(8)}$, $^1H_{(4)}$ $^1I_{(7)}$, $^1K_{(3)}$ $^1L_{(4)}$, $^1M_{(2)}$ $^1N_{(2)}$, ^1Q		$^3P_{(6)}$, $^3D_{(5)}$ $^3F_{(9)}$, $^3G_{(7)}$ $^3H_{(9)}$, $^3I_{(6)}$ $^3K_{(6)}$, $^3L_{(3)}$ $^3M_{(3)}$, ^3N ^3O		^5S , ^5P $^5D_{(3)}$, $^5F_{(2)}$ $^5G_{(3)}$, $^5H_{(2)}$ $^5I_{(2)}$, ^5K ^5L		^7F
f^7		$^2S^o_{(2)}$, $^2P^o_{(5)}$ $^2D^o_{(7)}$, $^2F^o_{(10)}$ $^2G^o_{(10)}$, $^2H^o_{(9)}$ $^2I^o_{(9)}$, $^2K^o_{(7)}$ $^2L^o_{(5)}$, $^2M^o_{(4)}$ $^2N^o_{(2)}$, $^2O^o$ $^2Q^o$		$^4S^o_{(2)}$, $^4P^o_{(2)}$ $^4D^o_{(6)}$, $^4F^o_{(5)}$ $^4G^o_{(7)}$, $^4H^o_{(5)}$ $^4I^o_{(5)}$, $^4K^o_{(3)}$ $^4L^o_{(3)}$, $^4M^o$ $^4N^o$		$^6P^o$, $^6D^o$ $^6F^o$, $^6G^o$ $^6H^o$, $^6I^o$	$^8S^o$
f^{14}	^1S						

since closed orbitals will only add a ^{1}S contribution (see section 6.2). The contribution from filled inner orbitals will change the energy, but only with an additive quantity that will be the same for all terms belonging to the configuration. It will not contribute to the complexity of the structure.

Rather few of the terms in tables 7.1 and 7.2 occur for atoms in pure, or even almost pure, form. To start with, the *LS*-coupling approximation will fail when the spin–orbit term of the total Hamiltonian gains importance, which will increasingly be the case as we go towards heavier atoms. For example, this will affect essentially all atoms with 4f- and 5f-configuration ground states — the lanthanides and the actinides. Moreover, there are many examples for which different configurations are almost degenerate, which will in a strict sense invalidate the CFA. Notwithstanding, it is still common to identify terms in the *LS*-coupling scheme, and to describe real observed energy levels as superpositions of these. Some examples of superpositions of *LS*-coupling configuration will be appraised in section 9.5.

In tables 7.1 and 7.2, the numbers in brackets after some terms indicate that the term appears more than once for the configuration. This can only occur with at least three valence electrons. For such atoms, the quantum numbers L, S, M_L and M_S are not sufficient to uniquely cover all permutations of individual m_{li} and m_{si} (see also section 6.3.4). As an example, we give the term chart for a d^3-configuration in table 7.3. If we pick out the terms from this, with the technique described in section 6.3, we first identify ^{2}H. When we proceed, we get ^{2}G, ^{4}F and ^{2}F, but we then find that we have to use ^{2}D twice in order to account for all table entries. Finally, we get single occurrences of ^{4}P and ^{2}P, in accordance with table 7.1.

For atoms with electrons in more than one open orbital, each contribution can be identified by help of tables 7.1 and 7.2. Thereafter, the different sets of terms (corresponding to intermediate values of L and S) can be added together, and none of the thus obtained vector sums will be inhibited by the Pauli principle.

An atomic ground state for which this occurs is Cr. We see in table 5.1 that the ground state configuration is 3d^{5}4s. Table 7.1 gives a large number of allowed *LS*-coupling terms — doublets, quartets and one sextet — for the 3d^5 part. The 4s electron will not change L, but its spin means that the multiplicity changes. We will have singlets, triplets, quintets and one septet. As evidenced by table 5.1, the ground state of Cr is indeed ^{7}S.

7.1.2 Genealogy of Atomic Terms

When a term occurs more than once for a configuration, the level can be specified further if it is possible to discriminate between different vectorial paths to the final vector coupling term. As an example, we consider three non-equivalent electrons with the l_i-quantum numbers 0, 1 and 2. An analysis of the possible terms that may emanate from this spd-configuration yields the result: ^{4}F, ^{2}F$_{(2)}$, ^{4}D, ^{2}D$_{(2)}$, ^{4}P and ^{2}P$_{(2)}$. That is, all three doublets occur twice. For this example, there is a way to differentiate between the terms that have the same notation, by identifying *parent terms*.

Table 7.3 Diagram for identification of *LS*-coupling terms not in conflict with the Pauli principle (see section 6.3.3), for a d^3 configuration. The entries are individual spin-orbitals, tabulated against M_L and M_S. The allowed terms are: ${}^2G$, ${}^4F$, ${}^2F$, ${}^2D_{(2)}$, ${}^4P$ and ${}^2P$, with ${}^2D$ occurring twice.

d^3		M_S	
		3/2	1/2
	6		
	5		$(2^+\ 2^-\ 1^+)$
	4		$(2^+\ 2^-\ 0^+)\ (2^+\ 1^+\ 1^-)$
	3	$(2^+\ 1^+\ 0^+)$	$(2^+\ 2^-\ {-1}^+)\ (2^+\ 1^+\ 0^-)$ $(2^+\ 1^-\ 0^+)\ (2^-\ 1^+\ 0^+)$
M_L	2	$(2^+\ 1^+\ {-1}^+)$	$(2^+\ 2^-\ {-2}^+)\ (2^+\ 1^+\ {-1}^-)$ $(2^+\ 1^-\ {-1}^+)\ (2^-\ 1^+\ {-1}^+)$ $(2^+\ 0^+\ 0^-)\ (1^+\ 1^-\ 0^+)$
	1	$(2^+\ 1^+\ {-2}^+)$ $(2^+\ 0^+\ {-1}^+)$	$(2^+\ 1^+\ {-2}^-)\ (2^+\ 1^-\ {-2}^+)$ $(2^-\ 1^+\ {-2}^+)\ (2^+\ 0^+\ {-1}^-)$ $(2^+\ 0^-\ {-1}^+)\ (2^-\ 0^+\ {-1}^+)$ $(1^+\ 1^-\ {-1}^+)\ (1^+\ 0^+\ 0^-)$
	0	$(2^+\ 0^+\ {-2}^+)$ $(1^+\ 0^+\ {-1}^+)$	$(2^+\ 0^+\ {-2}^-)\ (2^+\ 0^-\ {-2}^+)$ $(2^-\ 0^+\ {-2}^+)\ (2^+\ {-1}^+\ {-1}^-)$ $(1^+\ 1^-\ {-2}^+)\ (1^+\ 0^+\ {-1}^-)$ $(1^+\ 0^-\ {-1}^+)\ (1^-\ 0^+\ {-1}^+)$

We may first add the s- and p-electrons. This gives the two terms ${}^3P$ and ${}^1P$. If we then add the d-electron to the two terms above, we find the following possible total terms for the spd-configuration (all of odd parity, but we omit the superscript):

$$\begin{aligned} n_1 s n_2 p\,({}^3P)\,n_3 d &\Rightarrow {}^4F,\,{}^2F,\,{}^4D,\,{}^2D,\,{}^4P,\,{}^2P \\ n_1 s n_2 p\,({}^1P)\,n_3 d &\Rightarrow {}^2F,\,{}^2D,\,{}^2P \quad . \end{aligned} \tag{7.1}$$

The *LS*-coupling terms in the brackets are the parent terms. The states with the same final term, but with different parents, will for most examples have different energies, and the way the quantum numbers have been specified in (7.1) is a representation that renders H_{ee} diagonal.

We could instead have begun by adding the s- and d-electrons. That had resulted in the parent terms (${}^3D$) and (${}^1D$), and then the same set of final terms. The logical choice of order in which to add the electron angular momenta is to begin with the most tightly bound electrons. This means that the final electron is added to an ionic configuration, which is further specified by the parent term. In the example in table 7.3 — with a d^3-configuration — it is not possible to specify parentage in this way. The three electrons are identical, and we have no way to discriminate between the two ${}^2D$-terms.

A pure d^3-configuration is an example of a configuration which is relatively rare. Looking at the periodic system in figure 5.2, and the ground state configurations in table 5.1, it appears as if vanadium might have this configuration in its ground state. However, the tabulated ground state configuration is $3d^34s^2$, which is because the 3d and 4s electrons have very similar binding energies. It is therefore more proper to treat V as an atom with five valence electrons, even if, in the lowest energy state, two of the five electrons fill up an s-orbital. Indeed, the lowest energy *LS*-coupling term of the excited configuration $3d^44s$ lie lower than most terms in the ground state configuration $3d^34s^2$. The lower energy states in vanadium are a good example for illustrating almost degenerate orbitals. Therefore we will look at this atom more closely, and in figure 7.1 we show a partial Grotrian diagram for V (with fine-structure omitted).

Many things can be learned from figure 7.1. One is that the distribution of energy levels for terms of different configurations overlap. This indicates that here the CFA is a less than perfect approximation, when it is applied to the five outermost electrons. The angular part of the electrostatic repulsion between electrons is too strong to treat fully as a perturbation, with decent accuracy. The *LS*-coupling notation is still pertinent, but the assignment of quantum numbers should be seen as what the leading terms are in superposition states. Most terms in the diagram in figure 7.1 are strongly mixed. As pointed out earlier, we are not able to dedicate parent terms for the $3d^34s^2$ configuration. This holds true also for $3d^5$. For $3d^44s$, we have first given a parent term for the $3d^4$ configuration of the positive vanadium ion, and to that we have added a 4s electron, yielding the term for the neutral atom. For the configuration $3d^34s4p$ in figure 7.1, we have identified two *ancestral terms* — one for $3d^3$ and one for 4s4p. This is called for, because the final atomic terms can be related to either $3d^3(^4F)\,4s4p(^3P)$ or to $3d^3(^4F)\,4s4p(^1P)$.

7.2 Energies of *LS*-coupling Terms

Calculating absolute theoretical energies within the CFA is a daunting task, and for the radial part one frequently has to resort to heavy numerical calculations (see chapter 14 for a brief outline). This section specifically looks at energy contributions arising when we apply perturbation theory in order to compute *relative energies* of *LS*-coupling terms. In itself, this rarely gives accurate absolute state energies. However, it provides an understanding of the energy ordering of terms, and it gives insight into orders of magnitude of energy separations. Even though the calculations are quantitative, the main objective is to gain physical understanding.

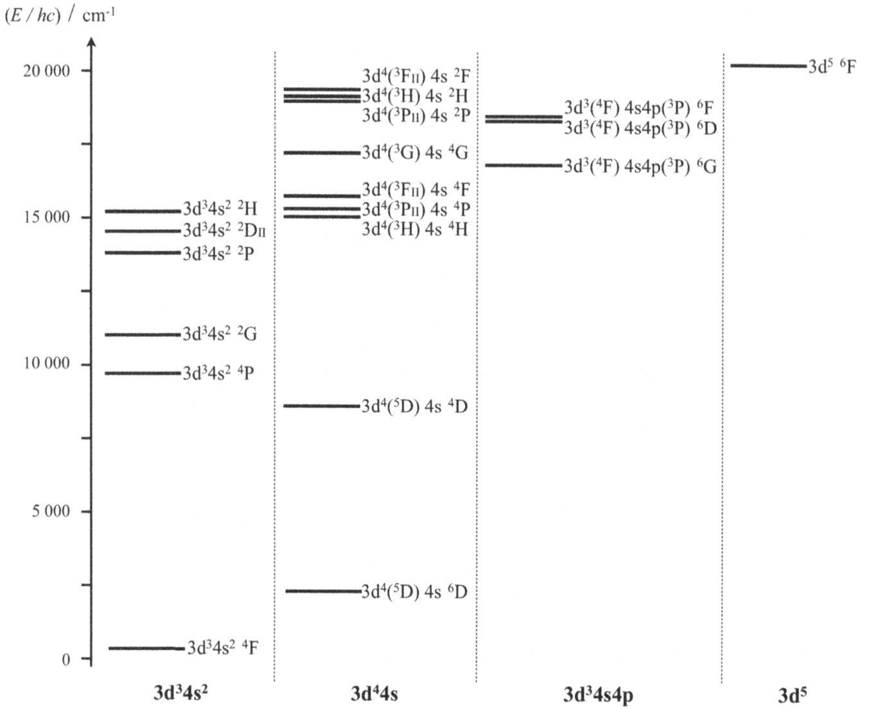

Fig. 7.1 Grotrian diagram for some of the lower energy levels in neutral vanadium [1]. The energy of the ground state, $3d^34s^2\ ^4F_{3/2}$ has been taken as zero. Fine-structure has been omitted in the diagram, and the energies indicated are the mean values of each respective multiplet term (which is why the lowest term is slightly above zero). Roman numeral subscripts under some terms appear when a term has a 'sibling', from which it cannot be discriminated with the quantum numbers available at the present level of approximation. Several terms belonging to one of the four configurations shown in the table have been excluded in the diagram (ones with energies above 20 000 cm^{-1}).

7.2.1 Reduction of the Problem with the Diagonal-Sum Rule

A configuration of electrons in open orbitals is specified by a set of quantum numbers n_i and l_i. This leaves a number of possibilities for all the different combinations of m_{li} and m_{si}. If there is only one populated orbital outside filled orbitals, it will have room for $N_O = 2(2l+1)$ electrons. If the number of (equivalent) electrons in the orbital is N_V, the number of ways that these can be arranged is:

$$\frac{N_O!}{N_V!\,(N_O - N_V)!}\ . \qquad (7.2)$$

These possibilities correspond to combinations of the different m_{li} and m_{si}, exactly as was shown in the tables of spin-orbitals in section 6.3, and also in table 7.3.

All entries in spin-orbital tables (or term charts) correspond to unique product wave functions, which are energy degenerate in the zero-order approximation — they all belong to the same configuration within the CFA. This means that the perturbation theory has to be of the degenerate kind, and at a first glance this seems to mean that to determine energy contributions, large secular problems have to be solved. As an example, for the d^3-configuration, shown in table 7.3, the first-order energy corrections could be found by first calculating the matrix elements of the perturbation Hamiltonian for the Slater determinant wave functions corresponding to each spin-orbital entry in table 7.3. That would leave us with the task of having to diagonalise a 120×120 matrix.

The problem can be greatly reduced by using basic algebra and inherent symmetries. We can take advantage of the fact that the perturbation Hamiltonian is diagonal in M_L and M_S, and then use the diagonal-sum rule for matrices, which states that the trace of a matrix equals the sum of its eigenvalues.

The degeneracy in M_L and M_S is a consequence of the fact that the operators L_z and S_z commute with the Hamiltonian, as long as the spin–orbit coupling is ignored. In the absence of H_{SO}, these two angular momentum projections are constants of the motion. To demonstrate this, we analyse the matrix element of the commutation relation for L_z and the Hamiltonian, between two spin-orbital wave functions, such as two entries in table 7.3. Consider the relation:

$$\left\langle M_L', M_S' \,|\, [L_z, H_{LS}] \,|\, M_L, M_S \right\rangle = 0\,. \tag{7.3}$$

H_{LS} is the non-relativistic Hamiltonian for the ensemble of valence electrons, in the LS-coupling scheme — it will be further defined in (7.5). The states $|M_L, M_S\rangle$ are eigenvectors to L_z, with eigenvalues M_L, so (7.3) reduces to:

$$(M_L' - M_L)\left\langle M_L', M_S' \,|\, H_{LS} \,|\, M_L, M_S \right\rangle = 0\,. \tag{7.4}$$

Thus, only elements diagonal in M_L will be non-zero. The same reasoning is valid for S_z and M_S.

This means that the large secular problem can be divided into many small ones. If we take again the example with the d^3 configuration, we will instead of a secular problem of dimension 120 have 36 separate ones, where no one is of higher dimension than eight — one matrix for each unique combination of M_L and M_S. In another example, a p^2 configuration (see table 6.3), the diagonalisation of a 15-dimensional matrix can be reduced to 11 separate secular equations, with no one being of higher dimension than three.

Determinants as large as 8×8, albeit better than 120×120, are still large. However, we can start with solving integrals for single determinantal wave functions, where there is a unique spin-orbital possible for a particular combination of M_L and M_S. We can then use the obtained results together with the diagonal-sum rule to simplify calculations for other terms.

As an example, we take the p^2-configuration. In table 7.4, we show a diagram very similar to that in table 6.3. The difference is that the spin-orbital entries have

Table 7.4 Diagram, analogous to table 6.3, connecting *LS*-coupling terms to combinations of the projection quantum numbers M_L and M_S, for a p^2 configuration. This table shows the *LS*-coupling terms that correspond to different combinations of the projection quantum numbers.

p^2		M_S 1	M_S 0	M_S −1
M_L	2		^{1}D	
	1	^{3}P	^{3}P , ^{1}D	^{3}P
	0	^{3}P	^{1}S , ^{3}P , ^{1}D	^{3}P
	−1	^{3}P	^{3}P , ^{1}D	^{3}P
	−2		^{1}D	

been replaced by the corresponding terms. The way to understand the table is that although we have 15 different spin-orbitals, the perturbation will not break the degeneracy more than by leaving three distinct energies — ones corresponding to ^{1}S, ^{3}P and ^{1}D. Nine eigenfunctions will be degenerate and they will all have the energy of the ^{3}P-term. Five states will have the energy of ^{1}D, and just one will have the non-degenerate energy corresponding to ^{1}S.

To proceed, as zero-order energy (a priori unknown), we take the CFA state of all electrons *in closed, inner orbitals*. We then calculate the perturbation energy eigenvalues for the ensemble of valence electrons. This will yield energy splittings of terms expressed as superpositions of two-electron integrals, with the latter being in the individual electron representations $|m_{li}, m_{si}, m_{lj}, m_{sj}\rangle$.

The *LS*-coupling valence electron perturbation Hamiltonian is:

$$H_{LS} = -\sum_{i=1}^{N} \frac{1}{2}\nabla^2_{r_i} - \sum_{i=1}^{N} \frac{Z_{\text{eff}}}{r_i} + \sum_{i=1}^{N}\sum_{j>i}^{N} \frac{1}{r_{ij}} = H_{\text{kin}} + H_{\text{nu}} + H_{\text{ee}} , \qquad (7.5)$$

with the summations only over the valence electrons. In the second part of the equation, we define shorthand notations for the two first terms of the Hamiltonian: the kinetic energy part and the electrostatic attraction between the valence electrons and the nucleus. The nuclear charge, Z_{eff}, in H_{nu} is an effective charge, taking into account the screening of inner electrons. Taking advantage of arguments from section 6.2, we rely on the fact that the sum of the interactions between closed orbitals and the valence electrons are, on average, spherically symmetric.

The first two terms in (7.5) are purely central. Within one valence electron configuration, they will add the same energy to all *LS*-coupling terms. We will postpone the analysis of that contribution, and instead concentrate on H_{ee}, which is responsible for the energy splitting between the terms.

Looking again at tables 7.4 and table 6.3, we can start by solving the matrix element for the combination $M_L = 2, M_S = 0$, which corresponds to the energy perturbation for ^{1}D. The sought energy for the term ^{1}D can be found from:

$$E^{(\mathrm{ee})}(^1\mathrm{D}) = \langle\, 2,0 \,|\, H_{\mathrm{ee}} \,|\, 2,0 \,\rangle_{M_L M_S} = \left\langle\, \mathbb{S}_{(1^+,1^-)} \,|\, H_{\mathrm{ee}} \,|\, \mathbb{S}_{(1^+,1^-)} \,\right\rangle_{m_l m_s} . \tag{7.6}$$

In this equation, the subscripts to the matrix elements indicate the respective representation used. The one in the Hilbert space of M_L and M_S corresponds to a unique energy (only one of the atomic terms have this combination of M_L and M_S). In the second matrix element, we instead use the spin-orbitals, represented as in table 6.3, but the eigenvector used in the expression must be the corresponding two-dimensional Slater determinant. The matrix element involving Slater determinants is defined in the same manner as in (2.29). For the present two-electron spin-orbital $(1^+,1^-)$, the energy is:

$$\begin{aligned} E^{(\mathrm{ee})}(^1\mathrm{D}) &= \left\langle\, \mathbb{S}_{(1^+,1^-)} \,|\, H_{\mathrm{ee}} \,|\, \mathbb{S}_{(1^+,1^-)} \,\right\rangle_{m_l m_s} \\ &= \iint \mathbb{S}^*_{(1^+,1^-)}\, H_{\mathrm{ee}}\, \mathbb{S}_{(1^+,1^-)}\, \mathbf{dq}_1 \mathbf{dq}_2 \,. \end{aligned} \tag{7.7}$$

To find the energy of the term $^3\mathrm{P}$, as expressed in two-electron integrals, we use the fact that there are six occurrences in table 7.4 where $^3\mathrm{P}$ appears alone. This means that also the energy $E_{^3\mathrm{P}}$ can be found by solving one single integral. Any of these six combinations of M_L and M_S can be chosen, for example:

$$E^{(\mathrm{ee})}(^3\mathrm{P}) = \langle\, 1,1 \,|\, H_{\mathrm{ee}} \,|\, 1,1 \,\rangle_{M_L M_S} = \left\langle\, \mathbb{S}^*_{(1^+,0^+)} \,|\, H_{\mathrm{ee}} \,|\, \mathbb{S}^*_{(1^+,0^+)} \,\right\rangle_{m_l m_s} . \tag{7.8}$$

The sole remaining term, $^1\mathrm{S}$ only occurs at one place in the table, for $M_L = 0$, $M_S = 0$. For the terms that are present for those values, we have a cubic secular problem, but two of the eigenvalues have already been found. The diagonal-sum rule then tells us that:

$$\begin{aligned} E^{(\mathrm{ee})}(^1\mathrm{S}) + E^{(\mathrm{ee})}(^3\mathrm{P}) + E^{(\mathrm{ee})}(^1\mathrm{D}) &= \left\langle\, \mathbb{S}_{(1^+,-1^-)} \,|\, H_{\mathrm{ee}} \,|\, \mathbb{S}_{(1^+,-1^-)} \,\right\rangle_{m_l m_s} \\ &+ \left\langle\, \mathbb{S}_{(0^+,0^-)} \,|\, H_{\mathrm{ee}} \,|\, \mathbb{S}_{(0^+,0^-)} \,\right\rangle_{m_l m_s} + \left\langle\, \mathbb{S}_{(1^-,-1^+)} \,|\, H_{\mathrm{ee}} \,|\, \mathbb{S}_{(1^-,-1^+)} \,\right\rangle_{m_l m_s} . \end{aligned} \tag{7.9}$$

This means that we get the energy perturbation for $^1\mathrm{S}$ in terms of three new two-electron integrals from which we subtract the previously attained results for $E^{(\mathrm{ee})}(^1\mathrm{D})$ and $E^{(\mathrm{ee})}(^3\mathrm{P})$. With that, we have managed to reduce all term energies to series of two-electron integrals (to be further analysed in the next section) without having to solve *any* secular problems, let alone one of dimension 15.

There are atomic configurations for which the diagonal-sum rule does not entirely save us from having to diagonalise matrices. As a further example, where one secular problem will remain, we will look again at the d^3-configuration, illustrated in table 7.3.

We know that there are seven different kinds of terms: $^2\mathrm{H}$, $^2\mathrm{G}$, $^4\mathrm{F}$, $^2\mathrm{F}$, $^2\mathrm{D}$, $^4\mathrm{P}$ and $^2\mathrm{P}$, with $^2\mathrm{D}$ appearing twice. We accordingly have eight energies to compute. The situation is illustrated in table 7.5. With matrix elements involving a single spin-orbital, we can calculate $E^{(\mathrm{ee})}(^2\mathrm{H})$ and $E^{(\mathrm{ee})}(^4\mathrm{F})$, since they are unique occur-

Table 7.5 Diagram, analogous to table 7.3, connecting *LS*-coupling terms to combinations of the projection quantum numbers M_L and M_S for a d^3 configuration. It shows *LS*-coupling terms corresponding to different combinations of the projection quantum numbers M_L and M_S. For this particular configuration, the term ^2D ($L=2$ and $S=1/2$) occurs for two distinct energies (a consequence of the *LS*-coupling term notation being incomplete for three valence electrons). The terms corresponding to the two different energies are notated by roman numeral subscripts.

d^3		M_S			
		3/2	1/2	−1/2	−3/2
	5		^2H	^2H	
	4		^2G, ^2H	^2G, ^2H	
	3	^4F	^2F, ^4F, ^2G, ^2H	^2F, ^4F, ^2G, ^2H	^4F
	2	^4F	^2D_I, $^2D_{II}$, ^2F ^4F, ^2G, ^2H	^2D_I, $^2D_{II}$, ^2F ^4F, ^2G, ^2H	^4F
	1	^4P ^4F	^2P, ^4P, ^2D_I, $^2D_{II}$ ^2F, ^4F, ^2G, ^2H	^2P, ^4P, ^2D_I, $^2D_{II}$ ^2F, ^4F, ^2G, ^2H	^4P ^4F
M_L	0	^4P ^4F	^2P, ^4P, ^2D_I, $^2D_{II}$ ^2F, ^4F, ^2G, ^2H	^2P, ^4P, ^2D_I, $^2D_{II}$ ^2F, ^4F, ^2G, ^2H	^4P ^4F
	−1	^4P ^4F	^2P, ^4P, ^2D_I, $^2D_{II}$ ^2F, ^4F, ^2G, ^2H	^2P, ^4P, ^2D_I, $^2D_{II}$ ^2F, ^4F, ^2G, ^2H	^4P ^4F
	−2	^4F	^2D_I, $^2D_{II}$, ^2F ^4F, ^2G, ^2H	^2D_I, $^2D_{II}$, ^2F ^4F, ^2G, ^2H	^4F
	−3	^4F	^2F, ^4F, ^2G, ^2H	^2F, ^4F, ^2G, ^2H	^4F
	−4		^2G, ^2H	^2G, ^2H	
	-5		^2H	^2H	

rences for specific combinations of M_L and M_S. The energies we thus obtain can then be used together with the diagonal-sum rule to facilitate a direct calculation of $E^{(ee)}(^2G)$. The three energies we then have means that it is straightforward to compute $E^{(ee)}(^2F)$.

However, we then run into a problem. When we get to the terms for $M_L=2$, $M_S=1/2$, we have a secular problem of dimension six, but we have only managed to predetermine four eigenvalues. For the remaining two terms, ^2D_I and $^2D_{II}$, there will be a few elements non-diagonal in L and S, and it will be unavoidable to diagonalise the remaining 2×2 matrix.

7.2.2 Energy Contributions from One- and Two-Electron Integrals

The next step in the determination of term energies is to express the two-electron integrals in the preceding section in a more synthesised form. Since we lack precise functional expressions for the individual electron wave functions in a multielectron atom, the end result will still fall short of level splittings expressed in absolute energy units. However, we will get results in terms of sets of standardised integral expressions, and we will be able to compare term energies. In a broad sense, the analysis that will follow is an extension of the perturbation theory calculations that were done for Li and He in chapter 2.

The perturbation Hamiltonian is the non-relativistic multielectron one of (7.5), with the spin–orbit interaction neglected. The zero-order states are the CFA configurations of all electrons in the inner orbitals. The valence electron wave functions are Slater determinants, with dimensions that correspond to the number of valence electrons. Each component in these determinants will be different permutations of electron configuration product functions, separable in radial, angular and spin parts, and specified by quantum numbers $\{n_i, l_i, m_{li}, m_{si}\}$.

When matrix elements are calculated for the components of the Hamiltonian in (7.5), using Slater determinants, H_{kin} and H_{nu} will yield sums of single-electron integrals, of the types:

$$\left\langle \Psi_{\mathrm{a}}(\mathbf{q}) \left| -\frac{1}{2}\nabla^2 \right| \Psi_{\mathrm{b}}(\mathbf{q}) \right\rangle , \tag{7.10}$$

and

$$\left\langle \Psi_{\mathrm{a}}(\mathbf{q}) \left| -\frac{Z_{\mathrm{eff}}}{r} \right| \Psi_{\mathrm{b}}(\mathbf{q}) \right\rangle , \tag{7.11}$$

where the variable $\mathbf{q}$ includes spin as well as the spatial coordinates. The two-electron component, H_{ee}, will give matrix elements that are functions of two-electron integrals, such as:

$$\left\langle \Psi_{\mathrm{a}}(\mathbf{q}_1)\Psi_{\mathrm{b}}(\mathbf{q}_2) \left| \frac{1}{r_{12}} \right| \Psi_{\mathrm{c}}(\mathbf{q}_1)\Psi_{\mathrm{d}}(\mathbf{q}_2) \right\rangle . \tag{7.12}$$

The objective of this section, and the ones that follow, is to derive explicit numerical factors for all angular and spin parts for all sums of matrix elements of the types in (7.10), (7.11) and (7.12). This should leave us with series of purely radial integrals preceded by numerical coefficients.

We begin with the kinetic energy term, that is, the operator $(-\nabla^2/2)$. The angular part of the wave function — consisting of spherical harmonics — and the spin function are both eigenvectors to the Laplacian. This gives us the constraints $l_{\mathrm{a}} = l_{\mathrm{b}}$, $m_{l\mathrm{a}} = m_{l\mathrm{b}}$ and $m_{s\mathrm{a}} = m_{s\mathrm{b}}$ for any non-zero contribution, and thus delta functions for the angular momentum quantum numbers. For a pair of electrons, indexed a and b, we get:

$$\left\langle \Psi_{n_a l_a m_{la} m_{sa}}(\mathbf{q}) \left| -\frac{1}{2}\nabla^2 \right| \Psi_{n_b l_b m_{lb} m_{sb}}(\mathbf{q}) \right\rangle$$

$$= \delta(m_{sa}, m_{sb}) \int_0^\infty \int_0^\pi \int_0^{2\pi} R^*_{n_a l_a}(r)\, Y^*_{l_a m_{la}}(\theta, \varphi)$$

$$\times \left\{ -\frac{1}{2}\left[\frac{1}{r^2}\frac{\partial}{\partial r}\left(r^2 \frac{\partial}{\partial r}\right) + \frac{\mathbf{L}^2}{r^2} \right] \right\} R_{n_b l_b}(r)\, Y_{l_b m_{lb}}(\theta, \varphi)\, r^2 \sin\theta \,\mathrm{d}r \mathrm{d}\theta \mathrm{d}\varphi$$

$$= \delta(l_a, l_b)\, \delta(m_{la}, m_{lb})\, \delta(m_{sa}, m_{sb})$$

$$\times \frac{1}{2}\int_0^\infty R^*_{n_a l}(r) \left[-\frac{\mathrm{d}}{\mathrm{d}r}\left(r^2 \frac{\mathrm{d}}{\mathrm{d}r}\right) - l(l+1) \right] R_{n_b l}(r)\, \mathrm{d}r \quad , \tag{7.13}$$

where, inside the last integral, we have taken advantage of the first delta function, and defined $l_a = l_b = l$. It should be noted that the radial functions $R_{nl}(r)$ are not hydrogenic ones. They are radial solutions of the valence electron Schrödinger equation in the CFA, in a screened Coulomb potential. We do not have access to their exact analytical shape, but it is reasonable to assume that they will be similar in form to hydrogenic radial functions, at least for large r.

The first term in the integrand in the last line of (7.13) can be developed by integration by parts. This gives the result:

$$\left\langle \Psi_{n_a l_a m_{la} m_{sa}}(\mathbf{q}) \left| -\frac{1}{2}\nabla^2 \right| \Psi_{n_b l_b m_{lb} m_{sb}}(\mathbf{q}) \right\rangle = \delta(l_a, l_b)\, \delta(m_{la}, m_{lb})\, \delta(m_{sa}, m_{sb})$$

$$\times \frac{1}{2}\int_0^\infty \left[r^2 \frac{\mathrm{d}R^*_{n_a l}(r)}{\mathrm{d}r} \frac{\mathrm{d}R_{n_b l}(r)}{\mathrm{d}r} + l(l+1)\, R^*_{n_a l}(r)\, R_{n_b l}(r) \right] \mathrm{d}r \,, \tag{7.14}$$

for the kinetic energy integral.

The matrix elements of the potential energy associated with the central potential, represented by H_{nu}, can be put directly into a form of a purely radial integral. Yet again we get delta functions in l, m_l and m_s, and the only thing remaining is an integral over $1/r$:

$$\left\langle \Psi_{n_a l_a m_{la} m_{sa}}(\mathbf{q}) \left| -\frac{Z_{\mathrm{eff}}}{r} \right| \Psi_{n_b l_b m_{lb} m_{sb}}(\mathbf{q}) \right\rangle$$

$$= \delta(l_a, l_b)\, \delta(m_{la}, m_{lb})\, \delta(m_{sa}, m_{sb})\ (-Z_{\mathrm{eff}}) \int_0^\infty r\, R^*_{n_a l}(r)\, R_{n_b l}(r)\, \mathrm{d}r \ . \tag{7.15}$$

We restrict the calculation to configurations with identical valence electrons. Within that limitation, we can disregard terms non-diagonal in n and l. For diagonal terms, we shorten the notation by introducing the following notation for the sum of the two single-electron integrals (7.14) and (7.15):

$$I(nl) \equiv \left\langle \Psi_{nlm_lm_s}(\mathbf{q}) \left| \left(-\frac{1}{2}\nabla^2 - \frac{Z}{r} \right) \right| \Psi_{nlm_lm_s}(\mathbf{q}) \right\rangle$$
$$= \int_0^\infty \left\{ \frac{r^2}{2} \frac{\mathrm{d}R^*_{nl}(r)}{\mathrm{d}r} \frac{\mathrm{d}R_{nl}(r)}{\mathrm{d}r} + \left[\frac{l(l+1)}{2} - Zr \right] R^*_{nl}(r)\, R_{nl}(r) \right\} \mathrm{d}r\,. \qquad (7.16)$$

The energy in (7.16) is that of one valence electron. For the ensemble of valence electrons, $\{n_i l_1 \ldots n_N l_N\}$, the total energy contribution from the sum of H_{kin} and H_{nu} is:

$$\langle n_i l_1 \ldots n_N l_N \,|H_{\mathrm{kin}} + H_{\mathrm{nu}}|\, n_i l_1 \ldots n_N l_N \rangle = \sum_{i=1}^{N} I(n_i l_i) \ . \qquad (7.17)$$

7.2.3 Slater Integrals

For the treatment of the final term in (7.5), with the exchange interaction Hamiltonian H_{ee}, we will begin with a general case, which includes four different electronic wave functions, $\Psi_{\mathrm{a}}(\mathbf{q})$, $\Psi_{\mathrm{b}}(\mathbf{q})$, $\Psi_{\mathrm{c}}(\mathbf{q})$ and $\Psi_{\mathrm{d}}(\mathbf{q})$. Since the operator acts on pairs of wave functions, the above four functions will have to be integrated, with H_{ee}, over two different sets of combined spin and spatial coordinates, $\mathbf{q}_1$ and $\mathbf{q}_2$. The general expression covers the special cases of the Coulomb integral and the exchange integral in the electron–electron repulsion (as defined in chapters 2 and 3). These will correspond to permutations of just two of the above four functions.

A more thorough description of how to treat Coulomb and exchange integrals is given in appendix D, see also [2]. In the following, we will refer to that appendix for most of the details of the calculations. The main differences between the analysis made here and that in appendix D, apart from the level of mathematical rigour, are two: that we here begin with a general example with four different orbitals, and the fact that we include the spin wave functions.

As mentioned previously, mixing continuous variables with a discrete one, such as the spin, in one vector parameter is mathematically frivolous. We here stick to this notation, due to its convenience. Moreover, the inclusion of the spin wave function will merely contribute to the conditions $m_{s\mathrm{a}} = m_{s\mathrm{c}}$ and $m_{s\mathrm{b}} = m_{s\mathrm{d}}$, that is to two delta functions.

In order to express everything in spherical coordinates, the function $1/r_{12}$ should be developed in terms of spherical harmonics. This expansion is demonstrated in appendix C.3, and we can use expression (C.56). We get the following equation for the full two-electron integral:

$$\left\langle \Psi_{n_\mathrm{a}l_\mathrm{a}m_{l\mathrm{a}}m_{s\mathrm{a}}}(\mathbf{q}_1)\, \Psi_{n_\mathrm{b}l_\mathrm{b}m_{l\mathrm{b}}m_{s\mathrm{b}}}(\mathbf{q}_2) \left| \frac{1}{r_{12}} \right| \Psi_{n_\mathrm{c}l_\mathrm{c}m_{l\mathrm{c}}m_{s\mathrm{c}}}(\mathbf{q}_1)\, \Psi_{n_\mathrm{d}l_\mathrm{d}m_{l\mathrm{d}}m_{s\mathrm{d}}}(\mathbf{q}_2) \right\rangle$$
$$= \delta(m_{s\mathrm{a}}, m_{s\mathrm{c}})\, \delta(m_{s\mathrm{b}}, m_{s\mathrm{d}}) \sum_{k=0}^{\infty} \sum_{q=-k}^{k} \frac{4\pi}{2k+1}$$

$$\times \int_0^\infty \int_0^\infty R^*_{n_a l_a}(r_1)\, R^*_{n_b l_b}(r_2)\, R_{n_c l_c}(r_1)\, R_{n_d l_d}(r_2)\, \frac{(r_<)^k}{(r_>)^{k+1}}\, r_1^2 r_2^2\, \mathrm{d}r_1 \mathrm{d}r_2$$

$$\times \int_0^{2\pi} \int_0^{\pi} Y^*_{l_a m_{la}}(\theta_1,\varphi_1)\, Y_{l_c m_{lc}}(\theta_1,\varphi_1)\, Y^*_{kq}(\theta_1,\varphi_1)\, \sin\theta_1\, \mathrm{d}\theta_1 \mathrm{d}\varphi_1$$

$$\times \int_0^{2\pi} \int_0^{\pi} Y^*_{l_b m_{lb}}(\theta_2,\varphi_2)\, Y_{l_d m_{ld}}(\theta_2,\varphi_2)\, Y_{kq}(\theta_2,\varphi_2)\, \sin\theta_2\, \mathrm{d}\theta_2 \mathrm{d}\varphi_2 \ . \tag{7.18}$$

In (7.18), k and q are summation variables. The parameters $r_<$ and $r_>$ symbolise, respectively, the smaller and the greater of the two radial coordinates, while the radial double integral is taken over both r_1 and r_2.

The explicit form of a spherical harmonic is that of a product of an associated Legendre polynomial of $\cos\theta$, an exponential in φ, and a normalisation factor. This means that the angular integrals in (7.18) can be factorised into ones over zenith angles and azimuthal angles — see (C.48). The two integrands containing φ_1 and φ_2 will be exponentials with the respective exponents $(-m_{la}+m_{lc}-q)$ and $(-m_{lb}+m_{ld}+q)$. If these are zero, each of the two azimuthal integrals will return 2π. For all other possibilities, they cancel. This means that those integrals can be replaced by a constant and a constraint for the m_l quantum numbers:

$$m_{la}+m_{lb}=m_{lc}+m_{ld} \ . \tag{7.19}$$

This condition will appear in the form of a delta function in our final expression. Physically, it means that the total angular momentum is conserved under the electrostatic interaction between the electrons. Mathematically, it means that the summation over q in (7.18) can be removed, since only one term will remain: $q=m_{lc}-m_{la}=m_{lb}-m_{ld}$.

The zenith integrals will each be over products of three associated Legendre polynomials. The general solution to this is known as Gaunt's formula, and it is presented in (D.7) and (D.11). With purely central potentials, as in the present case, it is possible to describe the angular integral entirely in products of *Gaunt coefficients*, defined as:

$$c^{(k)}[l_i m_{li} : l_j m_{lj}] \equiv \sqrt{\frac{4\pi}{2k+1}} \times \int_0^{2\pi} \int_0^{\pi} Y_{l_i,m_{li}}(\theta,\varphi)\, Y_{l_j,m_{lj}}(\theta,\varphi)\, Y_{k,m_{li}-m_{lj}}(\theta,\varphi)\, \sin\theta\, \mathrm{d}\theta \mathrm{d}\varphi \ . \tag{7.20}$$

The Gaunt coefficients are functions of two pairs of angular momentum quantum numbers — $\{l_i, m_{li}\}$ and $\{l_j, m_{lj}\}$ — and an index k, which corresponds to the summation variable k in (7.18). The general calculation of Gaunt coefficients is complex, but a generating formula has been derived, first by Gaunt (see [3] or [2]). Gaunt's original formula has subsequently been put into a more manageable form by Racah (see [4] or [5]), who expressed the coefficients in terms of Wigner $3j$-symbols (see appendix C.4):

$$c^{(k)}[l_i m_{li} : l_j m_{lj}] = (-1)^{m_{li}} \sqrt{2l_i+1}\,\sqrt{2l_j+1} \times \begin{pmatrix} l_i & k & l_j \\ 0 & 0 & 0 \end{pmatrix} \begin{pmatrix} l_i & k & l_j \\ -m_{li} & m_{li}-m_{lj} & m_{lj} \end{pmatrix} . \quad (7.21)$$

Algorithms that calculate Gaunt coefficients are included in many mathematical software packages, and they are tabulated in many volumes. A limited version of such a table is presented in table D.1.

The formulation of Gaunt coefficients in terms of $3j$-symbols reveals a triangular condition:

$$|l_i - l_j| \leq k \leq l_i + l_j \; . \quad (7.22)$$

A consequence of this constraint is that in the infinite sum in (7.18), only a few terms will remain non-zero. Putting all things together, the matrix element for the Coulomb electron–electron repulsion Hamiltonian for *one* electron pair becomes:

$$\left\langle \Psi_{n_a l_a m_{la} m_{sa}}(\mathbf{q}_1)\, \Psi_{n_b l_b m_{lb} m_{sb}}(\mathbf{q}_2) \left| \frac{1}{r_{12}} \right| \Psi_{n_c l_c m_{lc} m_{sc}}(\mathbf{q}_1)\, \Psi_{n_d l_d m_{ld} m_{sd}}(\mathbf{q}_2) \right\rangle$$
$$= \delta(m_{sa}, m_{sc})\, \delta(m_{sb}, m_{sd})\, \delta(m_{la}+m_{lb}, m_{lc}+m_{ld}) \times \sum_{k=0}^{\infty} c^{(k)}[l_a m_{la} : l_c m_{lc}]\, c^{(k)}[l_b m_{lb} : l_d m_{ld}]\; R^{(k)}(\mathrm{ab}:\mathrm{cd}) \; . \quad (7.23)$$

The final factor in this expression is a radial integral defined as:

$$R^{(k)}(\mathrm{ab}:\mathrm{cd}) \equiv \int_0^\infty \int_0^\infty R^*_{n_a l_a}(r_1)\, R^*_{n_b l_b}(r_2)\, R_{n_c l_c}(r_1)\, R_{n_d l_d}(r_2) \times \frac{(r_<)^k}{(r_>)^{k+1}}\, r_1^2 r_2^2\, \mathrm{d}r_1 \mathrm{d}r_2 \; . \quad (7.24)$$

Equation (7.24) is the general formulation of a *Slater integral*. Two special cases of the Slater integral are the *F-integral* — for a = c and b = d — and the *G-integral* — for a = d and b = c. That is, these two special integrals can be written as:

$$F^{(k)}(n_a l_a : n_b l_b) \equiv R^{(k)}(\mathrm{ab}:\mathrm{ab}) = \int_0^\infty \int_0^\infty |R_{n_a l_a}(r_1)|^2\, |R_{n_b l_b}(r_2)|^2\, \frac{(r_<)^k}{(r_>)^{k+1}}\, r_1^2 r_2^2\, \mathrm{d}r_1 \mathrm{d}r_2 \quad (7.25)$$

$$G^{(k)}(n_a l_a : n_b l_b) \equiv R^{(k)}(\mathrm{ab}:\mathrm{ba}) = \int_0^\infty \int_0^\infty R^*_{n_a l_a}(r_1)\, R^*_{n_b l_b}(r_2)\, R_{n_b l_b}(r_1)\, R_{n_a l_a}(r_2)\, \frac{(r_<)^k}{(r_>)^{k+1}}\, r_1^2 r_2^2\, \mathrm{d}r_1 \mathrm{d}r_2 \; . \quad (7.26)$$

In section 7.2.1 we concluded that the valence electron Hamiltonian is diagonal in M_L and M_S, at the current level of approximation. A consequence of that is that the F and G forms of the Slater integral will be the only ones that we will have use for when calculating term energies in the *LS*-coupling approximation (this will no longer hold for states with significant configuration mixing, see section 9.5). For each term in our final expressions for energies, there will be a sum of one-electron integrals, $I(nl)$ — one for each electron within the configuration. Added to that will appear a series of F- and G-integrals, corresponding to pairs of electrons in the configuration. The latter will be Coulomb and exchange integrals multiplied by Gaunt coefficients, as described in (D.17) and (D.18). Since the operator in the Slater integrals is a two-electron one, we have to keep in mind the rules for calculating matrix elements between Slater determinant functions for such operators (see section 2.3.4). It should be noted that for integrals with equivalent electrons ($n_\mathrm{a} = n_\mathrm{b}$ and $l_\mathrm{a} = l_\mathrm{b}$), the F- and G-integrals will be identical.

7.2.4 Examples of Term Energies

We shall demonstrate how to put the preceding analysis into practice with a few concrete examples. To start with, we consider the $2\mathrm{p}^2$ configuration (the ground state configuration of C), which has been described in tables 6.3 and 7.4, and following that we will look at $2\mathrm{p}^3$, illustrated in table 6.5.

In the general case, each valence electron configuration consists of product states specified by quantum numbers $\{n_1, l_1, m_{l1}, m_{s1}, \ldots, n_N, l_N, m_{lN}, m_{sN}\}$, many of them degenerate in energy. For electrons exclusively in one valence orbital, all n_i and all l_i are the same. Each product state corresponds to a unique combination of $\{M_L, M_S\}$, whereas each atomic term, ${}^{2S+1}L$, will typically be pertinent for a set of different states.

What we first need to calculate is the matrix elements of the Hamiltonian (7.5) for every product state:

$$E_\sigma = \langle \mathbb{S}_\sigma \,|\, H_\mathrm{kin} + H_\mathrm{nu} \,|\, \mathbb{S}_\sigma \rangle + \langle \mathbb{S}_\sigma \,|\, H_\mathrm{ee} \,|\, \mathbb{S}_\sigma \rangle \ , \tag{7.27}$$

with the definition:

$$\sigma \equiv \{n_1, l_1, m_{l1}, m_{s1}, \ldots, n_N, l_N, m_{lN}, m_{sN}\}_i \ . \tag{7.28}$$

$\mathbb{S}_\sigma$ is the Slater determinant corresponding to all permutations of the N electron coordinates in σ and the N sets of quantum numbers — that is, it is of dimension N.

The configuration $2\mathrm{p}^2$ is 15-fold degenerate, before any angular perturbations have been taken into account. We know from previous analyses that these 15 states are split up into three different energies: those for the *LS*-coupling terms ${}^1\mathrm{S}$, ${}^1\mathrm{D}$ and ${}^3\mathrm{P}$. Of these, we also know that ${}^3\mathrm{P}$ is nine-fold, and ${}^1\mathrm{D}$ five-fold degenerate.

Looking at tables 6.3 and 7.4 we see that one of the $|M_L, M_s\rangle$ states that belongs to ^{1}D is $|2,0\rangle$, and also that $|2,0\rangle$ cannot encompass any of the other two terms. This particular state corresponds to a unique spin-orbital:

$$\{m_{l1}=1\,,\, m_{s1}={}^1\!/\!_2\,,\, m_{l2}=1\,,\, m_{l2}=-{}^1\!/\!_2\} \longleftrightarrow (1^+\,1^-)\ , \tag{7.29}$$

see also (7.7).

This is enough information for determining the energy of ^{1}D in terms of two-electron integrals. The Slater determinant is:

$$\mathbb{S}_{(1^+,1^-)} = \frac{1}{\sqrt{2}} \begin{vmatrix} \Psi_1(\mathbf{q}_1) & \Psi_1(\mathbf{q}_2) \\ \Psi_2(\mathbf{q}_1) & \Psi_2(\mathbf{q}_2) \end{vmatrix}\ , \tag{7.30}$$

where the indices to the wave functions indicate the one-electron states:

$$\begin{aligned} &\Psi_1 : \{n_1=2\,,\, l_1=1\,,\, m_{l1}=1\,,\, m_{s1}={}^1\!/\!_2\} \\ &\Psi_2 : \{n_2=2, l_2=1, m_{l2}=1, m_{l2}=-{}^1\!/\!_2\}\ . \end{aligned} \tag{7.31}$$

For the sum of the two single-electron operators, the matrix element is:

$$\left\langle \mathbb{S}_{(1^+,1^-)} \,|\, H_{\mathrm{kin}} + H_{\mathrm{nu}} \,|\, \mathbb{S}_{(1^+,1^-)} \right\rangle = I(n_{\mathrm{a}} l_{\mathrm{a}}) + I(n_{\mathrm{b}} l_{\mathrm{b}}) = 2I(2\mathrm{p})\ , \tag{7.32}$$

where we have used (2.34) and the definition in (7.16).

For the two-electron operator, the matrix element has to be calculated with more care. Since the operator acts on a pair of electrons in the product wave function, we must take into account (2.36). Doing this, and using (7.23), we get:

$$\begin{aligned} \left\langle \mathbb{S}_{(1^+,1^-)} \,|\, H_{\mathrm{ee}} \,|\, \mathbb{S}_{(1^+,1^-)} \right\rangle &= \left\langle \Psi_1(\mathbf{q}_1)\,\Psi_2(\mathbf{q}_2) \left| \frac{1}{r_{12}} \right| \Psi_1(\mathbf{q}_1)\,\Psi_2(\mathbf{q}_2) \right\rangle \\ &\quad - \left\langle \Psi_1(\mathbf{q}_1)\,\Psi_2(\mathbf{q}_2) \left| \frac{1}{r_{12}} \right| \Psi_2(\mathbf{q}_1)\,\Psi_1(\mathbf{q}_2) \right\rangle \\ &= \sum_k^{\infty} c^{(k)}[l_1 m_{l1} : l_1 m_{l1}]\; c^{(k)}[l_2 m_{l2} : l_2 m_{l2}]\; F^{(k)}(n_1 l_1 : n_2 l_2) \\ &\quad - \delta(m_{s1}, m_{s2}) \sum_k^{\infty} |\, c^{(k)}[l_1 m_{l1} : l_2 m_{l2}]\,|^2\; G^{(k)}(n_1 l_1 : n_2 l_2) \\ &= F^{(0)}_{2\mathrm{p}:2\mathrm{p}} + \frac{1}{25} F^{(2)}_{2\mathrm{p}:2\mathrm{p}}\ . \end{aligned} \tag{7.33}$$

The term with G-integrals vanishes due to the delta functions in (7.18). To identify the Gaunt coefficients in (7.33), we have used table D.1. We now write the total term energy of $(2\mathrm{p}^2\,{}^1\mathrm{D})$:

$$E^{(LS)}(2\mathrm{p}^2\,{}^1\mathrm{D}) = 2I(2\mathrm{p}) + F^{(0)}_{2\mathrm{p}:2\mathrm{p}} + \frac{1}{25} F^{(2)}_{2\mathrm{p}:2\mathrm{p}}\ , \tag{7.34}$$

and this will be the energy for all occurrences of the ^{1}D term, before spin–orbit interaction (fine-structure) has been taken into account. The individual valence electron contribution, $2I(2p)$, is defined in (7.16) and (7.17).

In the next step, we calculate the energy for ^{3}P. This is nine-fold degenerate, but there are places in table 6.3 where it is isolated. One such occurrence is:

$$\{m_{l1}=1\,,\,m_{s1}={}^1\!/_2\,,\,m_{l2}=0\,,\,m_{s2}={}^1\!/_2\} \longleftrightarrow (1^+,0^+) \quad , \tag{7.35}$$

see also (7.8). If we calculate the energy for this state in the same manner as above, we find:

$$\begin{aligned} E^{(LS)}(2p^2\,{}^3P) &= 2I(2p) + F^{(0)}_{2p:2p} - \frac{2}{25}F^{(2)}_{2p:2p} - \frac{3}{25}G^{(2)}_{2p:2p} \\ &= 2I(2p) + F^{(0)}_{2p:2p} - \frac{5}{25}F^{(2)}_{2p:2p} \ , \end{aligned} \tag{7.36}$$

where in the last step we have taken advantage of the fact that $F^{(k)}=G^{(k)}$ for equivalent electrons.

The remaining term, ^{1}S, is only present at one place in table 6.3, namely for $M_L=0$ and $M_S=0$. In the corresponding cell in the diagram, we have the following three spin-orbitals:

$$\begin{aligned} \{m_{l1}=1\,,\,m_{s1}={}^1\!/_2\,,\,m_{l2}=-1\,,\,m_{s2}=-{}^1\!/_2\} &\longleftrightarrow (1^+,-1^-) \\ \{m_{l1}=0\,,\,m_{s1}={}^1\!/_2\,,\,m_{l2}=0\,,\,m_{s2}=-{}^1\!/_2\} &\longleftrightarrow (0^+,0^-) \\ \{m_{l1}=-1\,,\,m_{s1}={}^1\!/_2\,,\,m_{l2}=1\,,\,m_{s2}=-{}^1\!/_2\} &\longleftrightarrow (-1^+,1^-) \ . \end{aligned} \tag{7.37}$$

Without the diagonal-sum rule, we would have had to calculate all three diagonal, and all six non-diagonal matrix elements, and then diagonalise the matrix. However, since we already know two eigenvalues, it is enough to calculate the three diagonal elements, and then use the diagonal-sum rule — see also (7.9).

The diagonal matrix elements for the $1/r_{12}$ operator are in this case:

$$\begin{aligned} \left\langle \mathbb{S}_{(1^+,-1^-)}\,|H_{ee}|\,\mathbb{S}_{(1^+,-1^-)}\right\rangle &= F^{(0)}_{2p:2p} + \frac{1}{25}F^{(2)}_{2p:2p} \\ \left\langle \mathbb{S}_{(0^+,0^-)}\,|H_{ee}|\,\mathbb{S}_{(0^+,0^-)}\right\rangle &= F^{(0)}_{2p:2p} + \frac{4}{25}F^{(2)}_{2p:2p} \\ \left\langle \mathbb{S}_{(0^+,0^-)}\,|H_{ee}|\,\mathbb{S}_{(0^+,0^-)}\right\rangle &= F^{(0)}_{2p:2p} + \frac{1}{25}F^{(2)}_{2p:2p} \ . \end{aligned} \tag{7.38}$$

From the diagonal-sum rule, we can then extract the term energy for ^{1}S as:

$$\begin{aligned} E^{(LS)}(2p^2\,{}^1S) = 2I(2p) &+ \left\langle \mathbb{S}_{(1^+,-1^-)}\,|H_{ee}|\,\mathbb{S}_{(1^+,-1^-)}\right\rangle \\ &+ \left\langle \mathbb{S}_{(0^+,0^-)}\,|H_{ee}|\,\mathbb{S}_{(0^+,0^-)}\right\rangle + \left\langle \mathbb{S}_{(0^+,0^-)}\,|H_{ee}|\,\mathbb{S}_{(0^+,0^-)}\right\rangle \\ &- E^{(LS)}(2p^2\,{}^1D) - E^{(LS)}(2p^2\,{}^3P) \\ &= 2I(2p) + F^{(0)}_{2p:2p} + \frac{10}{25}F^{(2)}_{2p:2p} \ . \end{aligned} \tag{7.39}$$

If we compare (7.34), (7.36) and (7.39), we find that the energy ordering between the *LS*-coupling terms of the configuration $2p^2$ is:

$$E^{(LS)}(2p^{2\ ^1}S) > E^{(LS)}(2p^{2\ ^1}D) > E^{(LS)}(2p^{2\ ^3}P)\,. \tag{7.40}$$

What is not included is the energy contribution from the inner orbitals. However, this will only add a constant term to the three term energies. Moreover, we have not determined any energies on an absolute scale, since we do not have access to the radial functions, but we have still gained important knowledge. We have found that for the three *LS*-coupling terms in the $2p^2$-configuration, we can expect ^{3}P to lie lowest in energy, followed by ^{1}D and ^{1}S. This agrees with experimental data for C. We can also predict that the energy separations between the terms should be of the order of:

$$\begin{aligned}
E^{(LS)}(2p^{2\ ^1}S) - E^{(LS)}(2p^{2\ ^1}D) &= \frac{9}{25}F^{(2)}_{2p:2p} \\
E^{(LS)}(2p^{2\ ^1}D) - E^{(LS)}(2p^{2\ ^3}P) &= \frac{6}{25}F^{(2)}_{2p:2p} \\
\Rightarrow \quad \frac{E^{(LS)}(2p^{2\ ^1}S) - E^{(LS)}(2p^{2\ ^1}D)}{E^{(LS)}(2p^{2\ ^1}D) - E^{(LS)}(2p^{2\ ^3}P)} &= \frac{E^{(ee)}(^1S) - E^{(ee)}(^1D)}{E^{(ee)}(^1D) - E^{(ee)}(^3P)} = \frac{3}{2}\,.
\end{aligned} \tag{7.41}$$

Experimentally, the value of this ratio for C is about 1.13. The quantitative discrepancy is not surprising, given the level of approximation. One contributing factor to the deviation is an interaction of states within $2p^2$ with ones in the excited configuration 2p3p — that is a departure from the CFA.

Our second example is the three-electron configuration $2p^3$ (the ground state of N). We know from tables 7.2 and 6.5 that the allowed terms are ^{4}S, ^{2}P and ^{2}D. Table 6.5 also tells us that we can determine the energy for ^{4}S uniquely from the spin-orbital $(1^+, 0^+, -1^+)$. The difference here is that we have three electrons, and according to (2.36), we have to sum the contributions from all permutations of pairs. This means that the energy is:

$$\begin{aligned}
E^{(LS)}(2p^{3\ ^4}S) &= 3I(2p) + \left\langle \mathbb{S}_{(1^+,0^+)}\,|H_{ee}\,|\,\mathbb{S}_{(1^+,0^+)}\right\rangle + \left\langle \mathbb{S}_{(1^+,-1^+)}\,|H_{ee}\,|\,\mathbb{S}_{(1^+,-1^+)}\right\rangle \\
&\quad + \left\langle \mathbb{S}_{(0^+,-1^+)}\,|H_{ee}\,|\,\mathbb{S}_{(0^+,-1^+)}\right\rangle \\
&= 3I(2p) + F^{(0)}_{2p:2p} - \frac{2}{25}F^{(2)}_{2p:2p} - \frac{3}{25}G^{(2)}_{2p:2p} + F^{(0)}_{2p:2p} + \frac{1}{25}F^{(2)}_{2p:2p} \\
&\quad - \frac{6}{25}G^{(2)}_{2p:2p} + F^{(0)}_{2p:2p} - \frac{2}{25}F^{(2)}_{2p:2p} - \frac{3}{25}G^{(2)}_{2p:2p} \\
&= 3I(2p) + 3F^{(0)}_{2p:2p} - \frac{15}{25}F^{(2)}_{2p:2p}\,.
\end{aligned} \tag{7.42}$$

In the same manner, we can calculate:

$$E^{(LS)}(2\text{p}^3\ {}^2\text{D}) = 3I(2\text{p}) + 3F^{(0)}_{2\text{p:2p}} - \frac{6}{25}F^{(2)}_{2\text{p:2p}}\ . \tag{7.43}$$

Finally, we can apply the diagonal-sum rule on the two states in the table for $M_L = 1$ and $M_S = 1/2$, and we find:

$$E^{(LS)}(2\text{p}^3\ {}^2\text{P}) = 3I(2\text{p}) + 3F^{(0)}_{2\text{p:2p}}\ . \tag{7.44}$$

The conclusion is that for the 2p^3-configuration, the term ^{4}S should be the ground state, and the lowest energy excited term should be ^{2}D. This agrees with experimental data. The theoretical energy separations between the terms are:

$$\begin{aligned} E^{(LS)}(2\text{p}^3\ {}^2\text{P}) - E^{(LS)}(2\text{p}^3\ {}^2\text{D}) &= \frac{6}{25}F^{(2)}_{2\text{p:2p}} \\ E^{(LS)}(2\text{p}^3\ {}^2\text{D}) - E^{(LS)}(2\text{p}^3\ {}^4\text{S}) &= \frac{9}{25}F^{(2)}_{2\text{p:2p}} \\ \Rightarrow \quad \frac{E^{(LS)}(2\text{p}^3\ {}^2\text{P}) - E^{(LS)}(2\text{p}^3\ {}^2\text{D})}{E^{(LS)}(2\text{p}^3\ {}^2\text{D}) - E^{(LS)}(2\text{p}^3\ {}^4\text{S})} &= \frac{2}{3}\ . \end{aligned} \tag{7.45}$$

The corresponding experimental value for N is 0.50.

7.3 Fine-Structure

It is now time to include the spin–orbit interaction for the valence electrons in the analysis. This will give us the fine-structure of the *LS*-coupling atomic terms. The total Hamiltonian for the ensemble of valence electrons is now:

$$H_{\text{val}} = H_{\text{kin}} + H_{\text{nu}} + H_{\text{ee}} + H_{\text{SO}} = H_{LS} + H_{\text{SO}}\ . \tag{7.46}$$

That is, the spin–orbit Hamiltonian has been added as a perturbation to (7.5), and we have introduced the notation H_{val} for the total valence electron Hamiltonian.

The basis of the inclusion of the spin–orbit interaction as a perturbation to the *LS*-coupling terms was covered in section 6.3.5. The perturbation Hamiltonian is that of (6.31):

$$H_{\text{SO}} = \left[\sum_i^N \xi(r_i)\right] \mathbf{L}\cdot\mathbf{S}\ . \tag{7.47}$$

The radial factor in (7.47) will be approximated as being a function of the electronic configuration, and of L and S. It constitutes an operator and it commutes with $\mathbf{J}^2$ and J_z, as does the factor $\mathbf{L}\cdot\mathbf{S}$. As a consequence, the Hamiltonian (7.47) is diagonal in J and M_J, and we can expect a good state representation to be $|\gamma LSJM_J\rangle$. Each occurrence within a term, with a distinct value of J will be one fine-structure energy level, having $2J+1$ degenerate M_J states.

The energy contribution to a level in the representation above is found by taking the expectation value of the Hamiltonian:

$$\begin{aligned} E^{(\mathrm{SO})}(\gamma\, ^{2S+1}L_J) &= \langle \gamma LSJM_J \,|H_{\mathrm{SO}}|\, \gamma LSJM_J \rangle \\ &= \frac{1}{2} \left\langle \gamma LS \left| \sum_i^N \xi(r_i) \right| \gamma LS \right\rangle \langle LSJ \,|\, \mathbf{L}\cdot\mathbf{S} \,|\, LSJ \rangle \\ &= \frac{A(\gamma LS)}{2} \left[J(J+1) - L(L+1) - S(S+1) \right] . \end{aligned} \tag{7.48}$$

In the last line, we have substituted the *fine-structure factor* $A(\gamma LS)$, introduced in (6.34). The exact value of this is not easily accessible analytically, but as we shall see in the following, it can be expressed in terms of orbital integrals. This will provide ample insights, even without exact quantitative data. Empirical values of $A(\gamma LS)$ can also be deduced from spectroscopic data.

The fine-structure factor is specific for every atomic term but does not depend on J. This means that it does not contribute to the energy splitting of a term. The latter — the *fine-structure* — will be determined by the integral over $\mathbf{L}\cdot\mathbf{S}$. However, the value of $A(\gamma LS)$ will provide the scaling of this splitting, and thus the total energy width of a multiplet.

The treatment of the Hamiltonian (7.47) as diagonal with respect to *LS*-coupling terms works because we are taking advantage of the assumption that its energy contribution is small compared to the electron–electron interaction H_{ee}. If a complete Hamiltonian matrix was to be constructed, in the representation $|\gamma\, LSJM_J\rangle$, one would find finite non-diagonal elements. In fact, such elements are sometimes responsible for a breakdown of the approximation. A few examples of this will be shown in section 9.5. Here, we will work with states for which it is a sound approximation to take the perturbation Hamiltonian as diagonal (that is 'good *LS*-coupling').

7.3.1 Multiplet Intervals

The fine-structure energy in (7.48) leads to Landé's rule for energy splittings within a multiplet, stated in (6.37). For pure or almost pure *LS*-coupling, the energy interval between two neighbouring fine-structure levels is proportional to the fine-structure factor, and to the highest J-value of the pair. This property makes the Landé rule an excellent tool for empirical analysis of spectra. Observing the ratio between energy intervals, as in (6.38), for a group of spectral lines can help with identification of terms and provides a measure of the extent to which the *LS*-coupling approximation is tenable for the element, and the group of energy levels studied.

The *centre of gravity* of a *LS*-coupling term is the mean value of all the states of the multiplet. This is unchanged under the spin–orbit interaction. Mathematically, the above statement reads:

$$\sum_{J=|L-S|}^{L+S} [J(J+1)-L(L+1)-S(S+1)]\,(2J+1)=0 \ , \tag{7.49}$$

where the last factor on the left side is the degeneracy. This can be proven by using the polynomial expressions [2]:

$$\begin{aligned}\sum_{i=0}^{n} i &= \frac{1}{2}n(n+1)\\ \sum_{i=0}^{n} i^2 &= \frac{1}{6}n(n+1)(2n+1)\\ \sum_{i=0}^{n} i^3 &= \frac{1}{4}n^2(n+1)^2 \ .\end{aligned} \tag{7.50}$$

Physically, this is a consequence of conservation of total angular momentum, in the absence of any external interactions.

Equation (7.48) shows that whether the energy increases or decreases for higher J depends on the sign of $A(\gamma LS)$. If we consult the Grotrian diagrams in figures 6.4 and 6.5, we see that for those two examples we have $A(\gamma LS) > 0$. This is referred to as *normal order*, whereas the opposite — *inverted order* — means a negative $A(\gamma LS)$. For that eventuality the fine-structure level with the highest J has the lowest energy.

An empirical observation is that atoms with a less than half-filled valence orbital obey normal order, while in contrast, one gets inverted order if it is more than half full. Atoms that have their valence orbital exactly half-filled (such as N, P, Mn, As, Tc, Sb, Eu, Re and Bi) all have a $L = 0$ terms as their ground state, which means that they should have no ground state fine-structure splitting in a first-order approximation. The connection between the sign of $A(\gamma LS)$ and the filling ratio of a valence orbital can be put on a more robust theoretical footing, as will be shown in the following section.

7.3.2 Calculation of Fine-Structure Energy Intervals

In section 7.2.1, we computed energies — or rather perturbative energy corrections — for *LS*-coupling terms. We can make a similar calculation for multiplet intervals; that is the fine-structure perturbations to the term energies. When doing this, we will again make use of the diagonal-sum rule for matrices. A difference from the analysis in section 7.2.1, however, is that the spin–orbit coupling Hamiltonian is a single-electron operator, which means that we can avoid some of the complications that we had when estimating term energies.

The present calculation amounts to finding values for the fine-structure factor $A(\gamma LS)$ in terms of single-electrons integrals. Albeit we will not get absolute energy values (with the exact functional form of the radial wave function remaining unknown), the calculation will give a better qualitative understanding of the fine-structure. It will provide a basis for empirically observed rules, and it may facilitate numerical calculations, by furnishing better initial trial functions.

We have seen that the fine-structure Hamiltonian is diagonal in the representation $|\gamma LSJM_J\rangle$. However, in order to calculate $A(\gamma LS)$, we will apply the diagonal-sum rule for two different representations, namely $|\gamma LSM_LM_S\rangle$ and one specifying the individual quantum numbers of all N valence electrons: $|\gamma\, m_{l1}m_{s1},\ldots,m_{lN}m_{sN}\rangle$ (γ is a shorthand notation for $n_1l_1\ldots n_Nl_N$). In the individual electron representation, we use the more explicit form of the Hamiltonian given in (4.19):

$$H_{\mathrm{SO}}=\sum_{i=1}^{N}\xi(r_i)\,\mathbf{L}_i\cdot\mathbf{S}_i=\sum_{i=1}^{N}\xi(r_i)\,(L_{xi}S_{xi}+L_{yi}S_{yi}+L_{zi}S_{zi})\quad, \tag{7.51}$$

with the summation taken over the valence electrons. Since we are aiming for a solution via the diagonal-sum rule, we focus solely on the diagonal elements of the matrix.

The radial factor depends only on the configuration quantum numbers, γ. The terms in (7.51) that contain L_{xi}, S_{xi}, L_{yi} and S_{yi} will only have non-diagonal contributions in a representation based on the projection along $\hat{e}_z$ (since Cartesian angular momentum components do not commute), and accordingly, we will disregard them. The diagonal matrix elements are:

$$\begin{aligned}&\langle\gamma\, m_{l1}m_{s1},\ldots,m_{lN}m_{sN}\,|H_{\mathrm{SO}}|\,\gamma\, m_{l1}m_{s1},\ldots,m_{lN}m_{sN}\rangle\\&\qquad=\sum_{i=1}^{N}\langle\gamma_i\,|\,\xi(r_i)\,|\,\gamma_i\rangle\langle m_{li}m_{si}\,|L_{zi}S_{zi}|\,m_{li}m_{si}\rangle=\sum_{i=1}^{N}\xi_{n_il_i}\,m_{li}m_{si}\quad.\end{aligned} \tag{7.52}$$

In the last step, we have defined the integral form of the fine-structure factor. Note that one sum such as that in (7.52) will be needed for each spin-orbital (a combination of vales for $m_{l1}m_{s1}\ldots m_{lN}m_{sN}$). For a given fine-structure energy, there will typically be more than one possible spin-orbital that can result in the corresponding J, and accordingly several such sums will be needed in order to find the perturbation energy.

Next step is to correlate this with the $|\gamma LSM_LM_S\rangle$ representation. To get expressions for these matrix elements, we make use of the form of the spin–orbit Hamiltonian that is written in terms of the vector operators **L** and **S** for the entire atom. This can be taken from (7.47):

$$H_{\mathrm{SO}}=A(\gamma LS)\,\mathbf{L}\cdot\mathbf{S}\quad. \tag{7.53}$$

To compute the corresponding matrix elements, we expand the scalar product as — see (C.4):

$$\mathbf{L}\cdot\mathbf{S}=L_xS_x+L_yS_y+L_zS_z=L_zS_z+\frac{1}{2}\,(L_+S_-+L_-S_+)\quad. \tag{7.54}$$

For the same reason as earlier, we need not bother about the non-diagonal elements. For the diagonal ones, only one term in (7.54) will yield a non-zero contribution. We thus have:

$$\langle \gamma LSM_LM_S \,|\, H_{SO} \,|\, \gamma LSM_LM_S \rangle = A(\gamma LS)\, M_LM_S \,. \tag{7.55}$$

$A(\gamma LS)$ can be found by taking a given combination of values for M_L and M_S for a certain configuration (for example, by using diagrams such as in tables 6.3, 6.5 and 7.3) and identifying all *LS*-coupling terms that are represented for that combination. The sum of the diagonal elements in the representation given in (7.55) will equal the sum over the valence electrons for all represented spin-orbital combinations, in accordance with (7.52). We will illustrate the procedure by initially calculating $A(\gamma LS)$ for the two configurations in the examples in tables 6.3 and 6.5, that is $n\mathrm{p}^2$ and $n\mathrm{p}^3$, and also for $n\mathrm{p}n'\mathrm{p}$ and $n\mathrm{p}^4$.

For a $n\mathrm{p}^2$ configuration, the three *LS*-coupling terms are $^3\mathrm{P}$, $^1\mathrm{S}$ and $^1\mathrm{D}$. The two singlets do not have fine-structure splittings ($S=0$), so calculating a fine-structure factor for them does not have any meaning. However, we need to evaluate $A(n\mathrm{p}^2\,{}^3\mathrm{P})$ in order to find the energy splitting between $^3\mathrm{P}_2$, $^3\mathrm{P}_1$ and $^3\mathrm{P}_0$.

We see in table 6.3, and the discussion that follows it, that for the combination $M_L=+1$ and $M_S=+1$, $^3\mathrm{P}$ is the only feasible term. We also see that the only spin-orbital for that M_LM_S combination is (1^+0^+). Using (7.52) and (7.55), we find:

$$A(n\mathrm{p}^2\,{}^3\mathrm{P}) = \frac{1}{2}\,\xi_{n\mathrm{p}} \;. \tag{7.56}$$

The integral $\xi_{n\mathrm{p}}$ is proportional to the expectation value of r^{-3}, and must therefore be positive — see (4.13) and (B.39). This means that the fine-structure of $^3\mathrm{P}$ will follow normal order, which agrees with the empirical observation for less than half-filled orbitals.

The total valence electron energy perturbation to the CFA configuration of the fine-structure states for $n\mathrm{p}\,{}^3\mathrm{P}$ is thus:

$$\begin{aligned}
E^{(\mathrm{val})}(n\mathrm{p}^2\,{}^3\mathrm{P}_2) &= 2I(n\mathrm{p}) + F^{(0)}_{n\mathrm{p}:n\mathrm{p}} - \frac{5}{25}\,F^{(2)}_{n\mathrm{p}:n\mathrm{p}} + \frac{1}{2}\,\xi_{n\mathrm{p}} \\
E^{(\mathrm{val})}(n\mathrm{p}^2\,{}^3\mathrm{P}_1) &= 2I(n\mathrm{p}) + F^{(0)}_{n\mathrm{p}:n\mathrm{p}} - \frac{5}{25}\,F^{(2)}_{n\mathrm{p}:n\mathrm{p}} - \frac{1}{2}\,\xi_{n\mathrm{p}} \\
E^{(\mathrm{val})}(n\mathrm{p}^2\,{}^3\mathrm{P}_0) &= 2I(n\mathrm{p}) + F^{(0)}_{n\mathrm{p}:n\mathrm{p}} - \frac{5}{25}\,F^{(2)}_{n\mathrm{p}:n\mathrm{p}} - \xi_{n\mathrm{p}} \quad ,
\end{aligned} \tag{7.57}$$

where we have used (7.36), (7.56) and (7.48).

For $n\mathrm{p}n'\mathrm{p}$, there are two terms that have fine-structure splittings: $^3\mathrm{P}$ and $^3\mathrm{D}$. We now refer to table 6.2. We again begin with a M_LM_S cell with only one term. We can take $M_L=2$ and $M_S=1$, where the sole represented term is $^3\mathrm{D}$, and where the only spin-orbital is (1^+1^+). From this, we find the fine-structure factor for $^3\mathrm{D}$:

$$\begin{aligned} 2A(npn'\mathrm{p}\,^3\mathrm{D}) &= \frac{1}{2}\,\xi_{n\mathrm{p}} + \frac{1}{2}\,\xi_{n'\mathrm{p}} \\ \Rightarrow \quad A(npn'\mathrm{p}\,^3\mathrm{D}) &= \frac{1}{4}\left(\xi_{n\mathrm{p}} + \xi_{n'\mathrm{p}}\right) . \end{aligned} \tag{7.58}$$

To calculate $A(npn'\mathrm{p}\,^3\mathrm{P})$, we use the combination of the values $M_L = 1$ and $M_S = 1$, for which both $^3\mathrm{D}$ and $^3\mathrm{P}$ are represented, and where the two spin-orbitals are $(1^+\,0^+)$ and $(0^+\,1^+)$. The equality between the two different diagonal sums becomes:

$$\begin{aligned} A(npn'\mathrm{p}\,^3\mathrm{D}) + A(npn'\mathrm{p}\,^3\mathrm{P}) &= \frac{1}{2}\,\xi_{n\mathrm{p}} + \frac{1}{2}\,\xi_{n'\mathrm{p}} \\ \Rightarrow \quad A(npn'\mathrm{p}\,^3\mathrm{P}) &= \frac{1}{4}\left(\xi_{n\mathrm{p}} + \xi_{n'\mathrm{p}}\right) = A(npn'\mathrm{p}\,^3\mathrm{D}) . \end{aligned} \tag{7.59}$$

We conclude that for this configuration, the fine-structure factors are positive and the same for both triplets. If we combine this result with the Landé interval rule (6.37), we also learn that for this configuration, the ratio of the width of the multiplet $^3\mathrm{D}$ to that of $^3\mathrm{P}$ is five to three.

As yet another example, we study a np^3 configuration, as illustrated in table 6.5. The allowed terms are $^4\mathrm{S}$, $^2\mathrm{P}$ and $^2\mathrm{D}$. The ground state term $^4\mathrm{S}$ does not have fine-structure, since $L=0$, so we only need to calculate the factors for the two doublets. The cell ($M_L=2$, $M_S=1/2$) in table 6.5 gives:

$$A(n\mathrm{p}^3\,^2\mathrm{D}) = \frac{1}{2}\,\xi_{n\mathrm{p}} - \frac{1}{2}\,\xi_{n\mathrm{p}} = 0 . \tag{7.60}$$

From the cell ($M_L=1$, $M_S=1/2$), we have:

$$\frac{1}{2}\left[A(n\mathrm{p}^3\,^2\mathrm{D}) + A(n\mathrm{p}^3\,^2\mathrm{P})\right] = \frac{1}{2}\,\xi_{n\mathrm{p}} - \frac{1}{2}\,\xi_{n\mathrm{p}} - \frac{1}{2}\,\xi_{n\mathrm{p}} + \frac{1}{2}\,\xi_{n\mathrm{p}} = 0 . \tag{7.61}$$

The somewhat surprising result is that these fine-structure factors are both zero. This agrees with the hypothesis stated earlier that less than half-filled orbitals produce normal order, those more than half full displays inverted fine-structure order, and exactly half full orbitals have no fine-structure. However, the final part of this statement does not quite agree with experiment.

To give a concrete example, we can take the ground configuration of N: $2\mathrm{p}^3$. The experimental value for the fine-structure splitting for $^2\mathrm{D}$ is about 8.7 cm^{-1}, which would correspond to a fine-structure factor of 3.5 cm^{-1}, and the order is inverted. For $^2\mathrm{P}$ the splitting is 0.39 cm^{-1}, which makes the empirical fine-structure factor 0.26 cm^{-1}, and in that case, the ordering is normal. These numbers are non-zero, but they are very small compared to other typical fine-structure factors in the 2p-series of atoms.

As a comparison, we can take the $2\mathrm{p}^2\,^3\mathrm{P}$ ground state in C. The two fine-structure separations in the triplet are 16 cm^{-1} and 27 cm^{-1}, respectively. If we take the ratio between those numbers, we find that this does not fully agree with the Landé rule, but if we compute an average empirical fine-structure factor, the result is about 15 cm^{-1}. In another example, the $2p^4\,{}^3P$ ground state in O, the splittings in the triplet are 158 cm^{-1} and 69 cm^{-1}, yielding an empirical fine-structure constant of 74 cm^{-1}. Compared with these numbers, we see that the fine-structure in N, with an exactly half-filled orbital, is *almost* zero. The very small, but non-zero, fine-structure in the $2p^3$ configuration in N is a signature of a small departure from *LS*-coupling, and in general, we will experimentally find very small splittings in multiplets for configurations corresponding to exactly half full orbitals.

As a final example, we calculate the fine-structure factor for the already mentioned ground state triplet in O, with the configuration $2p^4$. We have not shown a spin-orbital diagram for this configuration, but as explained in section 6.3.3, it will be the same as that in table 6.3 for $2p^2$. The difference is that the two single-electron symbols, in this case, represent the *missing* electrons, and we have to mirror the table both horizontally and vertically.

Taking this aboard, we find that for ${}^3P$, and the cell ($M_L = 1$, $M_S = 1$), the only possible spin-orbital must be the four-electron function $(1^+\,1^-\,0^+\,-1^+)$. With the diagonal-sum rule, we can then find that the fine-structure is given by:

$$A(np^4\,{}^3\mathrm{P}) = \frac{1}{2}\,\xi_{np} - \frac{1}{2}\,\xi_{np} - \frac{1}{2}\,\xi_{np} = -\frac{1}{2}\,\xi_{np}\,. \tag{7.62}$$

We get a negative value, agreeing with the empiric observation of inverted order. In the same manner as above, estimates of fine-structure factors can be calculated for any configuration within the *LS*-coupling approximation.

7.4 Wave Functions

So far, the emphasis of the study in this chapter has been the energy perturbations beyond the electronic configurations, within the CFA. In particular, we have looked at the energy structure emanating from interactions between angular momenta, but we have not divulged into the mathematical form of the eigenfunctions corresponding to the respective energies. In this section, we will give a brief account also of how partial wave functions can be constructed in the *LS*-coupling scheme. This will be either approximations obtained with methods such as the variational method, or superpositions of Slater determinant functions. The latter may themselves not be explicitly known, nevertheless, an overview of the basic anatomy of the wave functions is useful.

The CFA and the various vector coupling schemes that spring from it represent methods that lend themselves excellently to the disentanglement of atomic spectra and to an understanding of the qualitative features of atomic structure. Many quantitative results are impressive, such as calculations of energy splittings, and this makes the methodology into a powerful tool in the analysis of atomic spectra. However, as was seen in the preceding sections, also quantitative calculations of energies suffer from the lack of knowledge of the functional form of zero-order wave functions. The term energies that we calculated in section 7.2.2 were stated in terms of special integrals, which we have not been able to further quantify, or put in explicit mathematical form.

To find functional forms for the full atomic wave functions through the CFA, and subsequent perturbation theory, is hard to do analytically. In principle, perturbation theory does provide means also to calculate perturbed wave functions, but a prerequisite is that explicit forms of the zero-order functions are known, and this is not the case in the CFA. However, we can get some approximative results.

The general methods of matrix algebra used in sections 7.2.2 and 7.3 can be built on also to express wave functions of *LS*-coupling levels in terms of superpositions of Slater determinants of single-electron spin-orbitals. This will be described in section 7.4.1. Following that, we will demonstrate that for small atoms, the variational principle can be used to approximate radial wave functions, which then can be used for quantitative determinations of Slater F- and G-integrals.

7.4.1 Eigenfunctions of Atomic Terms

How to construct eigenfunctions of *LS*-coupling terms as superpositions of spin-orbitals is thoroughly described in [2]. The working methods are based on the same type of diagrams as those used in tables 6.3 and 7.4. Such tables show that the *LS*-coupling terms are superpositions of spin-orbitals, and we know that the latter should be represented by Slater integrals.

The discussion will here to a large extent mirror that pursued in sections 7.1 and 7.2. The Hamiltonian H_{ee} is diagonal in L, S, M_L and M_S, and what we are looking for are superposition states that properly describe states of the form $|LSM_LM_S\rangle$ — or put in another way $|\,^{2S+1}L;(M_L,M_S)\,\rangle$. To identify the eigenfunctions, we will again use tables 6.3 and 7.4. Thereafter, we will use secular equations and angular momentum ladder operators to get the remaining functions.

In order to calculate the composite eigenvalues of L and S, we will use (C.5). For a general angular momentum, we can write:

$$\mathbf{J}^2 = J_+J_- + J_z^2 + J_z \ . \tag{7.63}$$

For all Cartesian components of $\mathbf{L}$ and $\mathbf{S}$, we have $L_x = l_{x1} + \ldots + l_{xN}$, and so on. Here, N is the number of valence electrons. Following from this is that also the ladder operators are additive:

$$L_+ = L_{1+} + \cdots + L_{N+} \ , \tag{7.64}$$

and the analogue operator relations for L_-, S_+ and S_-.

This means that we can formulate equations for the action of composite ladder operators on Slater determinants, using (C.20) and (C.22). For the Slater determinant corresponding to the spin-orbital $(m_{l1}^{+/-},\ldots,m_{lN}^{+/-})$ we use the same notation as in section 7.2:

$$\mathbb{S}_{(m_{l1}^{+/-},\ldots,m_{lN}^{+/-})} \quad . \tag{7.65}$$

For example, for a two-electron atom with $m_{l1}=1$, $m_{s1}=-1/2$, $m_{l2}=0$ and $m_{s2}=1/2$, we write $(1^{-}0^{+})$.

The relations for the action of the projection and step operators on Slater determinants are:

$$\begin{aligned}
L_{+}\,\mathbb{S}_{(m_{l1}^{+/-},\ldots,m_{lN}^{+/-})} &= \sum_{i=1}^{N}\sqrt{(l_i+m_{li}+1)(l_i-m_{li})}\;\mathbb{S}_{(m_{l1}^{+/-},\ldots,[m_{li}+1]^{+/-},\ldots,m_{lN}^{+/-})}\\
L_{-}\,\mathbb{S}_{(m_{l1}^{+/-},\ldots,m_{lN}^{+/-})} &= \sum_{i=1}^{N}\sqrt{(l_i-m_{li}+1)(l_i+m_{li})}\;\mathbb{S}_{(m_{l1}^{+/-},\ldots,[m_{li}-1]^{+/-},\ldots,m_{lN}^{+/-})}\\
L_{z}\,\mathbb{S}_{(m_{l1}^{+/-},\ldots,m_{lN}^{+/-})} &= \left[\sum_{i=1}^{N}m_{li}\right]\mathbb{S}_{(m_{l1}^{+/-},\ldots,m_{lN}^{+/-})}\\
S_{+}\,\mathbb{S}_{(m_{l1}^{+/-},\ldots,m_{lN}^{+/-})} &= \sum_{i=1}^{N}\delta(m_{si},-1/2)\;\mathbb{S}_{(m_{l1}^{+/-},\ldots,m_{li}^{+},\ldots,m_{lN}^{+/-})}\\
S_{-}\,\mathbb{S}_{(m_{l1}^{+/-},\ldots,m_{lN}^{+/-})} &= \sum_{i=1}^{N}\delta(m_{si},1/2)\;\mathbb{S}_{(m_{l1}^{+/-},\ldots,m_{li}^{-},\ldots,m_{lN}^{+/-})}\\
S_{z}\,\mathbb{S}_{(m_{l1}^{+/-},\ldots,m_{lN}^{+/-})} &= \left[\sum_{i=1}^{N}m_{si}\right]\mathbb{S}_{(m_{l1}^{+/-},\ldots,m_{lN}^{+/-})} \quad .
\end{aligned} \tag{7.66}$$

A core caveat is that any term in the sums above where the quantum numbers m_{li} or m_{si} are at their upper or lower extreme, for the raising and lowering operators, respectively, will return zero.

This procedure can be illustrated by a few examples of two-electron configurations, with successively increasing complexity. To begin with, we take a 1s2s configuration, which corresponds to the lowest excited configuration in He — previously analysed in section 2.2.3. Here, we have $l_1=l_2=0$, and thus no other options for the total orbital angular momentum than $L=M_L=0$. This means that we can restrict the analysis to **S**.

The four possible spin-orbitals are $(0^{+}0^{+})$, $(0^{+}0^{-})$, $(0^{-}0^{+})$ and $(0^{-}0^{-})$. Using (7.63) and (7.66), we can compute the action of $\mathbf{S}^2$ on the corresponding Slater determinant functions:

$$\begin{aligned}
\mathbf{S}^2\,\mathbb{S}_{(0^{+}0^{+})} &= 2\,\mathbb{S}_{(0^{+}0^{+})}\\
\mathbf{S}^2\,\mathbb{S}_{(0^{+}0^{-})} &= \mathbf{S}^2\,\mathbb{S}_{(0^{-}0^{+})} = \mathbb{S}_{(0^{+}0^{-})}+\mathbb{S}_{(0^{-}0^{+})}\\
\mathbf{S}^2\,\mathbb{S}_{(0^{-}0^{-})} &= 2\,\mathbb{S}_{(0^{-}0^{-})} \quad .
\end{aligned} \tag{7.67}$$

The two functions with aligned spins are eigenstates to the operator, corresponding to $S=1$. These will be identical to the wave functions for two of the terms. For the remaining two, the *LS*-coupling states must be superpositions. The secular equation is simple and gives the eigenvalues 2 and 0. In other words, the eigenstates will be one triplet and one singlet, exactly as was found with simpler means in chapter 2.

In this case, the normalisation of the superposition states is trivial, but for the sake of giving an example, we shall demonstrate how it can be obtained with step operators. The wave functions for the kets $|\,^3\mathrm{S};(0,1)\,\rangle$ and $|\,^3\mathrm{S};(0,-1)\,\rangle$ correspond, respectively, to $\mathbb{S}_{(0^+0^+)}$ and $\mathbb{S}_{(0^-0^-)}$. The operation of the lowering operator on the first of these (with $M_S = +1$) can be derived from (C.20) and (C.22), and if we use the Slater determinant notation, (7.66) is still valid. This gives us the following two relations:

$$S_-\ |^3\mathrm{S};(0,1)\,\rangle = \sqrt{(S - M_S + 1)(S + M)}\ \ |^3\mathrm{S};(0,0)\,\rangle = \sqrt{2}\,|^3\mathrm{S};(0,0)\,\rangle$$
$$S_-\ \mathbb{S}_{(0^+0^+)} = \mathbb{S}_{(0^+0^-)} + \mathbb{S}_{(0^-0^+)}\ . \tag{7.68}$$

Since the two expressions to the left correspond to the same state, the two on the right do also. Thus, we find the correct normalised superposition wave function for $|\,^3\mathrm{S};(0,0)\,\rangle$.

The only state remaining to specify is the singlet $|\,^1\mathrm{S};(0,0)\,\rangle$, and we can find this by using the requirement that all the wave functions must be mutually orthogonal. The result is an anti-symmetric superposition. The normalised *LS*-coupling wave functions, expressed in Slater determinant functions of spin-orbitals, for the 1s2s configuration are as shown in table 7.6.

Table 7.6 Decomposition of the *LS*-coupling wave functions for the electronic configuration 1s2s, in Slater determinants of composite spin-orbitals. These superposition states are eigenfunctions of $\mathbf{L}^2$, L_z, $\mathbf{S}^2$ and S_z.

Term	M_L	M_S	eigenstate
		1	$\mathbb{S}_{(0^+0^+)}$
^{3}S	0	0	$\frac{1}{\sqrt{2}}\left[\mathbb{S}_{(0^+0^-)} + \mathbb{S}_{(0^-0^+)}\right]$
		-1	$\mathbb{S}_{(0^-0^-)}$
^{1}S	0	0	$\frac{1}{\sqrt{2}}\left[\mathbb{S}_{(0^+0^-)} - \mathbb{S}_{(0^-0^+)}\right]$

For the simplest two-electron atom with a non-zero orbital angular momentum, we turn to the 1s2p. This has twelve possible spin-orbitals, for which we have to diagonalise $\mathbf{L}^2$ and $\mathbf{S}^2$. That will produce twelve *LS*-coupling states (with specific values of M_L and M_S), distributed over the terms ^{3}P and ^{1}P. We will do this by beginning with a spin-orbital that is unique for one combination of M_L and M_S, and from there, we will use step operators to find the other superposition states.

The spin-orbital $(0^+\,1^+)$ is the only one that can give the values $M_L = 1$ and $M_S - 1$ at the same time. This means that the corresponding Slater determinant must be the good wave function for $|\,^3\text{P};(1,1)\,\rangle$. We can confirm that this is an eigenstate of $\mathbf{L}^2$ and $\mathbf{S}^2$ by using (7.63) and (7.66). This gives the expected results of $L=1$ and $S=1$.

From there, we may use the operator S_- to find the correct, normalised superposition for $|\,^3\text{P};(1,0)\,\rangle$. This is one of two states with $M_L=1$ and $M_S=0$. The other one is $|\,^1\text{P};(1,0)\,\rangle$, and with one eigenfunction found for a secular equation of dimension two, the other one can be obtained from the orthogonality criterion.

We can use S_- also on $|\,^3\text{P};(1,0)\rangle$ to find $|\,^3\text{P};(1,-1)\,\rangle$, and L_- on $|\,^3\text{P};(1,1)\rangle$ for $|\,^3\text{P};(0,1)\,\rangle$. Then, we can continue to use the condition of orthogonality and step operators for all states, and we can for each state confirm that it is an eigenfunction to $\mathbf{L}^2$ and $\mathbf{S}^2$. The results for the are compiled in table 7.7.

Table 7.7 Decomposition of the *LS*-coupling wave functions for the electronic configuration 1s2p, in Slater determinants of spin-orbitals

Term	M_L	M_S	eigenstate
^{3}P	1	1	$\mathbb{S}_{(0^+\,1^+)}$
		0	$\frac{1}{\sqrt{2}}\left[\mathbb{S}_{(0^+\,1^-)}+\mathbb{S}_{(0^-\,1^+)}\right]$
		-1	$\mathbb{S}_{(0^-\,1^-)}$
	0	1	$\mathbb{S}_{(0^+\,0^+)}$
		0	$\frac{1}{\sqrt{2}}\left[\mathbb{S}_{(0^+\,0^-)}+\mathbb{S}_{(0^-\,0^+)}\right]$
		-1	$\mathbb{S}_{(0^-\,0^-)}$
	-1	1	$\mathbb{S}_{(0^+\,-1^+)}$
		0	$\frac{1}{\sqrt{2}}\left[\mathbb{S}_{(0^+\,-1^-)}+\mathbb{S}_{(0^-\,-1^+)}\right]$
		-1	$\mathbb{S}_{(0^-\,-1^-)}$
^{1}P	1	0	$\frac{1}{\sqrt{2}}\left[\mathbb{S}_{(0^+\,1^-)}-\mathbb{S}_{(0^-\,1^+)}\right]$
	0	0	$\frac{1}{\sqrt{2}}\left[\mathbb{S}_{(0^+\,0^-)}-\mathbb{S}_{(0^-\,0^+)}\right]$
	-1	0	$\frac{1}{\sqrt{2}}\left[\mathbb{S}_{(0^+\,-1^-)}-\mathbb{S}_{(0^-\,-1^+)}\right]$

For a two-electron atom with non-equivalent p-electrons, such as, for example, 2p3p, a diagram of the possible spin-orbitals is given in table 6.2. The methods outlined above for the simpler configurations still work, and while they get more tedious to carry through, they do not get conceptually more difficult. A natural place to begin is the orbital $(1^+\,1^+)$, which is the good wave function for $|\,^3\text{D};(2,1)\,\rangle$. With the operators, S_- and L_-, we can then pick out all fifteen states belonging to ^{3}D, and they will all be automatically normalised (albeit with an arbitrary phase).

Carrying on, the orthogonality requirement for the pair of spin-orbitals corresponding to $(M_L=1, M_S=1)$ provides us with a function for $|\,^3\text{P};(1,1)\,\rangle$, and then all nine states belonging to that triplet can be recovered by lowering operators. In this way, we may continue. When we eventually get to $(M_L=0, M_S=0)$, we will have superpositions of six spin-orbital Slater determinants, corresponding to one

LS-coupling state for each of the six terms present. Five of those will already have been determined by step operators, and the remaining — $|{}^1\mathrm{S};0,0\rangle$ — must be orthogonal to all of these.

In tables 7.8 and 7.9, we show the result of this exercise. Because of the size of the tables, we have divided them into one for the triplets and one for the singlets.

Table 7.8 Decomposition of the *LS*-coupling wave functions for the electronic 2p3p, in Slater determinants of spin-orbitals. This table shows the terms ^{3}D, ^{3}P and ^{3}S. The singlet terms are given in table 7.9.

Term	M_L	M_S	eigenstate
		1	$\mathbb{S}_{(1^+1^+)}$
	2	0	$\frac{1}{\sqrt{2}}\left[\mathbb{S}_{(1^+1^-)}+\mathbb{S}_{(1^-1^+)}\right]$
		−1	$\mathbb{S}_{(1^-1^-)}$
		1	$\frac{1}{\sqrt{2}}\left[\mathbb{S}_{(1^+0^+)}+\mathbb{S}_{(0^+1^+)}\right]$
	1	0	$\frac{1}{2}\left[\mathbb{S}_{(1^+0^-)}+\mathbb{S}_{(1^-0^+)}+\mathbb{S}_{(0^+1^-)}+\mathbb{S}_{(0^-1^+)}\right]$
		−1	$\frac{1}{\sqrt{2}}\left[\mathbb{S}_{(1^-0^-)}+\mathbb{S}_{(0^-1^-)}\right]$
		1	$\frac{1}{\sqrt{6}}\left[\mathbb{S}_{(1^+-1^+)}+2\mathbb{S}_{(0^+0^+)}+\mathbb{S}_{(-1^+1^+)}\right]$
^{3}D	0	0	$\frac{1}{\sqrt{12}}\left[\mathbb{S}_{(1^+-1^-)}+\mathbb{S}_{(1^--1^+)}+2\mathbb{S}_{(0^+0^-)}+2\mathbb{S}_{(0^-0^+)}+\mathbb{S}_{(-1^+1^-)}+\mathbb{S}_{(-1^-1^+)}\right]$
		−1	$\frac{1}{\sqrt{6}}\left[\mathbb{S}_{(1^--1^-)}+2\mathbb{S}_{(0^-0^-)}+\mathbb{S}_{(-1^-1^-)}\right]$
		1	$\frac{1}{\sqrt{2}}\left[\mathbb{S}_{(-1^+0^+)}+\mathbb{S}_{(0^+-1^+)}\right]$
	−1	0	$\frac{1}{2}\left[\mathbb{S}_{(-1^+0^-)}+\mathbb{S}_{(-1^-0^+)}+\mathbb{S}_{(0^+-1^-)}+\mathbb{S}_{(0^--1^+)}\right]$
		−1	$\frac{1}{\sqrt{2}}\left[\mathbb{S}_{(-1^-0^-)}+\mathbb{S}_{(0^--1^-)}\right]$
		1	$\mathbb{S}_{(-1^+-1^+)}$
	−2	0	$\frac{1}{\sqrt{2}}\left[\mathbb{S}_{(-1^+-1^-)}+\mathbb{S}_{(-1^--1^+)}\right]$
		−1	$\mathbb{S}_{(-1^--1^-)}$
		1	$\frac{1}{\sqrt{2}}\left[\mathbb{S}_{(1^+0^+)}-\mathbb{S}_{(0^+1^+)}\right]$
	1	0	$\frac{1}{2}\left[\mathbb{S}_{(1^+0^-)}+\mathbb{S}_{(1^-0^+)}-\mathbb{S}_{(0^+1^-)}-\mathbb{S}_{(0^-1^+)}\right]$
		−1	$\frac{1}{\sqrt{2}}\left[\mathbb{S}_{(1^-0^-)}-\mathbb{S}_{(0^-1^-)}\right]$
		1	$\frac{1}{\sqrt{2}}\left[\mathbb{S}_{(1^+-1^+)}-\mathbb{S}_{(-1^+1^+)}\right]$
^{3}P	0	0	$\frac{1}{2}\left[\mathbb{S}_{(1^+-1^-)}+\mathbb{S}_{(1^--1^+)}-\mathbb{S}_{(-1^+1^-)}-\mathbb{S}_{(-1^-1^+)}\right]$
		−1	$\frac{1}{\sqrt{2}}\left[\mathbb{S}_{(1^--1^-)}-\mathbb{S}_{(-1^-1^-)}\right]$
		1	$\frac{1}{\sqrt{2}}\left[\mathbb{S}_{(0^+-1^+)}-\mathbb{S}_{(-1^+0^+)}\right]$
	−1	0	$\frac{1}{2}\left[\mathbb{S}_{(0^+-1^-)}+\mathbb{S}_{(0^--1^+)}-\mathbb{S}_{(-1^+0^-)}-\mathbb{S}_{(-1^-0^+)}\right]$
		−1	$\frac{1}{\sqrt{2}}\left[\mathbb{S}_{(0^--1^-)}-\mathbb{S}_{(-1^-0^-)}\right]$
		1	$\frac{1}{\sqrt{3}}\left[\mathbb{S}_{(1^+-1^+)}-\mathbb{S}_{(0^+0^+)}+\mathbb{S}_{(-1^+1^+)}\right]$
^{3}S	0	0	$\frac{1}{\sqrt{6}}\left[\mathbb{S}_{(1^+-1^-)}+\mathbb{S}_{(1^--1^+)}-\mathbb{S}_{(0^+0^-)}-\mathbb{S}_{(0^-0^+)}+\mathbb{S}_{(-1^+1^-)}+\mathbb{S}_{(-1^-1^+)}\right]$
		−1	$\frac{1}{\sqrt{3}}\left[\mathbb{S}_{(1^--1^-)}-\mathbb{S}_{(0^-0^-)}+\mathbb{S}_{(-1^-1^-)}\right]$

Table 7.9 Decomposition of the *LS*-coupling wave functions for the electronic configuration 2p3p, in Slater determinants of spin-orbitals. This table shows the terms ^{1}D, ^{1}P and ^{1}S. The triplet terms are given in table 7.8.

Term	M_L	M_S	eigenstate
	2	0	$\frac{1}{\sqrt{2}}\left[\mathbb{S}_{(1^+1^-)} - \mathbb{S}_{(1^-1^+)}\right]$
	1	0	$\frac{1}{2}\left[\mathbb{S}_{(1^+0^-)} - \mathbb{S}_{(1^-0^+)} + \mathbb{S}_{(0^+1^-)} - \mathbb{S}_{(0^-1^+)}\right]$
^{1}D	0	0	$\frac{1}{\sqrt{12}}\left[\mathbb{S}_{(1^+-1^-)} - \mathbb{S}_{(1^--1^+)} + 2\mathbb{S}_{(0^+0^-)} - 2\mathbb{S}_{(0^-0^+)} + \mathbb{S}_{(-1^+1^-)} - \mathbb{S}_{(-1^-1^+)}\right]$
	−1	0	$\frac{1}{2}\left[\mathbb{S}_{(-1^+0^-)} - \mathbb{S}_{(-1^-0^+)} + \mathbb{S}_{(0^+-1^-)} - \mathbb{S}_{(0^--1^+)}\right]$
	−2	0	$\frac{1}{\sqrt{2}}\left[\mathbb{S}_{(-1^+-1^-)} - \mathbb{S}_{(-1^--1^+)}\right]$
	1	0	$\frac{1}{2}\left[\mathbb{S}_{(1^+0^-)} - \mathbb{S}_{(1^-0^+)} - \mathbb{S}_{(0^+1^-)} + \mathbb{S}_{(0^-1^+)}\right]$
^{1}P	0	0	$\frac{1}{2}\left[\mathbb{S}_{(1^+-1^-)} - \mathbb{S}_{(1^--1^+)} - \mathbb{S}_{(-1^+1^-)} + \mathbb{S}_{(-1^-1^+)}\right]$
	−1	0	$\frac{1}{2}\left[\mathbb{S}_{(0^+-1^-)} - \mathbb{S}_{(0^--1^+)} - \mathbb{S}_{(-1^+0^-)} + \mathbb{S}_{(-1^-0^+)}\right]$
^{1}S	0	0	$\frac{1}{\sqrt{6}}\left[\mathbb{S}_{(1^+-1^-)} - \mathbb{S}_{(1^--1^+)} - \mathbb{S}_{(0^+0^-)} + \mathbb{S}_{(0^-0^+)} + \mathbb{S}_{(-1^+1^-)} - \mathbb{S}_{(-1^-1^+)}\right]$

The preceding study is for two p-electrons in different orbitals. For atomic ground states, it is more relevant to take an example with equivalent electrons, such as $2p^2$. For this, the spin-orbital diagram was shown in table 6.3.

Finding eigenfunctions with step operators here involves less work than for the example 2p3p because we have 15 states instead of 36. We can see in tables 7.8 and 7.9 which functions that need to be removed. There are six instances where the two electrons are identical and these must be eliminated. Then we can, for example, look at the state $|^3\mathrm{D};(2,0)\rangle$. The two Slater determinants $\mathbb{S}_{(1^+1^-)}$ and $\mathbb{S}_{(1^-1^+)}$ are identical, except for an interchange of two columns. Swapping two columns in a determinant means a change of sign, and thus the wave function for $|^3\mathrm{D};(2,0)\rangle$, shown in table 7.8, must cancel. With the same argument, we get rid of another 15 states. Note that these arguments provide further justification to the one about indistinguishable electrons, made in section 6.3.3 when identifying terms allowed by the Pauli principle.

We are left with only the states belonging to ^{1}D, ^{3}P and ^{1}S. We can pick these out from tables 7.8 and 7.9, but we have to be careful with the normalisations and with the signs. For an illustration of the importance of the choice of sign, take the example with $|^3\mathrm{P};(1,1)\rangle$ from table 7.8. As argued above, the two Slater determinants in the superposition are the same except for the sign, and since they are here separated by a minus sign, the correct answer will be one of them multiplied by two. For a single function, we can choose the sign as we wish, but as we go through the whole table, we have to be consistent in how we decide which orbital to keep and which to throw away.

What this amounts to is to decide upon an order in which to write electrons, based on their individual quantum numbers, and to then follow that order throughout. There is a standard for this, introduced by Condon & Shortley [6], and we shall adhere to this. It consists in ordering first the orbitals after *increasing* values of n, and thereafter l (1s, 2s, 2p, 3s, 3p, 3d, 4s and so on). Next, we order after *decreasing* values of m_l, and finally we take spin-up before spin-down. Albeit not mentioned earlier, this standard has been used when setting up the spin-orbital diagrams for equivalent electrons in tables 6.3, 6.5 and 7.3.

From the example above, with $|{}^3\text{P};(1,1)\rangle$, it is also evident that we have to renormalise the wave functions when we take them from tables 7.8 and 7.9. The end result for the 2p^2 configuration is presented in table 7.10.

Table 7.10 Decomposition of the *LS*-coupling wave functions for the electronic configuration 2p^2, in Slater determinants of spin-orbitals

Term	M_L	M_S	eigenstate
	2	0	$\mathbb{S}_{(1^+,1^-)}$
	1	0	$\frac{1}{\sqrt{2}}\left[\mathbb{S}_{(1^+0^-)} - \mathbb{S}_{(1^-0^+)}\right]$
^{1}D	0	0	$\frac{1}{\sqrt{6}}\left[\mathbb{S}_{(1^+-1^-)} - \mathbb{S}_{(1^--1^+)} + 2\mathbb{S}_{(0^+0^-)}\right]$
	−1	0	$\frac{1}{\sqrt{2}}\left[\mathbb{S}_{(0^+-1^-)} - \mathbb{S}_{(0^--1^+)}\right]$
	−2	0	$\mathbb{S}_{(-1^+,-1^-)}$
		1	$\mathbb{S}_{(1^+0^+)}$
	1	0	$\frac{1}{\sqrt{2}}\left[\mathbb{S}_{(1^+0^-)} + \mathbb{S}_{(1^-0^+)}\right]$
		−1	$\mathbb{S}_{(1^-,0^-)}$
		1	$\mathbb{S}_{(1^+-1^+)}$
^{3}P	0	0	$\frac{1}{\sqrt{2}}\left[\mathbb{S}_{(1^+-1^-)} + \mathbb{S}_{(1^--1^+)}\right]$
		−1	$\mathbb{S}_{(1^--1^-)}$
		1	$\mathbb{S}_{(0^+-1^+)}$
	−1	0	$\frac{1}{\sqrt{2}}\left[\mathbb{S}_{(0^+-1^-)} + \mathbb{S}_{(0^--1^+)}\right]$
		−1	$\mathbb{S}_{(0^--1^-)}$
^{1}S	0	0	$\frac{1}{\sqrt{3}}\left[\mathbb{S}_{(1^+-1^-)} - \mathbb{S}_{(1^--1^+)} - \mathbb{S}_{(0^+0^-)}\right]$

It should be noted that the tables 7.6, 7.7, 7.8, 7.9 and 7.10 show wave functions in $|\gamma LSM_LM_S\rangle$ representations. If we want to go one step further and calculate the wave functions with fine-structure included, in the representation $|\gamma LSJM_J\rangle$, we must apply procedures for a change of the basis, using Clebsch–Gordan coefficients (see appendix C.5).

The above described procedure for determining superposition states, using step-operators and the orthogonality criterion, is not the only method. Another route is to use projection operators (see [2]). This is conceptually more complex, but may still be preferable for configurations with a large number of states (for instance, the example above with 2p3p). However, we refer to the suggested reading for a description of how that method can be applied.

7.4.2 Radial Functions for Light Atoms

The wave functions found in section 7.4.1 still suffer from the lack of having explicit mathematical forms. They merely express eigenfunctions of $\mathbf{L}^2$ and $\mathbf{S}^2$ (that is, *LS*-coupling terms) as superpositions of spin-orbitals. The latter are still left undetermined in section 7.4.1. However, for small atoms, functional forms may be estimated with the variational principle.

The variational principle method was explained in section 3.2, where it was applied to He and Li. For some of the slightly larger atoms — notably the 2p elements — the variational principle can be used in order to find approximate expressions for the radial wave functions. Here, we give but a brief example, and we refer to [2, 7] and [8] for more details.

To express eigenstates for the ground states of the 2p atoms, we need CFA radial wave functions for the orbitals 1s, 2s and 2p. An early example of variational functions used in such an analysis [8] is:

$$\begin{aligned} R_{1s}^{(a,\mu)}(r_i) &= \sqrt{4\mu^3 a^3}\ \mathrm{e}^{-\mu a r_i} \\ R_{2s}^{(b,\mu)}(r_i) &= \sqrt{\frac{4\mu^5}{3B}} \left(r_i \mathrm{e}^{-\mu r_i} - \frac{3A}{\mu} \mathrm{e}^{-\mu b r_i} \right) \\ R_{2p}^{(c,\mu)}(r_i) &= \sqrt{\frac{4\mu^5 c^5}{3}}\ r_i \mathrm{e}^{-\mu c r_i} \quad , \end{aligned} \tag{7.69}$$

where A and B are defined as:

$$\begin{aligned} A &\equiv \frac{(a+b)^3}{(a+1)^4} \\ B &\equiv 1 - \frac{48A}{(b+1)^4} + \frac{3A^2}{b^3} \quad . \end{aligned} \tag{7.70}$$

The variational parameters are μ, a, b and c, and the definitions of A and B assure that the functions are orthonormal.

With the functions in (7.69), one- and two-electron integrals can be computed using (7.27), with the Hamiltonian (7.5) and the methods described in section 7.2.2. For every atom and every configuration, the parameters μ, a, b and c must be varied for a minimisation of the energy expectation values.

In order to demonstrate the strength of the method, we show in table 7.11 some computed energies for a few neutral atoms, found from a variational calculation of wave functions. The theoretical energies are compared to experimental values.

Table 7.11 Energies of some low lying levels, in eV, for the atoms B, C, N and O, calculated with the variational principle (from [9]). These are compared to experimental values from [1].

Atom		term	$E_{\mathrm{var/eV}}$	$E_{\mathrm{exp/eV}}$
B	$1s^2\,2s^2\,2p$	^{2}P	−666.7	−670.98
		^{3}P	−1023.9	−1030.10
C	$1s^2 2s^2\,2p^2$	^{1}D	−1022.1	−1028.84
		^{1}S	−1019.5	−1027.42
		^{4}S	1476.9	−1486.07
N	$1s^2\,2s^2\,2p^3$	^{2}D	−1471.5	−1483.69
		^{2}P	−1473.7	−1482.50
		^{3}P	−2028.6	−2043.82
O	$1s^2\,2s^2\,2p^4$	^{1}D	−2026.1	−2041.88
		^{1}S	−2022.7	−2039.65

As can be seen, all the calculated values are higher than the respective measured ones, which should always be the case with the variational method. Nevertheless, the agreement is rather good even for this early computation, taken from [9]. With contemporary variational calculations, this can be substantially improved [7].

7.5 Energy Levels in the *LS*-coupling Scheme

In this chapter, we have taken great advantage of the spherically symmetric contribution of closed orbital electrons, explained in section 6.2. The Hamiltonians we have used, and the corresponding energy perturbations in the chapter, refer solely to the contribution from the individual valence electrons and the interactions between these. Figure 7.2 schematises the valence electron Hamiltonians that we have introduced, and how we couple them. This is further outlined in table 7.12, where we include the notation used for the energy contributions, both for *LS*-coupling (left column) and for *jj*-coupling (right column).

The angular momentum coupling scheme covered in this chapter is based on the assumption that the torque introduced by the electrostatic interaction between valence electrons has an energy contribution much greater than that of the spin–orbit interaction. This gives us the possibility to split up the overabundance of angular momentum interactions into distinct parts, which we can then apply as successive perturbations to the Hamiltonian in a mathematically convenient order. The emphasis on angular effects at this stage of the analysis is made possible by the central-field approximation, which allows the separation of radial and angular dependencies. The CFA presents us with the possibility to assume a zero-order Hamiltonian, which is represented by the electronic configuration. It enables us to limit the analyses of

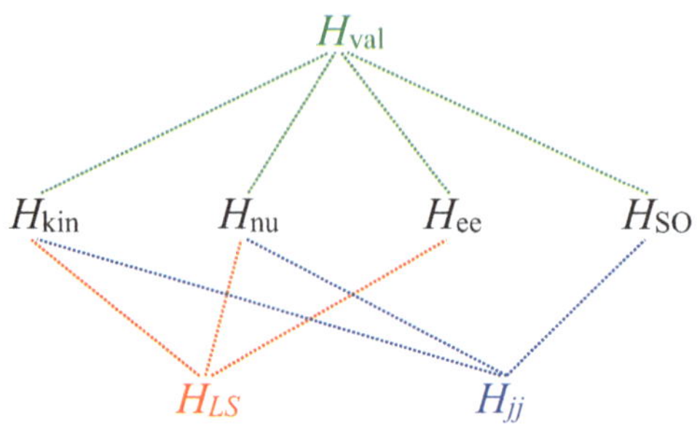

Fig. 7.2 The total Hamiltonian for all valence electrons is H_{val}. This is the sum of H_{kin}, H_{nu}, H_{ee} and H_{SO}. In the *LS*-coupling approximation, we first apply three of them (as indicated in red) as a perturbation, and then the final remaining Hamiltonian (H_{SO}) is a small perturbation to that. In *jj*-coupling, as we shall see in chapter 8, the roles of H_{ee} and H_{SO} have to be reversed (indicated in blue).

Table 7.12 Table describing the notation of energies corresponding to different perturbation Hamiltonians used in chapters 7 and 8. The left column is for *LS*-coupling and the right for *jj*-coupling. The part $H_{kin}+H_{nu}$ is common for the schemes.

$H_{kin}+H_{nu} \Rightarrow \sum_{i=1}^{N} I(nl)$	
$H_{ee} \Rightarrow E^{(ee)}(^{2S+1}L)$	$H_{SO} \Rightarrow E^{(SO)}([\{nl_j\}])$
$H_{LS} \Rightarrow E^{(LS)}(\gamma\, ^{2S+1}L)$	$H_{jj} \Rightarrow E^{(jj)}([\{nl_j\}])$
$H_{SO} \Rightarrow E^{(SO)}(\gamma\, ^{2S+1}L_J)$	$H_{ee} \Rightarrow E^{(ee)}([\{nl_j\}]_J)$
$H_{val} \Rightarrow E^{(val)}(\gamma\, ^{2S+1}L_J)$	$H_{val} \Rightarrow E^{(val)}([\{nl_j\}]_J)$

interactions between angular momenta to the valence electrons, and taking the latter as perturbations works because we assume that the radial part of the Coulomb interactions gives the dominating energy scale in the atom.

The terminology *LS*-coupling can be deceptive, since it does not signify a coupling between the vectors **L** and **S**, in the way that spin–orbit coupling really is a coupling between two angular momenta. The name *LS*-coupling rather signifies that it is a scheme where the first, and most important, angular coupling actions are to *form* the vectors **L** and **S**. The vector operators $\mathbf{L}^2$, L_z, $\mathbf{S}^2$ and S_z are then assumed to commute with the Hamiltonian. Another issue with the nomenclature is the order of L and S in the naming of the coupling scheme. It has been argued that a more proper name should be *SL*-coupling. Indeed, the order in which **L** and **S** are summed does introduce a difference in phase. This is briefly mentioned and demonstrated in appendix C.5. In this book, we have simply taken the name as a semantic question, and we have followed the overwhelmingly dominating standard. For the examples studied in this volume, the phase ambiguity is of no consequence. On the subject of semantics, one should note that some literature, in particular older works, refer to *LS*-coupling as Russell–Sunders coupling.

We have shown that the spin–orbit interaction scales as the fourth power of the nuclear charge. This means that *LS*-coupling is generally an excellent approximation for light atoms. Also for atoms in the middle of the periodic chart, the approximation remains pretty good. Only for the heaviest atoms in the stable part of the periodic table can one see very significant departures from *LS*-coupling behaviour. However, also for these atoms, *LS*-coupling notation is often used to label energy levels (see, for example, [1]). In such cases, the true energy levels are actually superpositions of *LS*-coupling coupling terms, since the representation $|LSJM_J\rangle$ leaves us with non-diagonal terms that cannot be neglected. In energy level tables, a common standard is therefore to give a *leading term*, described in *LS*-coupling notation, while assigning also other involved terms, when the contributions of these are known. This will be further discussed in section 9.6.

The application of *LS*-coupling notation also for atomic states where the approximation is invalid may seem surprising, but it is justified by the clear and standardised form of the notation. The *atomic term*,

$$^{2S+1}L_J \ , \tag{7.71}$$

is such a widely accepted nomenclature for atomic energy levels that its usage transcends the body of energy eigenstates for which the underlying approximation is actually valid. It also helps that selection rules for radiative transitions can be clearly formulated in the *LS*-coupling scheme. Thereby, the notation facilitates the analysis of atomic spectra and the identification of levels involved in spectral lines.

In the earlier sections of this chapter, we have provided means to calculate approximate energies for states described by *LS*-coupling terms. Even if these are somewhat approximate, they provide a good qualitative picture, and they can help with the identification of the energy order of terms within a configuration. Empirical rules for the energy order were established early on [10], by a thorough analysis of experimental spectra. These are known as *Hund's rules* and they are still useful, in particular for identifying ground states. However, they must be applied with some caution.

Hund's rules are stated slightly differently in different books. However, a relatively standard formulation is the following:

1. Within a given electron configuration, the lowest energy state is the one with the highest value of S (the greatest multiplicity).
2. In the case of several states having the maximum S, the lowest one is that among them with the highest L (largest orbital angular momentum).
3. The lowest fine-structure state is the one with minimum J, if the valence orbital is less than half filled (normal order). If the valence orbital is more than half full, the highest value of J gives the lowest energy (inverted order).

There are many examples for which Hund's rules give the wrong answer, in particular if there are electrons in more than one open orbital. The states for which the rules can be safely applied are those with just one non-filled orbital and those with one open orbital plus *one* single s-electron, and when they are applied solely for identifying the ground state.

If we consider the results from the analyses presented in sections 7.2 and 7.3, we find that in general the energy order of *LS*-coupling terms and of fine-structure levels agree quite well with Hund's empirical rules. This may also provide a qualitative understanding of some of the rules. For example, the reason why a higher multiplicity tends to give lower energy is that electrons with parallel spins (ones that add up to a greater $|\mathbf{S}|$) are forbidden to occupy the same volume in space, due to the Pauli principle. This means that they are more likely to be far apart, which leads to a smaller Coulomb repulsion, and a lower total energy. For the second of Hund's rules, a hand-waving argument is that electrons with high orbital angular momenta tend to interact less with the other electrons, because of a smaller spatial overlap.

In the next chapter, we shall study the opposite limiting case of angular momentum coupling: *jj*-coupling. In chapter 9, we will look at intermediate situations, and examples of atomic structures where different configurations interact, the latter being one major reason for breakdowns of Hund's rules.

Further Reading

The theory of atomic spectra, by Condon & Shortley [6]
Quantum theory of atomic structure, by Slater [2]
Atomic spectra, by Kuhn [11]
Atomfysik, by Lindgren & Svanberg [12]
The theory of atomic structure and spectra, by Cowan [5]
Atomic Many-Body Theory, by Lindgren & Morrison [13]
Springer Handbook of Atomic, Molecular, and Optical Physics, by Drake [14]

References

1. A. Kramida, Y. Ralchenko, J. Reader, and NIST ASD Team. NIST Atomic Spectra Database (ver. 5.3). [Online]. Available: http://physics.nist.gov/asd (2018). Accessed: 2019-07-14
2. J.C. Slater, *Quantum theory of atomic structure* (McGraw-Hill, New York, 1960)
3. J.A. Gaunt, Phil. Trans. Roy. Soc. (London) **A228**, 151 (1929)
4. G. Racah, Phys. Rev. **61**, 186 (1942)
5. R.D. Cowan, *The theory of atomic structure and spectra* (University of California press, Berkeley, 1981)
6. E.U. Condon, G.H. Shortley, *The theory of atomic spectra* (Cambridge University Press, Cambridge, 1935)
7. G.W.F. Drake, in *Springer Handbook of Atomic, Molecular, and Optical Physics*, ed. by G.W.F. Drake (Springer-Verlag, New York, 2006), p. 199
8. P.M. Morse, L.A. Young, E.S. Haurwitz, Phys. Rev. **48**, 948 (1935)
9. A. Tubis, Phys. Rev. **102**, 1049 (1956)
10. F. Hund, *Linienspektren und periodisches System der Elemente* (Springer, Berlin, 1927)
11. H.G. Kuhn, *Atomic spectra* (Longmans, London, 1969)
12. I. Lindgren, S. Svanberg, *Atomfysik* (Universitetsförlaget, Uppsala, 1974)
13. I. Lindgren, J. Morrison, *Atomic Many-Body Theory*, 2nd edn. (Springer Verlag, Berlin, 1986)
14. G.W.F. Drake (ed.), *Springer Handbook of Atomic, Molecular, and Optical Physics* (Springer-Verlag, New York, 2006)

Chapter 8
jj-Coupling

In chapter 6, we introduced the other extreme of the CFA, besides *LS*-coupling — *jj*-coupling. In this limiting approximation, we again begin by determining the electronic configuration in the CFA, and the following step is this time to consider the Hamiltonian of (6.39), H_{SO}. This is pertinent for atomic states with a spin–orbit interaction so strong that it swamps the angular part of the electron–electron interactions for the valence electrons. It is rather rare in its pure (or even almost pure) form. However, detailed knowledge of both *jj*-coupling and *LS*-coupling will enable us to better analyse any general situation, where the two competing perturbation Hamiltonians are of comparable importance. Understanding of both *LS*-coupling and *jj*-coupling will also be useful when more subtle interactions — such as spin–spin interaction and spin-other-orbit interaction — have to be taken into account.

The present chapter begins by establishing which *jj*-coupling atomic terms that are possible for a certain configuration, including fine-structure — which here appears after adding the electrostatic repulsion Hamiltonian for the valence electrons. Following that, we look at theoretical estimates of level energies. The task of constructing wave functions for *jj*-coupling does not differ significantly from what was explained in chapter 7.4 for *LS*-coupling, except that the limitations from the Pauli principle are often less important. For the notations employed for Hamiltonians and for perturbative energies, we refer back to table 7.12. A schematic illustration of coupling schemes, including *jj*-coupling, is shown in figure 7.2. For a more detailed survey of *jj*-coupling energies, we refer to, for example, [1] or [2].

8.1 Allowed *jj*-coupling Terms

The number of allowed atomic terms does not depend on the choice of *LS*-coupling or *jj*-coupling. The total number of states is set by the electronic configuration, and the occurrences of the quantum number J are the same, regardless of what

A. Kastberg, *Structure of Multielectron Atoms*, Springer Series on Atomic, Optical, and Plasma Physics 112, https://doi.org/10.1007/978-3-030-36420-5_8

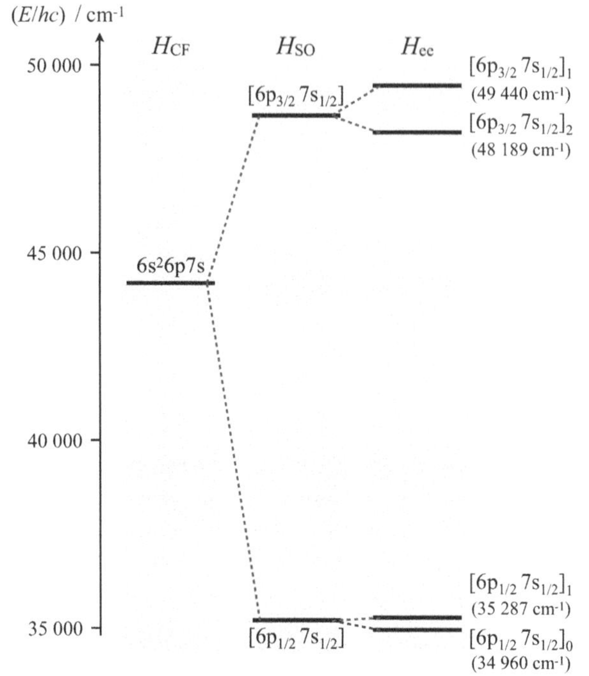

Fig. 8.1 Partial energy level diagram for Pb, showing the 6p7s-configuration. There are many other energy levels — for different configurations and with intermediate energies — which have been omitted in the figure. The four fine-structure levels in the figure group into two pairs, which is a signature of *jj*-coupling. In this diagram, zero energy has been set to the energy of the ground state $[6p^2_{1/2}]_0$. The empiric values are from [3].

path we take when we form the vector operator **J**. The difference lies in how the fine-structure levels group up, and in which quantum numbers we use for representing the Hilbert space.

We illustrate this with a diagram of a *jj*-coupled atom, showing how states group differently than they do in *LS*-coupling. Figure 8.1 depicts a partial Grotrian diagram for Pb, with the energy levels for the excited configuration 6p7s. The energies in the figure are ones experimentally obtained [3]. We see how a *jj*-coupling analysis will lead to two distinct pairs of energies. In a corresponding *LS*-coupling scheme with a nsn'p configuration, we would instead have one triplet with $J = 0, 1, 2$ and one singlet with $J = 1$.

When unravelling the allowed terms in *jj*-coupling, there is no need to delay the assignment of the fine-structure. In section 6.4, the *jj*-coupling terms of the equivalent electron configuration $6p^2$ were determined. There we saw that the values of J could be extracted directly from the table. Since the quantum numbers j_i and m_{ji} refer to the individual electrons, an antisymmetric wave function requires unique combinations $\{n_i l_i j_i m_{ji}\}$ for every electron, and these must be possible to decompose in unique sets of $\{n_i l_i m_{li} m_{si}\}$. One consequence of this, demonstrated in

Table 8.1 Diagram for identification of possible *jj*-coupling terms, not in conflict with the Pauli principle, for a np^3-configuration. The numbers within brackets are the projection quantum numbers m_{ji}. The bottom half of the table, with negative values of M_J is a mirror image and has been omitted. There are no allowed states for the j_i combination (1/2 1/2 1/2).

np^3		j_1 j_2 j_3		
		3/2 3/2 3/2	3/2 3/2 1/2	3/2 1/2 1/2
	9/2			
	7/2			
	5/2		(3/2 1/2 1/2)	
M_J	3/2	(3/2 1/2 −1/2)	(3/2 1/2 −1/2) (3/2 −1/2 1/2)	(3/2 1/2 −1/2)
	1/2	(3/2 1/2 −3/2)	(3/2 −1/2 −1/2) (3/2 −3/2 1/2) (1/2 1/2 1/2)	(1/2 1/2 −1/2)
possible J:		3/2	5/2 or 3/2 or 1/2	3/2

table 6.7 for the terms $[6p_{3/2}^2]$ and $[6p_{1/2}^2]$, is the fact that for two identical electrons with the same value of j_i, only even values of J are allowed.

In table 8.1, we give an example of a diagram that could be used for finding out the *jj*-coupling terms allowed for a state with three equivalent electrons, in this particular example a p^3-configuration. From this, we can pick out the possible terms, following the same methodology as that explained in connection to table 6.7. The conclusion from the table is that the allowed *jj*-coupling terms are:

$$\begin{array}{lll} [6p_{3/2}^3]_{3/2} & , & [6p_{3/2}^2\,6p_{1/2}]_{5/2} \\ [6p_{3/2}^2\,6p_{1/2}]_{3/2} & , & [6p_{3/2}^2\,6p_{1/2}]_{1/2} \\ [6p_{3/2}\,6p_{1/2}^2]_{3/2} & . & \end{array} \quad (8.1)$$

With the same method, allowed terms and J values can be determined for any configuration. In table 8.2, we present a table for possible *jj*-coupling states for configurations with equivalent electrons $[nl_j^w]$, with values of j up to 9/2 (w is here the number of equivalent electrons). This table is limited to showing the possible resulting J for configurations with identical j. For states such as the ones in (8.1), where there are more than one value of j occurring in the configuration, the total value of J is easily found by vector addition, then unrestricted by the Pauli principle.

A feature that we can recognise from *LS*-coupling (see tables 7.1 and 7.2) is that for an orbital that is more than half full, the possible states will be the same as for the case where we count the number of holes instead of the number of electrons.

Table 8.2 Possible values of the total angular momentum quantum number J for different $[nl_j^w]$ configurations of identical electrons. w is the number of equivalent electrons, and l is represented by its spectroscopic symbol. The subscripts appearing in brackets for some J symbolise that these states occur more than once for the configuration in question.

j	l	w	possible J
1/2	s , p	0 , 2	0
		1	1/2
3/2	p , d	0 , 4	0
		1 , 3	3/2
		2	0 , 2
5/2	d , f	0 , 6	0
		1 , 5	5/2
		2 , 4	0 , 2 , 4
		3	3/2 , 5/2 , 9/2
7/2	f , g	0 , 8	0
		1 , 7	7/2
		2 , 6	0 , 2 , 4 , 6
		3 , 5	3/2 , 5/2 , 7/2 , 9/2 , 11/2 , 15/2
		4	0 , $2_{(2)}$, $4_{(2)}$, 5 , 6 , 8
9/2	g , h	0 , 10	0
		1 , 9	9/2
		2 , 8	0 , 2 , 4 , 6 , 8
		3 , 7	3/2 , 5/2 , 7/2 , $9/2_{(2)}$, 11/2 , 13/2 , 15/2 , 17/2 , 21/2
		4 , 6	$0_{(2)}$, $2_{(2)}$, 3 , $4_{(3)}$, 5 , $6_{(3)}$, 7 , $8_{(2)}$, 9 , 10 , 12
		5	1/2 , 3/2 , $5/2_{(2)}$, $7/2_{(2)}$, $9/2_{(3)}$, $11/2_{(2)}$, $13/2_{(2)}$ $15/2_{(2)}$, $17/2_{(2)}$, 19/2 , 21/2 , 25/2

8.2 Theoretical Energies of *jj*-Coupled States

Calculating energies in terms of one- and two-electron integrals for *jj*-coupling is done in very much the same way as for *LS*-coupling (see section 7.2). The difference is that the two computations are made in the opposite order. This simplifies the calculation because it eliminates the need for the diagonal-sum rule for the first of the two perturbations — which is now the spin–orbit interaction energy (non-diagonal matrices will, however, return when we compute the *jj*-coupling fine-structure). Since the *jj*-coupling terms are specified with individual electron quantum numbers, and the spin–orbit Hamiltonian only involves one electron at the time in this representation, the first energy splitting in *jj*-coupling can be found simply by summing over the valence electrons.

8.2.1 The Spin–Orbit Energies

For an atom with N valence electrons, which is well described by *jj*-coupling, we use the notation $|\gamma, n_1 l_1 j_1 \ldots n_N l_N j_N\rangle \equiv |\gamma, \{nl_j\}\rangle$, with γ representing the electronic configuration of all closed orbitals. The projections of individual electron orbital angular momenta and spins, m_{li} and m_{si}, are not good quantum numbers, since L_z and S_z do not commute with the spin–orbit Hamiltonian, which is:

$$H_{\mathrm{SO}} = \sum_{i=1}^{N} \xi(r_i)\, \mathbf{L}_i \cdot \mathbf{S}_i \quad . \tag{8.2}$$

The radial fine-structure function $\xi(r_i)$ is defined in (4.19). The summation goes over the electrons in non-filled orbitals (that is, N is the number of valence electrons).

From (8.2) and (4.11), we can deduce the matrix elements of the Hamiltonian (8.2), and the associated energy displacements for a *jj*-coupling term as:

$$\begin{aligned} E^{(\mathrm{SO})}([\{nl_j\}]) &= \langle n_1 l_1 j_1 \ldots n_N l_N j_N \,|H_{\mathrm{SO}}|\, n_1 l_1 j_1 \ldots n_N l_N j_N\rangle \\ &= \sum_{i=1}^{N} \frac{\xi_{n_i l_i}}{2} \left[j_i(j_i+1) - l_i(l_i+1) - s_i(s_i+1) \right] . \end{aligned} \tag{8.3}$$

Here we have eliminated all contributions from inner orbitals, with the same rationale as previously (see sections 6.2 and 7.2), and s_i is always one half.

We will demonstrate how to put this into practice with a couple of examples, beginning with a *jj*-coupled $6\mathrm{p}^2$ configuration. From (6.43), we know that the terms we have to compute energies for are:

$$[6\mathrm{p}_{3/2}^2] \ , \ [6\mathrm{p}_{3/2}\, 6\mathrm{p}_{1/2}] \ , \ [6\mathrm{p}_{1/2}^2] \ . \tag{8.4}$$

For the first one, $[6\mathrm{p}_{3/2}^2]$, the two j_i values are the same, so the proper zero-order wave function should be a Slater determinant. However, we know from (2.34) that for a single-electron operator, it suffices to take the sum of the matrix elements of the diagonal product functions. The spin–orbit energy, from (8.3), is thus:

$$E^{(\mathrm{SO})}([6\mathrm{p}_{3/2}^2]) = \frac{\xi_{6\mathrm{p}}}{2} \left[\left(\frac{3}{2}\frac{5}{2} - 2 - \frac{1}{2}\frac{3}{2} \right) + \left(\frac{3}{2}\frac{5}{2} - 2 - \frac{1}{2}\frac{3}{2} \right) \right] = \xi_{6\mathrm{p}} \ , \tag{8.5}$$

and the total perturbative term energy is:

$$E^{(jj)}([6\mathrm{p}_{3/2}^2]) = 2I(6\mathrm{p}) + \xi_{6\mathrm{p}} \ , \tag{8.6}$$

with the single-electron contribution $I(6\mathrm{p})$ being defined as in (7.32). For the other two terms, we find in the same fashion that:

$$\begin{aligned} E^{(jj)}([6\mathrm{p}_{3/2}\,6\mathrm{p}_{1/2}]) &= 2I(6\mathrm{p}) - \frac{1}{2}\,\xi_{6\mathrm{p}} \\ E^{(jj)}([6\mathrm{p}_{1/2}^2]) &= 2I(6\mathrm{p}) - 2\,\xi_{6\mathrm{p}} \;. \end{aligned} \tag{8.7}$$

For two non-equivalent p-electrons, such as, for example, the 6p7p-configuration described in table 6.6, the calculation will be almost identical with the one for the $6\mathrm{p}^2$-case. The spin–orbit energies are:

$$\begin{aligned} E^{(jj)}([6\mathrm{p}_{3/2}\,7\mathrm{p}_{3/2}]) &= I(6\mathrm{p}) + I(7\mathrm{p}) + \frac{1}{2}\,\xi_{6\mathrm{p}} + \frac{1}{2}\,\xi_{7\mathrm{p}} \\ E^{(jj)}([6\mathrm{p}_{3/2}\,7\mathrm{p}_{1/2}]) &= I(6\mathrm{p}) + I(7\mathrm{p}) + \frac{1}{2}\,\xi_{6\mathrm{p}} - \xi_{7\mathrm{p}} \\ E^{(jj)}([6\mathrm{p}_{1/2}\,7\mathrm{p}_{3/2}]) &= I(6\mathrm{p}) + I(7\mathrm{p}) - \xi_{6\mathrm{p}} + \frac{1}{2}\,\xi_{7\mathrm{p}} \\ E^{(jj)}([6\mathrm{p}_{1/2}\,7\mathrm{p}_{1/2}]) &= I(6\mathrm{p}) + I(7\mathrm{p}) - \xi_{6\mathrm{p}} - \xi_{7\mathrm{p}} \;. \end{aligned} \tag{8.8}$$

After having calculated all this, the corresponding results for the allowed terms in a *jj*-coupled p^3-configuration, such as $6\mathrm{p}^3$, can be directly inferred, since we already know the relevant single-electron contributions from (8.6) and (8.7). The result is:

$$\begin{aligned} E^{(jj)}([6\mathrm{p}_{3/2}^3]) &= 3I(6\mathrm{p}) + \frac{3}{2}\,\xi_{6\mathrm{p}} \\ E^{(jj)}([6\mathrm{p}_{3/2}^2\,6\mathrm{p}_{1/2}]) &= 3I(6\mathrm{p}) \\ E^{(jj)}([6\mathrm{p}_{3/2}\,6\mathrm{p}_{1/2}^2]) &= 3I(6\mathrm{p}) - \frac{3}{2}\,\xi_{6\mathrm{p}} \;. \end{aligned} \tag{8.9}$$

The single-electron, radial spin–orbit coupling function ξ_{nl} is always positive. These calculations will thus give us the energy order and also a good estimate of the energy difference between *jj*-coupling terms.

8.2.2 Energy Contribution from the Electron–Electron Interaction

The *jj*-coupling fine-structure is the splitting of the *jj*-coupling term energies into states with different J. This we calculate by using the electrostatic repulsion Hamiltonian, H_{ee} — see (7.5) — as perturbation to the *jj*-coupling terms. The matrix elements to be calculated are:

$$\langle n_1 l_1 j_1 \ldots n_N l_n j_N\, J M_J \,|\, H_{\mathrm{ee}} \,|\, n_1 l_1 j_1 \ldots n_N l_n j_N\, J M_J \rangle \;. \tag{8.10}$$

The terms will spilt up in J, but we will retain a degeneracy in M_J.

A problem here is that the entries in *jj*-coupling term diagrams, such as tables 6.7 and 8.1, are not in the same representation as (8.10). They are instead in a basis of individual values of m_{ji}. To accommodate for the mismatch in representation we could diagonalise matrices for every entry in the term tables. However, as for the first perturbative contribution for *LS*-coupling, we can simplify the problem by using the diagonal-sum rule.

The wave functions going into each individual matrix element are product functions, which means that we must use Slater determinants. Thus, the rules developed in section 2.3.4 must be adhered to. The result will be energies that can be expressed in the same two-electron Slater *F*- and *G*-integrals as those used in section 7.2.2 (see also appendix D). However, electron wave functions that go into Slater integrals are typically formulated in the $|\, lsm_l m_s \rangle$-configuration, and in the *jj*-coupling scheme we are working with the representation $|\, lsjm_j \rangle$. As a consequence, we have to express each single-electron state found in diagrams such as table 6.7 as a superposition of $|\, lsm_l m_s \rangle$-states, using vector coupling coefficients (see appendix C.5).

To start with, for every product-function state in a configuration with N valence electrons, $\sigma = \{l_1 j_1 m_{j1} \ldots l_N j_N m_{jN}\}$, we need to calculate the matrix element:

$$\langle \mathbb{S}_\sigma \,|\, H_{\mathrm{ee}} \,|\, \mathbb{S}_\sigma \rangle \;, \tag{8.11}$$

where the Slater determinant is:

$$\mathbb{S}_\sigma = \frac{1}{\sqrt{N!}} \begin{vmatrix} \Psi_1(\mathbf{q}_1) & \ldots & \Psi_1(\mathbf{q}_N) \\ \vdots & \ddots & \vdots \\ \Psi_N(\mathbf{q}_1) & \ldots & \Psi_N(\mathbf{q}_N) \end{vmatrix} \;. \tag{8.12}$$

The indices i on the functions Ψ_i in (8.12) are shorthand for individual electron quantum numbers $(l_i j_i m_{ji})$, and the states must be expressed as superpositions in the $\{m_{li} m_{si}\}$-basis. However, since we always have $s_i = 1/2$, and the only two options for m_{si} are spin-up and spin-down, no superposition will have more than two terms (one with $m_{li} = m_{ji} + 1/2$ and $m_{si} = -1/2$, and the other with $m_{li} = m_{ji} - 1/2$ and $m_{si} = 1/2$). If we express the conversion between the representations as state vectors in Dirac notation, and use the formulae (C.85) and (C.95) in appendix C.5, it is:

$$\begin{aligned} |\, l, 1/2, j, m_j \rangle_{jm_j} &= (-1)^{l-1/2+m_j} \sqrt{2j+1} \\ &\times \left[\begin{pmatrix} l & 1/2 & j \\ m_j + 1/2 & -1/2 & -m_j \end{pmatrix} |\, l, 1/2, m_j + 1/2, -1/2 \rangle_{m_l m_s} \right. \\ &+ \left. \begin{pmatrix} l & 1/2 & j \\ m_j - 1/2 & 1/2 & -m_j \end{pmatrix} |\, l, 1/2, m_j - 1/2, 1/2 \rangle_{m_l m_s} \right] \;, \end{aligned} \tag{8.13}$$

where the subscripts to the kets are used to identify the respective bases.

Instead of doing the conversion in (8.13) for every single state, a generic table can be produced, given the symmetry of (8.13) and the relatively limited variation in possible values for the involved quantum numbers (excluding states with very high orbital angular momenta). Following a convention introduced by Condon and Shortley [1], we apply the following notation:

$$\begin{aligned}
E^{(\mathrm{ee})}([\{nl_j\}]_J) &= \langle \mathbb{S}_\sigma \,|\, H_{\mathrm{ee}} \,|\, \mathbb{S}_\sigma \rangle \\
&= \sum_{\alpha=1}^{N} \sum_{\beta>\alpha}^{N} \Bigg[\left\langle \Psi_{l_\alpha j_\alpha m_{j\alpha}}(\mathbf{q}_1)\, \Psi_{l_\beta j_\beta m_{j\beta}}(\mathbf{q}_2) \left| \frac{1}{r_{12}} \right| \Psi_{l_\alpha j_\alpha m_{j\alpha}}(\mathbf{q}_1)\, \Psi_{l_\beta j_\beta m_{j\beta}}(\mathbf{q}_2) \right\rangle \\
&\quad - \left\langle \Psi_{l_\alpha j_\alpha m_{j\alpha}}(\mathbf{q}_1)\, \Psi_{l_\beta j_\beta m_{j\beta}}(\mathbf{q}_2) \left| \frac{1}{r_{12}} \right| \Psi_{l_\beta j_\beta m_{j\beta}}(\mathbf{q}_1)\, \Psi_{l_\alpha j_\alpha m_{j\alpha}}(\mathbf{q}_2) \right\rangle \Bigg] \\
&\equiv \sum_{\alpha=1}^{N} \sum_{\beta>\alpha}^{N} T(l_\alpha j_\alpha m_{j\alpha} \!:\! l_\beta j_\beta m_{j\beta}) \,,
\end{aligned} \tag{8.14}$$

where we have used (2.36) for developing a matrix element involving Slater determinants and a two-electron operator (see also [4]). Note that the composite variables $\mathbf{q}_1$ and $\mathbf{q}_2$ are integration variables (strictly speaking, the expectation values will contain integrations over $\mathbf{r}_1$ and $\mathbf{r}_2$ and sums over the spins). The sums run over all valence electrons, and will involve all permutations of valence electron pairs.

By using (8.13), (D.1), (D.24), (D.25), (D.22), (D.23), and table D.1, one can compile a table with the *T-polynomials*, $T(l_\alpha j_\alpha m_{j\alpha} : l_\beta j_\beta m_{j\beta})$. Each $T(l_\alpha j_\alpha m_{j\alpha} \!:\! l_\beta j_\beta m_{j\beta})$ is in turn a linear superposition of Slater F- and G-integrals. Examples of such tables of *jj*-coupling fine-structure polynomials are shown in appendix D (tables D.3, D.4, D.5 and D.6).

We will continue to use p-configurations to exemplify calculations. For the heavy atom configuration $6p^2$, we now want to find the fine-structure energy corrections to the term energies that were calculated in (8.6) and (8.7). To do this, we consult table 6.7. Since there are only two valence electrons, the double sum over T-polynomials in (8.14) only contains one term.

If we start with $[6p^2_{3/2}]$, we have the two fine-structure levels $[6p^2_{3/2}]_2$ and $[6p^2_{3/2}]_0$. We can see in the left-most column in table 6.7 that the level with $J=2$ is the only one compatible with the individual quantum number combination $m_{j1}={}^3\!/2$ and $m_{j2}={}^1\!/2$. This means that for the fine-structure energy $E^{(\mathrm{ee})}([6p^2_{3/2}]_2)$, it suffices to take the matrix element of the product function with those two individual m_{ji}. We can use (8.14) and table D.4, and we find that the fine-structure energy is:

$$\begin{aligned}
E^{(\mathrm{ee})}([6p^2_{3/2}]_2) &= \langle \mathbb{S}_{(3/2,\,1/2)} \,|\, H_{\mathrm{ee}} \,|\, \mathbb{S}_{(3/2,\,1/2)} \rangle \\
&= F^{(0)}_{6p:6p} - F^{(2)}_{6p:6p} - 2\,G^{(2)}_{6p:6p} = F^{(0)}_{6p:6p} - 3\,F^{(2)}_{6p:6p} \quad ,
\end{aligned} \tag{8.15}$$

where in the last line we have used the fact that the Slater F- and G-integrals are identical for pairs of electrons with the same l_i. The total energy of the level (including single valence electron energies, term energy and fine-structure) is:

$$E^{(\mathrm{val})}([6\mathrm{p}^2_{3/2}]_2) = E^{(jj)}([6\mathrm{p}^2_{3/2}]) + E^{(\mathrm{ee})}([6\mathrm{p}^2_{3/2}]_2)$$
$$= 2I(6\mathrm{p}) + \xi_{6\mathrm{p}} + F^{(0)}_{6\mathrm{p:6p}} - 3F^{(2)}_{6\mathrm{p:6p}} , \qquad (8.16)$$

where the term energy has been taken from (8.6).

For the other fine-structure level, $[6\mathrm{p}^2_{3/2}]_0$, we can use the diagonal-sum rule, and the cell for $M_J = 0$ in the left column of table 6.7. Applying again (8.14) and table D.4, we find that this sum of fine-structure energies is:

$$E^{(\mathrm{ee})}([6\mathrm{p}^2_{3/2}]_2) + E^{(\mathrm{ee})}([6\mathrm{p}^2_{3/2}]_0)$$
$$= \langle \mathbb{S}_{(3/2,-3/2)} \,|H_{\mathrm{ee}}|\, \mathbb{S}_{(3/2,-3/2)} \rangle + \langle \mathbb{S}_{(1/2,-1/2)} \,|H_{\mathrm{ee}}|\, \mathbb{S}_{(1/2,-1/2)} \rangle . \qquad (8.17)$$

From that, we deduce:

$$E^{(\mathrm{ee})}([6\mathrm{p}^2_{3/2}]_0) = F^{(0)}_{6\mathrm{p:6p}} + F^{(2)}_{6\mathrm{p:6p}} + F^{(0)}_{6\mathrm{p:6p}} + F^{(2)}_{6\mathrm{p:6p}} - F^{(0)}_{6\mathrm{p:6p}} + 3F^{(2)}_{6\mathrm{p:6p}}$$
$$= F^{(0)}_{6\mathrm{p:6p}} + 5F^{(2)}_{6\mathrm{p:6p}} , \qquad (8.18)$$

and for the total energy:

$$E^{(\mathrm{val})}([6\mathrm{p}^2_{3/2}]_0) = 2I(6\mathrm{p}) + \xi_{6\mathrm{p}} + F^{(0)}_{6\mathrm{p:6p}} + 5F^{(2)}_{6\mathrm{p:6p}} . \qquad (8.19)$$

In the same vein, the energies of the remaining terms for the configuration, with fine-structure energy included, can also be calculated. The outcome is:

$$E^{(\mathrm{val})}([6\mathrm{p}_{3/2}6\mathrm{p}_{1/2}]_2) = 2I(6\mathrm{p}) - \frac{1}{2}\xi_{6\mathrm{p}} + F^{(0)}_{6\mathrm{p:6p}} - F^{(2)}_{6\mathrm{p:6p}}$$
$$E^{(\mathrm{val})}([6\mathrm{p}_{3/2}6\mathrm{p}_{1/2}]_1) = 2I(6\mathrm{p}) - \frac{1}{2}\xi_{6\mathrm{p}} + F^{(0)}_{6\mathrm{p:6p}} - 5F^{(2)}_{6\mathrm{p:6p}}$$
$$E^{(\mathrm{val})}([6\mathrm{p}^2_{1/2}]_0) = 2I(6\mathrm{p}) - 2\xi_{6\mathrm{p}} + F^{(0)}_{6\mathrm{p:6p}} , \qquad (8.20)$$

with term energies from (8.7).

8.3 Applicability of *jj*-coupling

For an atomic state to be truly well described by *jj*-coupling, the spin–orbit Hamiltonian has to have an energy contribution so much bigger than the mutual Coulomb repulsion between valence electrons that the latter can be safely treated as a perturbation to the former. From the analyses presented in chapters 6, 7 and the present one, it is evident that this is very far from being the case for the low energy states for any neutral atom in the first half of the periodic table.

The relative importance of *jj*-coupling increases dramatically, however, for heavier atoms. The reason for this is the Z^4-scaling of the spin–orbit interaction — see (4.14). In figure 8.2, we illustrate this with an empiric and qualitative comparison

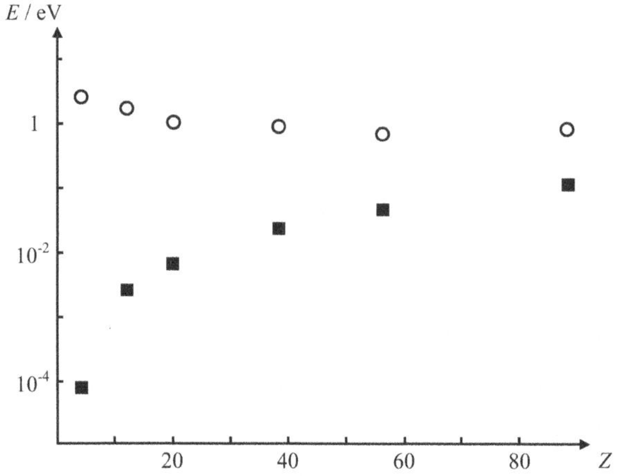

Fig. 8.2 Comparison between energy scales for term energies (shown by open circles — defined as in *LS*-coupling) and spin–orbit interaction fine-structure (filled squares), for the atoms Be, Mg, Ca, Sr, Ba and Ra. The spin–orbit energy is taken as the splitting between the levels $^3P_0$ and $^3P_1$ of the configurations nsnp (with n ranging from 2 to 7). The mutual Coulomb interaction energy has been chosen as the interval between $^3P_1$ and $^1P_1$ for the same configuration. Energies are in electronvolts and are shown as functions of Z (see also [5]). The empirical data are from [3].

of relative energy contributions from H_{ee} and H_{SO}. With a sequence of p^2-elements from Be to Ra, taken as an illustrative example, we plot energy contributions from H_{ee} and H_{SO}, on a logarithmic scale (in eV), as a function of Z. The important points are, first of all, that we see a progressive transformation from a situation where the spin–orbit coupling is indeed much smaller than electron–electron interaction from H_{ee}, to the situation for high Z, where it is essentially of the same order of magnitude. The second point is that pure *jj*-coupling is a poor approximation also for heavier atoms. H_{SO} never becomes quite large enough — in the sequence shown — not even with the Z^4-scaling.

Other systems for which *jj*-coupling must be taken seriously into account are highly charged ions and highly excited states. In the latter case, one can think of an atom with at least two valence electrons, whereof one is excited to a state corresponding to a classical orbital radius much larger than that of the electronic core (see also section 13.1). This will reduce the electrostatic interaction between the excited electron and the core (the expectation value of $1/r_{12}$ becomes small), while the spin–orbit interaction remains of the same order of magnitude. Vector coupling situations similar to this are partially treated in chapter 9. It should also be mentioned that *jj*-coupling formalism is of relevance for studies of nuclear structure [6].

Since pronounced *jj*-coupling is so rare, it may be a pertinent question why its study is of any importance. The answer lies rather in the non-universality of the *LS*-coupling scheme. As good as *LS*-coupling is for light atoms, one does not have to extend the analyses of atomic spectra very far in order to see that it is inadequate as a *general* model of atomic structure.

The preferred solution to this dilemma would be to apply H_{ee} and H_{SO} together, as a unified perturbation, to the electron configuration. However, regardless of which representation that would then be used to define the Hilbert space, one would get secular equations of such high orders that solutions would be untenable (which is, in essence, the raison d'être for the vector model). Therefore, the best practical route is to instead reduce the problem to simplified limiting cases, and from there one might, by interpolation or otherwise, be able to decipher also intermediate and general situations, with acceptable precision.

In that sense, *jj*-coupling represents the extreme of an axis on which *LS*-coupling is at the other end. It should be added that *LS*-coupling notation is widely used for labelling atomic energy levels, even when it is clear that the approximation on which it builds does not hold. This is due to the practicality and clarity of *LS*-coupling notation and of its universal acceptance. This practice is only tenable if we have tools with which we can make transformations between *LS*-coupling and other coupling schemes. To this end, a good understanding of *jj*-coupling is adamant.

Further Reading

The theory of atomic spectra, by Condon & Shortley [1]
Quantum theory of atomic structure, by Slater [4]
Atomic Many-Body Theory, by Lindgren & Morrison [2]
Atomic Physics, by Foot [5]

References

1. E.U. Condon, G.H. Shortley, *The theory of atomic spectra* (Cambridge University Press, Cambridge, 1935)
2. I. Lindgren, J. Morrison, *Atomic Many-Body Theory*, 2nd edn. (Springer Verlag, Berlin, 1986)
3. A. Kramida, Y. Ralchenko, J. Reader, and NIST ASD Team. NIST Atomic Spectra Database (ver. 5.3). [Online]. Available: http://physics.nist.gov/asd (2018). Accessed: 2019-07-14
4. J.C. Slater, *Quantum theory of atomic structure* (McGraw-Hill, New York, 1960)
5. C. Foot, *Atomic Physics* (Oxford University Press, Oxford, 2005)
6. L.D. Landau, E.M. Lifshitz, *Quantum Mechanics — Course of Theoretical Physics, volume 3*, 3rd edn. (Butterworth-Heinemann, Amsterdam, 1981)

Chapter 9
Other Coupling Schemes — Intermediate Cases

The angular momentum coupling schemes that have been covered in the two preceding chapters — LS and jj — are limiting cases. No real atom has pure LS-coupling or jj-coupling, albeit for many light atoms LS-coupling is a very good approximation. In many instances, the spin–orbit interaction and the angular part of the electron–electron repulsion are of similar order of importance, and adding one as a perturbative term before the other leads to erroneous results. There are also cases where, for example, one electron may have a spin–orbit interaction that is weaker than its electrostatic coupling to one of its neighbours, but stronger than that to another.

For an atom with many valence electrons, distributed over more than one orbital, the couplings within distinct groups of electrons may be strong, whereas the connections between them are weaker. For such an atom, a proper description is one where each group is first organised in whatever scheme that is internally most pertinent, thereby forming the respective J_1, J_2 and so on. Following that, these different group angular momenta may be coupled together, as the overall least important perturbation. That will then, as usual, result in overall quantum numbers J and M_J. These kind of coupling schemes are often referred to as J_1J_2, or in the case that the second group is a lone electron as J_1j.

It is important to note that even if two groups that form J_1 and J_2 are each well described by LS-coupling, this is not the same as an LS-coupling situation with ancestral terms, such as those described in section 7.1.2. In the latter case, *all* electrons are coupled together, before *any* spin–orbit interaction is taken into account. The parent terms merely describe different paths to a final LS-coupling term. For J_1J_2-coupling, some spin–orbit interaction take precedence over some interelectronic repulsions.

Examples of J_1j-coupling can be seen in table 5.1 for the first ionised levels of some of the lanthanides. One such is the ground state of singly ionised terbium. The neutral atom has eleven valence electrons, and the ground state electronic configuration is $4f^9\,6s^2$. For the ion, the energetically lowest configuration is that for which one of the 6s electrons has been removed. For the remaining electrons the lone tenth electron, 6s, has a small overlap with, and only couples weakly to, the nine 4f electrons. This means that for a proper perturbation treatment, one should first form

A. Kastberg, *Structure of Multielectron Atoms*, Springer Series on Atomic, Optical, and Plasma Physics 112, https://doi.org/10.1007/978-3-030-36420-5_9

a term for the $4f^9$, and then add the spin–orbit interactions within that term. That results in a quantum number that we refer to as J_1. The remaining interaction to add is that between J_1 and $j_{10} = 1/2$. It turns out that the value of J_1 that gives lowest energy is $15/2$, and the conclusion is that the ion ground state is:

$$4f^9\,(^6H)\,6s\,\left[\tfrac{15}{2},\tfrac{1}{2}\right]_8 \ . \tag{9.1}$$

This is also how this J_1j-coupled level is described in table 5.1.

This notation is far from unique. The values of J_1 and j are sometimes additionally given as subscript to, in this example, 'H' and 's'. General features of coupling schemes other than *LS*-coupling and *jj*-coupling are that their notations are not well standardised, they are not well known outside of a community of specialists, and they are inherently confusing. Whenever these kinds of states are to be tabulated, the adopted notation should each time be clarified.

The angular momentum vector coupling becomes even more complex if there are some interactions between different groups of electrons that are stronger than some of those inside the groups. Two such examples, J_1K-coupling and *LK*-coupling are described in sections 9.2 and 9.3 in this chapter. Before that, we will, in section 9.1, look into a couple of angular momentum interactions that we have hitherto mostly ignored. Towards the end of the chapter, we will in section 9.4 handle cases that are amenable for a description with a superposition of *LS*-coupling and *jj*-coupling. In section 9.5, we discuss departures from the CFA, where the angular effect perturbation Hamiltonian has non-diagonal terms between different electronic configurations.

9.1 Other Interactions Between Electronic Angular Momenta

An electron in a multielectron atom interacts electrically and magnetically with the spins and orbital angular momenta of *all* other electrons, and on top of that, it has its own internal spin–orbit interaction. Magnetic electron–electron interactions that we have not heretofore included in our analyses are the spin–spin interaction, the spin-other-orbit interaction, and the magnetic orbit–orbit interaction.

The meanings of these contributions are evident from their names. All of them are quite small, and in most instances, ignoring them is perfectly acceptable. For the lightest atoms, however, the internal spin–orbit interaction is also small, since it scales with the nuclear charge as Z^4. For these atoms, the aforementioned contributions may be of the same order as the spin–orbit term, or even bigger, and they will frequently be responsible for a breakdown of the Landé interval rule.

In appendix F, we derive a more complete magnetic interaction Hamiltonian, shown in (F.14). The terms corresponding to the three effects mentioned above are identified in section F.2.2. In the following, we will provide a qualitative treatment. For detailed studies of these effects, we direct the reader to other volumes (for example [1, 2] or [3]), or to some of the references at the end of this chapter.

The Hamiltonian describing electromagnetic interactions between one pair of electrons — derived in (F.14) — is:

$$H_{ij} = \frac{1}{r_{ij}} + \frac{\mathrm{i}\,\alpha^2\,[\boldsymbol{\sigma}_i \cdot (\mathbf{r}_{ij} \times \nabla_i) - \boldsymbol{\sigma}_j \cdot (\mathbf{r}_{ij} \times \nabla_j)]}{4\,r_{ij}^3} + H_{\mathrm{ss}} + H_{\mathrm{soo}} + H_{\mathrm{moo}}\,. \tag{9.2}$$

In this equation, $\boldsymbol{\sigma}_i$ and $\boldsymbol{\sigma}_j$ are the Pauli spin matrices for electrons i and j — defined in (E.15) — and r_{ij} is the separation between the two interacting electrons. The first term in (9.2) is the electrostatic interaction between the pair. The second is an internal spin–orbit effect in each of the two electrons, but it is just the contribution to this emanating from the orbital torque caused by the interelectronic electrostatic interaction. This is typically lumped together with the main intraelectronic spin–orbit interaction, which comes from the orbital motion around the nucleus. The three last terms — the ones we are concerned with in this section — the spin–spin, spin-other-orbit and magnetic orbit–orbit Hamiltonians, are:

$$H_{\mathrm{ss}} = \frac{\alpha^2}{4}\left[\frac{\boldsymbol{\sigma}_i \cdot \boldsymbol{\sigma}_j}{r_{ij}^3} - \frac{3\,(\boldsymbol{\sigma}_i \cdot \mathbf{r}_{ij})\,(\boldsymbol{\sigma}_j \cdot \mathbf{r}_{ij})}{r_{ij}^5}\right] \tag{9.3}$$

$$H_{\mathrm{soo}} = \mathrm{i}\,\alpha^2\,\frac{[\boldsymbol{\sigma}_j\,(\mathbf{r}_{ij} \times \nabla_i) - \boldsymbol{\sigma}_i\,(\mathbf{r}_{ij} \times \nabla_j)]}{2\,r_{ij}^3} \tag{9.4}$$

$$H_{\mathrm{moo}} = \frac{\alpha^2}{2}\left[\frac{\nabla_i \cdot \nabla_j}{r_{ij}} + \frac{(\mathbf{r}_{ij} \cdot \nabla_i)\,(\mathbf{r}_{ij} \cdot \nabla_j)}{r_{ij}^3} + \frac{\mathbf{r}_{ij} \cdot \nabla_i - \mathbf{r}_{ij} \cdot \nabla_j}{2\,r_{ij}^3}\right]. \tag{9.5}$$

9.1.1 Spin–Spin Interaction

The spin–spin interaction is the coupling between two magnetic dipoles. It is exactly analogous to the interaction between two spins in a magnetic material. The form of the Hamiltonian is that of (9.3) — and also of (F.21). If we write it in terms of the spin vector operators, rather than the Pauli matrices, the Hamiltonian for one pair of electrons becomes:

$$H_{\mathrm{ss}}(\mathbf{r}_i, \mathbf{r}_j) = \alpha^2\left[\frac{\mathbf{S}_i \cdot \mathbf{S}_j}{r_{ij}^3} - \frac{3\,(\mathbf{S}_i \cdot \mathbf{r}_{ij})(\mathbf{S}_j \cdot \mathbf{r}_{ij})}{r_{ij}^5}\right]. \tag{9.6}$$

This is valid for $r_{ij} \neq 0$.

Examples for which this interaction is pronounced are the triplet states of neutral He, where its influence changes the relative distance between fine-structure levels dramatically, and it explains the poor validity of the Landé rule for helium triplets. This is illustrated in the energy diagrams in figures 9.1 and 9.2, and commented on further in section 9.1.4.

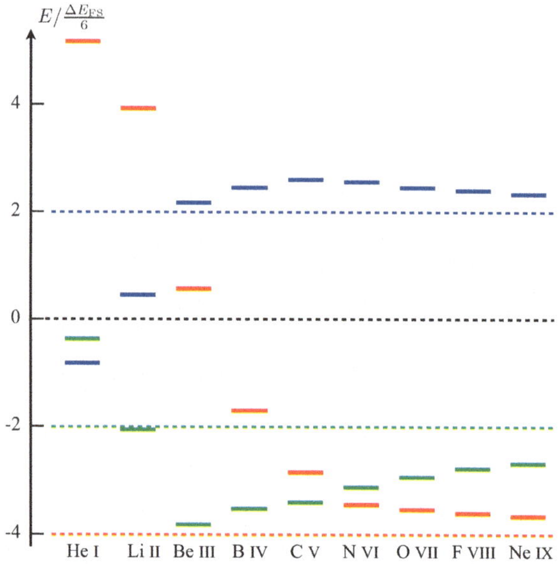

Fig. 9.1 Empiric energy fine-structure for the triplet 1s2p ^{3}P for He and for ions in its isoelectronic series up to Ne IX. For each element, the actual energy span of the triplet is different, but in this diagram it has been normalised as ΔE_{FS}. The zero-point for each triplet has been put at the geometric mean value of the partially degenerate levels, and the unit on the vertical axis is $\Delta E_{FS}/6$. In the diagram, the red levels correspond to the fine-structure $J=0$, the green to $J=1$ and the blue to $J=2$. The dashed red, green and blue lines indicate where the levels would be if the triplets followed ideal *LS*-coupling. The data are from [4].

9.1.2 Spin-Other-Orbit Interaction

The spin-other orbit interaction Hamiltonian has a form very similar to the spin–orbit Hamiltonian — it contains scalar products between spin matrices and angular momenta. From (9.4) we can deduce the following for two electrons with position vectors $\mathbf{r}_i$ and $\mathbf{r}_j$:

$$H_{\text{soo}}(\mathbf{r}_i, \mathbf{r}_j) = \xi'(\mathbf{r}_i, \mathbf{r}_j)\ [\mathbf{L}_{ij} \cdot \mathbf{S}_j + \mathbf{L}_{ji} \cdot \mathbf{S}_i]\ . \tag{9.7}$$

The angular momenta $\mathbf{L}_{ij}$ and $\mathbf{L}_{ji}$ are here defined as in (F.16). The equation is the same as (F.22), except that we have defined a radial integral $\xi'(\mathbf{r}_i, \mathbf{r}_j)$. This is analogous to the radial fine-structure function used in (4.19), albeit it is a function of two electron position vectors.

Taken on its own, the spin-other-orbit interaction will result in an interval rule similar to the normal Landé one, except that it will be quantified by angular momenta that are sums of the $\mathbf{L}_i$ and $\mathbf{S}_j$ (with $i \neq j$), instead of by J. The multiplicative factor $\xi'(\mathbf{r}_i, \mathbf{r}_j)$ may be negative, in which case it can result in a reversal of the internal ordering in a fine-structure multiplet for a light atom.

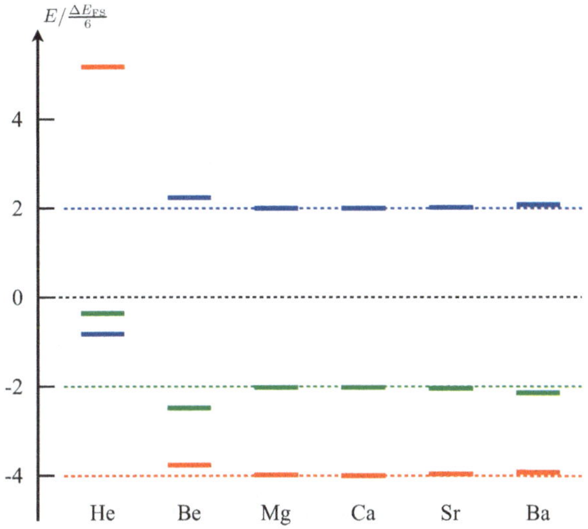

Fig. 9.2 Energy fine-structure of low lying triplets of neutral alkaline earth atoms: 1s2p ^{3}P for He, 2s2p ^{3}P for Be, 3s3p ^{3}P for Mg, 4s4p ^{3}P for Ca, 5s5p ^{3}P for Sr and 6s6p ^{3}P for Ba. For each element, the actual energy span of the triplet has been normalised as ΔE_{FS}, in the same fashion as in figure 9.1. The red levels correspond to the fine-structure $J=0$, the green to $J=1$ and the blue to $J=2$. The dashed red, green and blue lines indicate where the levels would be if the triplets followed ideal *LS*-coupling. The data are from [4].

9.1.3 Magnetic Orbit–Orbit Interaction

The magnetic interaction caused by the currents associated with the respective orbital angular momenta of two electrons, the magnetic orbit–orbit interaction, is typically smaller still than the two previously discussed magnetic interelectronic interactions. It can be traced to several terms in (F.14), leading to the expression in (9.5) — and in (F.23).

9.1.4 An Example of Magnetic Electron–Electron Interactions

We end section 9.1 by analysing, as a concrete example, the aforementioned lowest excited triplet, 1s2p^{3}P, in neutral helium. In figure 9.1, we show the relative energies of the three fine-structure states $J=0$, $J=1$ and $J=2$ for the isoelectronic series from He I to Ne IX. The energy span of each triplet has been normalised to be the same. The dashed red, green and blue lines indicate where the levels would be for a perfectly *LS*-coupled case (that is, one in keeping with the Landé rule).

As can be seen, the Landé rule is widely off the mark for He I, which is because the spin–spin coupling is of the same order as the spin–orbit interaction. With increasing Z, the relative importance of the spin–orbit contribution grows rapidly, and we recover something more akin to a standard *LS*-coupling triplet.

We can also compare neutral He with other neutral alkaline earths. This is illustrated in figure 9.2. We see that already for Be, we have good *LS*-coupling, and for the thereafter following elements it becomes almost perfect. For the heaviest element in the figure, Ba, a slight deviation caused by relativistic effects starts to be discernible.

9.2 J_1K-Coupling

The coupling schemes that differ from *LS*-coupling and *jj*-coupling have not been given entirely consistent names or definitions. The exact notation and description of in which ways the angular momentum vectors are added is often adapted to the situation at hand.

If one single electron is highly excited, its spin may contribute very little to the energy, compared to the couplings between all the other $\mathbf{L}_i$ and $\mathbf{S}_i$ among the valence electrons. As a result, the energy levels of such an atom tend to appear in pairs, corresponding to the two possible orientations of the last spin. This is particularly prevalent if l_i is high for the excited electron; partly because an electron with high orbital angular momentum penetrates the core electrons very little, and partly because for each individual electron, the spin–orbit coupling diminishes rapidly with increasing l_i — roughly proportionally to l_i^{-3}, see (4.14). The situation is sometimes referred to as pair coupling. However, that name may be less than ideal, since it does not actually describe what couples to what, but instead a phenomenological consequence.

One type of pair coupling is J_1K-coupling (or jK-coupling), where all valence electrons except a lone highly excited one intercouple to an intermediate, *almost* total, angular momentum. This vector is given the symbol $\mathbf{J}_1$, with associated quantum number J_1 (strictly speaking, J_1 is the quantum number that relates to the operator $\mathbf{J}_1^2$). This angular momentum then couples to the orbital angular momentum of the last electron, $\mathbf{L}_N$, electrostatically, and as a symbol for the resultant, we introduce the angular momentum vector $\mathbf{K}$. As a final stage, this $\mathbf{K}$ interacts with the remaining electron spin, $\mathbf{S}_N$, through the spin–orbit effect. That gives us the overall $\mathbf{J}$.

The situation is most readily described by an example with just two valence electrons, of which one is more excited than the other. The order in which we couple the four angular momenta $\mathbf{L}_1$, $\mathbf{S}_1$, $\mathbf{L}_2$ and $\mathbf{S}_2$ is:

$$\begin{aligned} \mathbf{L}_1 + \mathbf{S}_1 &= \mathbf{J}_1 \\ \mathbf{J}_1 + \mathbf{L}_2 &= \mathbf{K} \\ \mathbf{K} + \mathbf{S}_2 &= \mathbf{J} \ . \end{aligned} \tag{9.8}$$

The notation is not well standardised, but one way to write a J_1K-coupling term for a state with two valence electrons is:

$$n_1 l_1 \left({}^{2S_1+1}L_{1J_1} \right) n_2 l_2 \, [K]_J \ . \tag{9.9}$$

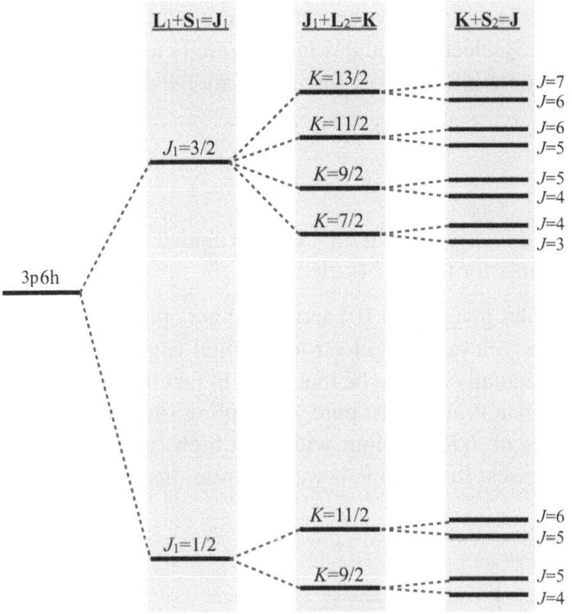

Fig. 9.3 Schematic energy level diagram (not to scale) for a hypothetical atom with two valence electrons, excited to the configuration 3p6h. The most important angular momentum coupling is that intrinsic in the 3p electron, which forms the vector $\mathbf{J}_1$. Thereafter, the greatest contribution to the energy is the coupling $\mathbf{J}_1 + \mathbf{L}_2 = \mathbf{K}$, with the orbital angular momentum of the 6h electron. Finally, the spin of the outermost electron couple to $\mathbf{K}$, as the last perturbation. The final states appear in pairs.

We will give a more concrete example. Consider the two-electron configuration 3p6h. If J_1K-coupling is a good description for the atom in question, the possible states are:

$$
\begin{array}{lll}
3\mathrm{p}\left({}^2\mathrm{P}_{3/2}\right)6\mathrm{h}\,[{}^{13}\!/_2]_7 & , & 3\mathrm{p}\left({}^2\mathrm{P}_{3/2}\right)6\mathrm{h}\,[{}^{13}\!/_2]_6 \\
3\mathrm{p}\left({}^2\mathrm{P}_{3/2}\right)6\mathrm{h}\,[{}^{11}\!/_2]_6 & , & 3\mathrm{p}\left({}^2\mathrm{P}_{3/2}\right)6\mathrm{h}\,[{}^{11}\!/_2]_5 \\
3\mathrm{p}\left({}^2\mathrm{P}_{3/2}\right)6\mathrm{h}\,[{}^{9}\!/_2]_5 & , & 3\mathrm{p}\left({}^2\mathrm{P}_{3/2}\right)6\mathrm{h}\,[{}^{9}\!/_2]_4 \\
3\mathrm{p}\left({}^2\mathrm{P}_{3/2}\right)6\mathrm{h}\,[{}^{7}\!/_2]_4 & , & 3\mathrm{p}\left({}^2\mathrm{P}_{3/2}\right)6\mathrm{h}\,[{}^{7}\!/_2]_3 \\
3\mathrm{p}\left({}^2\mathrm{P}_{1/2}\right)6\mathrm{h}\,[{}^{11}\!/_2]_6 & , & 3\mathrm{p}\left({}^2\mathrm{P}_{1/2}\right)6\mathrm{h}\,[{}^{11}\!/_2]_5 \\
3\mathrm{p}\left({}^2\mathrm{P}_{1/2}\right)6\mathrm{h}\,[{}^{9}\!/_2]_5 & , & 3\mathrm{p}\left({}^2\mathrm{P}_{1/2}\right)6\mathrm{h}\,[{}^{9}\!/_2]_4 \quad .
\end{array}
\tag{9.10}
$$

In figure 9.3, we show a schematic energy level diagram of this configuration.

In the case of a two-electron atom, as above, this coupling scheme is also referred to as jK-coupling. However, the description J_1K-coupling is more general, since it caters also for atoms with three or more electrons. To give an example with a three-electron atom, we can take the configuration $4\mathrm{s}^2\,5\mathrm{g}$. This could be assumed to couple as a single 5g valence electron, but that would be erroneous if the angular momentum of the 5g electron couples more strongly to the $4s^2$ core than it does to its own spin. The possible J_1K-coupling states are the ones in the following pair:

$$4s^2\left({}^1S_0\right)5g\,[4]_{9/2}$$
$$4s^2\left({}^1S_0\right)5g\,[4]_{7/2}\,. \tag{9.11}$$

The levels with the intermediate term ^3S are excluded due to the restrictions posed by the Pauli principle for the two 4s-electrons.

The two examples given in (9.10) and (9.11) are ones with g- and h-orbitals. For J_1K-coupling to be relevant, the electronic orbital angular momentum for the last electron does not actually have to be that high. In fact, for real atoms, J_1K-coupling is more common than is an almost pure jj-coupling situation.

Good examples of J_1K-coupling without a high l_N are some excited configurations of noble gases. In figure 9.4, we illustrate this by showing some levels in

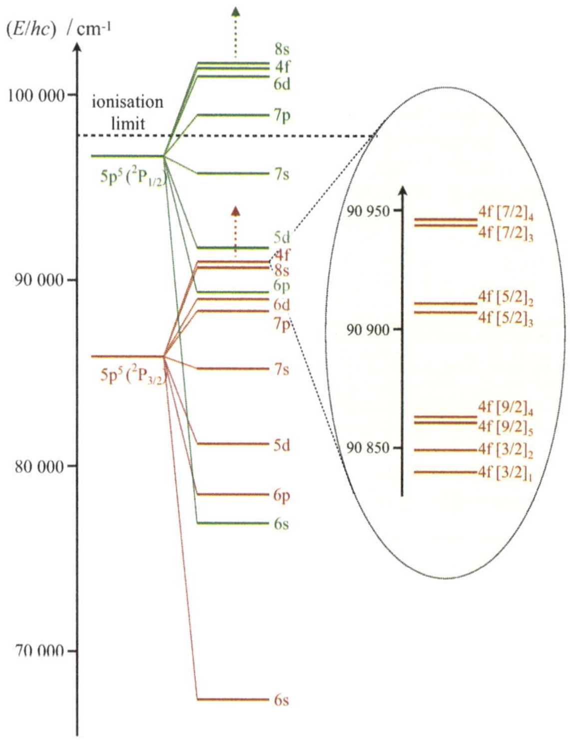

Fig. 9.4 Partial energy level diagram for some excited states in neutral Xe, with the data taken from [4]. The zero of the energy scale has been set at the Xe ground state. The two J_1 levels $5p^5\left({}^2P_{3/2}\right)$ (red in the figure) and $5p^5\left({}^2P_{1/2}\right)$ (green) have been placed at somewhat arbitrarily weighted averages. Both multiplets have many higher lying excited orbitals than the ones shown in the figure, and these converge towards respective series limits. Several of the states belonging to $5p^5\left({}^2P_{1/2}\right)$ are auto-ionising (see section 13.2). In the yellow blow-up, we have magnified the splitting of the 4f excited orbital into different quantum numbers K, and subsequently pairs of J.

neutral Xe. The ground state of Xe is $5p^6\,{}^1S_0$. When one of the electrons in the 5p orbital is excited, the most important energy difference between different excited states is that between the two parent terms $5p^5(^2P_{3/2})$ and $5p^5(^2P_{1/2})$. The energy splitting between the different excited electron orbitals comes first thereafter. For the state $5p^5(^2P_{3/2})4f$, we show in the figure the splitting between four different pairs, each with a different value of K. This represents the coupling between $l_6 = 3$ and $J_1 = 3/2$. Finally, the spin–orbit interaction with the remaining spin will be the weakest interaction, and this gives us the pairs.

It is not obvious how to unambiguously write the state vector for a J_1K-coupled atom in a compact form. One way to do so is the following:

$$|\,\gamma\delta J_1 KJM_J\rangle_{J_1K}\,. \tag{9.12}$$

In this notation, γ is the electron configuration of all the filled orbitals, and δ describes both the valence electron configuration and the parent term. If we are dealing with a two-electron atom, we can use the following more explicit notation:

$$|\,\gamma\,\{[(l_1,s_1)j_1\,,\,l_2]\,K\}\,JM_J\rangle_{J_1K}\,. \tag{9.13}$$

9.3 *LK*-Coupling

LK-coupling is, like J_1K-coupling, a form of pair coupling, and the two schemes have many similarities. The common feature is that there is one valence electron with a high l_i, which has the consequence that the coupling of its spin is weaker than all other angular momentum couplings.

The difference here is the order of all the other couplings. In the *LK*-coupling approximation, we assume that *all* electrostatic couplings are stronger than any spin–orbit interaction. This means that it is rational to first couple all l_i and to thus form an overall L. Then, this L is coupled to the sum of all s_i *except* the one for the single electron with high l_i. This sum, we call S_1, and this time the angular momentum K is formed by the coupling of L to S_1. The final step is to include the weak spin–orbit coupling to the last spin, which again results in pairs of levels.

One way to notate an *LK*-coupled state, for an atom with N valence electrons, is as follows:

$$n_1\,l_1\,\ldots\,n_{N-1}\,l_{N-1}\,\left(^{2S_1+1}L_1\right)\,n_N\,l_N\,L\,[K]_J\,. \tag{9.14}$$

Here, we first write all the valence electrons — from 1 to $(N-1)$ — except the one which is very weakly spin–orbit coupled. This *valence core* makes up an initial parent term, notated according to the standard *LS*-coupling convention, but without a J_1. We then include the final valence electron. When the orbital angular momentum

l_N is added to L_1, we get the total L. The penultimate step is to add the spin–orbit interaction with the valence core. This is the vector addition of L and S_1, which results in K. Finally, the last spin–orbit interaction is included, and hence we get a pair structure, with two values of J.

We can also illustrate the order in which the vector coupling should be made, using a two-electron atom as an example. The proper vector addition is:

$$\begin{aligned} \mathbf{L}_1 + \mathbf{L}_2 &= \mathbf{L} \\ \mathbf{L} \ + \mathbf{S}_1 &= \mathbf{K} \\ \mathbf{K} + \mathbf{S}_2 &= \mathbf{J} \ . \end{aligned} \tag{9.15}$$

There is also a form of LK-coupling with two or more electrons having a weak spin coupling. For that eventuality, the discussion above has to be adjusted accordingly.

LK-coupling is less common than its J_1K sibling, but it does occur. One such example is the 2p 4f configuration in the singly ionised species N II. The possible LK-coupling terms for this configuration are:

$$\begin{aligned}
&2p(^2P)\,4f\ \ D\,[3/2]_1 \quad , \quad && 2p(^2P)\,4f\ \ D\,[3/2]_2 \\
&2p(^2P)\,4f\ \ D\,[5/2]_2 \quad , \quad && 2p(^2P)\,4f\ \ D\,[5/2]_3 \\
&2p(^2P)\,4f\ \ F\,[5/2]_2 \quad , \quad && 2p(^2P)\,4f\ \ F\,[5/2]_3 \\
&2p(^2P)\,4f\ \ F\,[7/2]_3 \quad , \quad && 2p(^2P)\,4f\ \ F\,[7/2]_4 \\
&2p(^2P)\,4f\ \ G\,[7/2]_3 \quad , \quad && 2p(^2P)\,4f\ \ G\,[7/2]_4 \\
&2p(^2P)\,4f\ \ G\,[9/2]_4 \quad , \quad && 2p(^2P)\,4f\ \ G\,[9/2]_5 \ .
\end{aligned} \tag{9.16}$$

In figure 9.5, we show a partial energy level diagram for this ion.

As was the case for J_1K-coupling, writing a state vector for LK-coupling is somewhat awkward. If we use the same philosophy as for (9.12) and (9.13), we can write the ket as:

$$|\,\gamma\delta\, LKJM_J\,\rangle_{LK}\ , \tag{9.17}$$

and if we have a two-electron atom, an explicit type of notation is:

$$|\,\gamma\,\{[(l_1, l_2)L\,,\, s_1]\,K\,,\, s_2\}\,JM_J\,\rangle_{LK}\ . \tag{9.18}$$

9.4 Intermediate Coupling

There are atomic energy levels for which neither the electrostatic repulsion between electrons nor the spin–orbit interaction dominates over the other. The variants of pair coupling, such as J_1K-coupling and LK-coupling — described in the previous sections — may not be pertinent either. In such cases, the true energy levels are not

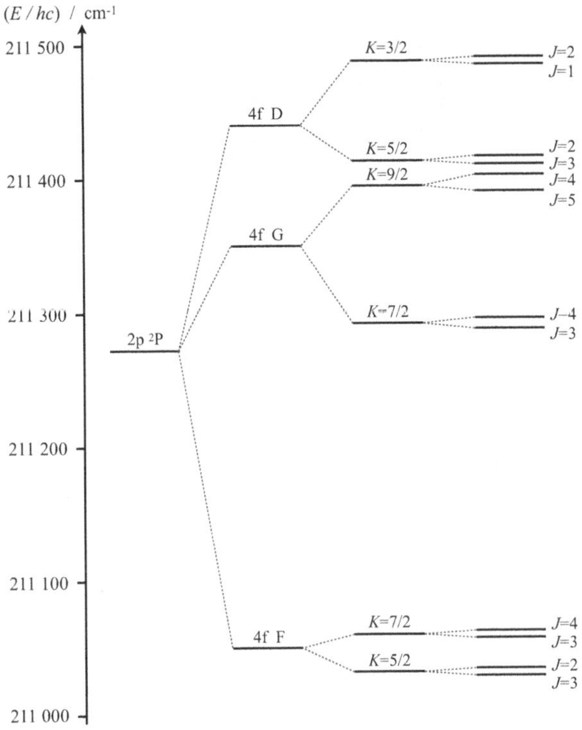

Fig. 9.5 Experimentally obtained energy levels for the *LK*-coupled configuration 2p4f in singly ionised atomic nitrogen [4]. The first vector coupling is that of $l_1 = 1$ and $l_2 = 3$, which gives the three alternatives of 2, 3 and 4 for L_1. That L_1 then couples with $s_1 = 1/2$ to set up K, and eventually the coupling of K with s_2 results in our final J. This is an example of *LK*-coupling for two valence electrons.

well described by any of the angular momentum coupling schemes that we have heretofore explained. This means that for a theoretical analysis, one is typically stuck with having to treat the sum of the Hamiltonians beyond the CFA, $H_{\mathrm{ee}} + H_{\mathrm{SO}}$ (summed over all valence electrons) as *one* perturbation to the electronic configuration.

Taking all the angular effects as one single correction is cumbersome because the perturbation Hamiltonian will not be close to diagonal in any of the representations used for *LS*-coupling or *jj*-coupling. One has to choose one representation, formulate the entire matrix for the Hamiltonian within that and then solve the resulting secular equation in order to find the energy levels. For relatively simple cases, analytic solutions may be feasible, but often the perturbation matrix has to be diagonalised numerically.

Most common is to choose the representation $|\gamma LSJM_J\rangle$ for the calculation, and here we will restrict the analysis to this choice. This means that we have to find the non-diagonal matrix element of H_{SO} in this representation, using transformations between coupling schemes.

The complete secular equation will typically be quite large, but it can always be broken up into smaller chunks, due to the fact that the full perturbation Hamiltonian is diagonal in J and M_J in the absence of an external field. The total angular momentum is conserved through all intra-atomic interactions. For example, this means that if there is one unique fine-structure level with a particular value of J for a configuration, this J-state will be represented by the same wave function in the LS-coupling and jj-coupling schemes, as well as for all intermediate situations. If a value of J occurs twice in the ensemble of allowed terms, we need to diagonalise a two-by-two matrix for that J, and so on.

9.4.1 Transformations Between Coupling Schemes

The mathematics of the coupling of quantum mechanical angular momenta is reviewed in appendix C.5. Therein, expressions for vector coupling factors, *Clebsch–Gordan coefficient*, are derived, and expressed in terms of Wigner $3j$-symbols (see appendix C.4). In section C.5.3 we demonstrate transformations between different ways to couple *three* angular momenta. A repeated application of the transformation rules for three vectors makes it possible to find expressions also for states with four or more angular momenta.

In the following, we will show examples of how this can be applied to an atom with two valence electrons. In other words, we have four angular momenta: $\mathbf{L}_1$, $\mathbf{S}_1$, $\mathbf{L}_2$ and $\mathbf{S}_2$, and depending on the relative strengths of the interactions between these, the summation (really the successive application of perturbation Hamiltonians) needs to be done in a certain order. That, in turn, leads to different sets of quantum numbers that render the Hamiltonian approximately diagonal, and such sets of quantum numbers are the same thing as a representation.

We begin by a transformation between LS-coupling and LK-coupling (see section 9.3). In order to demonstrate the task at hand, we first write the state vectors for both coupling schemes in an explicit form, such as is shown for LK-coupling in (9.18). Ignoring closed orbital electrons, the configuration is $(n_1 l_1, n_2 l_2)$ and we want to transform between the two representations:

$$\left| \, [(l_1,l_2)L\,,\,(s_1,s_2)S]\,J \right\rangle_{LS} \longleftrightarrow \left| \, \{[(l_1,l_2)L\,,\,s_1]\,K\,,\,s_2\}\,J \right\rangle_{LK} . \tag{9.19}$$

We have left out the labels for the projection quantum number M_J.

In order to make the notation more readable, we will shorten (9.19), in two steps, into:

$$\left| \, [\varepsilon\, LS]\,J \right\rangle_{LS} \longleftrightarrow \left| \, \{[\varepsilon\, L]\,K\}\,J \right\rangle_{LK} , \tag{9.20}$$

and

$$\left| LSJ \right\rangle_{LS} \longleftrightarrow \left| LKJ \right\rangle_{LK} . \tag{9.21}$$

In (9.20), we have eliminated the redundant s_1 and s_2, and ε is shorthand for the electron configuration of the valence electrons.

The transformation matrix that we want to calculate is:

$$\mathbb{T}_{LS,LK}(\varepsilon) = \sum_{i=1}^{N_{\rm FS}} \sum_{j=1}^{N_{\rm FS}} \big|(LSJ)_i\big\rangle_{LS} \; \big\langle (LKJ)_j\big|_{LK} \; . \tag{9.22}$$

The summation limit $N_{\rm FS}$ is the total number of allowed fine-structure terms (independent of coupling scheme). The square matrix in (9.22) will have the possible *LS*-coupling terms as columns, and the *LK*-coupling ones as rows, and the matrix elements will be:

$$T_{LS,LK}(\varepsilon, [LSJ]_{LS}, [LKJ]_{LK}) = {}_{LK}\langle LKJ \mid LSJ\rangle_{LS} \; . \tag{9.23}$$

Both schemes begin by adding $\mathbf{L}_1$ and $\mathbf{L}_2$ in order to get $\mathbf{L}$. Thus, we are in effect comparing two ways to couple the *three* vectors $\mathbf{L}$, $\mathbf{S}_1$ and $\mathbf{S}_2$. This means that we can use (C.102) and (C.96), and we find that the matrix elements will be:

$$\begin{aligned} &T_{LS,LK}(\varepsilon, [LSJ]_{LS}, [LKJ]_{LK}) \\ &\quad = \delta(L_{LS}, L_{LK})\,(-1)^{L+1+J}\,\sqrt{2K+1}\sqrt{2S+1}\begin{Bmatrix} L & 1/2 & K \\ 1/2 & J & S \end{Bmatrix} . \end{aligned} \tag{9.24}$$

The delta factor occurs because L is coupled in the same way in the two coupling schemes, or to put it a more technical way, the partially coupled state vectors in the sub-basis $|(l_1,l_2)LM_L\rangle$ are mutually orthogonal.

From *LK*-coupling we can transform to J_1K-coupling (section 9.3), by again using an equation for the coupling of three angular momenta. This time, the two schemes concerned have as common final step the coupling of $\mathbf{K}$ with $\mathbf{S}_2$. The thing that differs is the order in which the vectors $\mathbf{L}_1$, $\mathbf{L}_2$ and $\mathbf{S}_1$ are added in the formation of $\mathbf{K}$. Also, this can be derived with the above method, and the result is that the elements of the transformation matrix are — this time with a delta factor in K — as follows:

$$\begin{aligned} &T_{LK,J_1K}(\varepsilon, [LKJ]_{LK}, [J_1KJ]_{J_1K}) = {}_{J_1K}\langle J_1KJ \mid LKJ\rangle_{LK} \\ &\quad = \delta(K_{LK}, K_{J_1K})\,(-1)^{l_2-1/2+L+J_1}\,\sqrt{2J_1+1}\sqrt{2L+1}\begin{Bmatrix} 1/2 & l_1 & J_1 \\ l_2 & K & L \end{Bmatrix} . \end{aligned} \tag{9.25}$$

Thus, if a transformation is sought between *LS*-coupling and J_lK-coupling, one can apply (9.24) and (9.25) in succession. In principle, that would also require a summation over the possible quantum numbers $[LK]_{LK}$, but thanks to the delta factors in L and K in the two equations, this reduces to one single term.

For the transformation between J_1K-coupling and *jj*-coupling, we have to consider the two different ways to combine the vectors $\mathbf{J}_1$, $\mathbf{L}_2$ and $\mathbf{S}_2$ in order to arrive at $\mathbf{J}$. The matrix elements for this are:

$$T_{J_1K,jj}(\varepsilon,[J_1KJ]_{J_1K},[j_1j_2J]_{jj}) = {}_{jj}\langle j_1j_2J \mid J_1KJ\rangle_{J_1K}$$
$$= \delta(J_1,j_1)\,(-1)^{l_2-1/2-j_1+J}\,\sqrt{2j_2+1}\sqrt{2K+1}\begin{Bmatrix} 1/2 & l_2 & j_2 \\ j_1 & J & K \end{Bmatrix}. \qquad (9.26)$$

We can note that for this two-electron example, J_1 and j_1 is the same thing. If we need to transform from LK-coupling to jj-coupling, we take (9.25) and (9.26) and take advantage of the respective delta factors to avoid having to sum over all the $[J_1K]_{J_1K}$ terms.

The transformation between LS-coupling and jj-coupling is the one which is most commonly needed in intermediate coupling. The matrix elements of that can be obtained by applying the closure theorem twice, over the bases $|J_1KJ\rangle$ and $|LKJ\rangle$, to reach a formula with a product of all three of the transformations (9.24), (9.25) and (9.26). The delta functions will eliminate three of the four summations, but one will still be required. The resulting equation for the matrix elements is:

$$T_{LS,jj}(\varepsilon,[LSJ]_{LS},[j_1j_2J]_{jj}) = {}_{jj}\langle j_1j_2J \mid LSJ\rangle_{LS}$$
$$= \sum_K T_{LS,LK}\,T_{LK,J_1K}\,T_{J_1K,jj}$$
$$= (-1)^{2J}\sqrt{2L+1}\sqrt{2S+1}\sqrt{2j_1+1}\sqrt{2j_2+1}$$
$$\times \sum_{K=K_{min}}^{K_{max}} (2K+1)\begin{Bmatrix} L & 1/2 & K \\ 1/2 & J & S \end{Bmatrix}\begin{Bmatrix} 1/2 & l_1 & j_1 \\ l_2 & K & L \end{Bmatrix}\begin{Bmatrix} 1/2 & l_2 & j_2 \\ j_1 & J & K \end{Bmatrix}$$
$$= \sqrt{2L+1}\sqrt{2S+1}\sqrt{2j_1+1}\sqrt{2j_2+1}\begin{Bmatrix} l_2 & 1/2 & j_2 \\ l_1 & 1/2 & j_1 \\ L & S & J \end{Bmatrix}. \qquad (9.27)$$

In the last line, we have converted the sum of products of $6j$-symbols to a $9j$-symbol, using (C.69). The phase has been adjusted by observing that $2K$ will always be an odd integer for a two-electron atom.

For atoms with more than two valence electrons, the principle described above remains the same, but the level of complexity increases dramatically. In many instances, analytical approaches to atoms with three or more electrons, and intermediate coupling, are hardly feasible, and thus the derivation of explicit transformation matrices is not always a useful exercise.

9.4.2 Examples of Intermediate Coupling for Two-Electron Atoms

We will demonstrate transformations between LS-coupling and jj-coupling schemes with a few examples. As a first one, we take a configuration $ns\,n'$p. This gives four different states, which in the limiting representations have the following labels:

$$LS : \begin{cases} ns\,n'\mathrm{p}\ ^1\mathrm{P}_1 \\ ns\,n'\mathrm{p}\ ^3\mathrm{P}_2 \\ ns\,n'\mathrm{p}\ ^3\mathrm{P}_1 \\ ns\,n'\mathrm{p}\ ^3\mathrm{P}_0 \end{cases} , \quad jj : \begin{cases} [n\mathrm{s}_{1/2}\,n'\mathrm{p}_{3/2}]_1 \\ [n\mathrm{s}_{1/2}\,n'\mathrm{p}_{3/2}]_2 \\ [n\mathrm{s}_{1/2}\,n'\mathrm{p}_{1/2}]_1 \\ [n\mathrm{s}_{1/2}\,n'\mathrm{p}_{1/2}]_0 \end{cases} . \tag{9.28}$$

Using the formalism of section 7.2, we can find that in the *LS*-coupling limit, if we initially neglect the spin–orbit coupling, the two term energies are:

$$\begin{aligned} E_{^1\mathrm{P}} &= I(ns) + I(n'\mathrm{p}) + F^{(0)}_{ns:n'\mathrm{p}} + \frac{1}{3}\,G^{(1)}_{ns:n'\mathrm{p}} \\ E_{^3\mathrm{P}} &= I(ns) + I(n'\mathrm{p}) + F^{(0)}_{ns:n'\mathrm{p}} - \frac{1}{3}\,G^{(1)}_{ns:n'\mathrm{p}} \quad , \end{aligned} \tag{9.29}$$

with the definitions of, and notations for, the integrals I, F and G as explained in chapter 7. Thus, the span in energy for pure *LS*-coupling for the ensemble of states is $\frac{2}{3}\,G^{(1)}_{ns:n'\mathrm{p}}$.

For the opposite limit, we learn from section 8.2 that the energies of the *jj*-coupling terms are:

$$\begin{aligned} E_{[1/2\,,\,3/2]} &= I(ns) + I(n'\mathrm{p}) + \frac{1}{2}\,\xi_{n'\mathrm{p}} \\ E_{[1/2\,,\,1/2]} &= I(ns) + I(n'\mathrm{p}) - \xi_{n'\mathrm{p}} \quad , \end{aligned} \tag{9.30}$$

where the radial fine-structure function integral ξ is defined in (7.52). This gives a range in the energies of $\frac{3}{2}\,\xi_{n'\mathrm{p}}$.

From this analysis we can conclude that the respective virtues of the approximations *LS*-coupling and *jj*-coupling is directly dependent on the relative measures of the integrals $G^{(1)}_{ns:n'\mathrm{p}}$ and $\xi_{n'\mathrm{p}}$. A practical way to quantify the quality of the two approximations is therefore by a comparative parameter χ, which for the present example will be defined as follows:

$$\chi = \frac{\frac{3}{2}\,\xi_{n'\mathrm{p}}}{\frac{2}{3}\,G^{(1)}_{ns:n'\mathrm{p}} + \frac{3}{2}\,\xi_{n'\mathrm{p}}} . \tag{9.31}$$

The fact that the total perturbation is diagonal in J saves us from the task of having to diagonalise the entire 4×4-matrix. There is only one state with $J=2$, regardless of representation. This means that the *LS*-coupling state $ns\,n'\mathrm{p}\,^3\mathrm{P}_2$ is the same as the *jj*-coupling state $[n\mathrm{s}_{1/2}\,n'\mathrm{p}_{3/2}]_2$. The only thing that will vary with χ is its energy eigenvalue. In an analogue fashion, $ns\,n'\mathrm{p}\,^3\mathrm{P}_0$ is the same state as $[n\mathrm{s}_{1/2}\,n'\mathrm{p}_{1/2}]_0$. The only fine-structure quantum number represented by more than one state is $J=1$, and hence we will have just one secular equation, and it is of dimension two.

The full 4×4–matrix form of $H_{\mathrm{to}}+H_{\mathrm{SO}}$, we choose to formulate in the *LS*-coupling representation. The ordering of the rows and columns is arbitrary. We will use the following sequence: $^3\mathrm{P}_2$, $^3\mathrm{P}_1$, $^1\mathrm{P}_1$, $^3\mathrm{P}_0$; that is after decreasing J. We already have the expectation values of H_{ee} from (9.29), and the diagonal elements of H_{SO} can be had from the procedure described in section 7.3.

What remains is the determination of two off-diagonal elements. For that, we will need to transform between the two coupling schemes. The relevant transformation matrix can be found from (9.27). Its elements are:

$$T_{LS,jj}(ns\,n'\mathrm{p},[1S1]_{LS},[{}^1\!/\!{}_2\,j_2 1]_{jj}) = \sqrt{6}\sqrt{2S+1}\sqrt{2j_2+1}\begin{Bmatrix} 1 & 1/2 & j_2 \\ 0 & 1/2 & 1/2 \\ 1 & S & 1 \end{Bmatrix}, \tag{9.32}$$

and the $J=1$–matrix is:

$$T_{LS,jj}(ns\,n'\mathrm{p}) = \frac{1}{\sqrt{3}}\begin{pmatrix} 1 & \sqrt{2} \\ \sqrt{2} & -1 \end{pmatrix}, \tag{9.33}$$

where the rows have been ordered as $S=1$ followed by $S=0$, and the columns as $j_2 = 3/2$ before $j_2 = 1/2$ (S and j_2 are the only two quantum numbers that vary for the concerned states). This provides us with a way to write the LS-coupling states in terms of jj-coupling and, with (9.30) and the closure theorem, we then have all we need for writing down the full matrix. It is:

$$\begin{aligned} H_{\mathrm{ee}} + H_{\mathrm{SO}} = {} & [I(ns) + I(n'\mathrm{p}) + F^{(0)}_{ns:n'\mathrm{p}}]\,\mathbb{1} + \frac{G^{(1)}_{ns:n'\mathrm{p}}}{3}\begin{pmatrix} -1 & 0 & 0 & 0 \\ 0 & -1 & 0 & 0 \\ 0 & 0 & 1 & 0 \\ 0 & 0 & 0 & -1 \end{pmatrix} \\ & + \frac{\xi_{n'\mathrm{p}}}{2}\begin{pmatrix} 1 & 0 & 0 & 0 \\ 0 & -1 & \sqrt{2} & 0 \\ 0 & \sqrt{2} & 0 & 0 \\ 0 & 0 & 0 & -2 \end{pmatrix}, \end{aligned} \tag{9.34}$$

with $\mathbb{1}$ being the identity matrix. The corresponding eigenenergies can be computed as a function of χ. The result of such a calculation is shown in figure 9.6.

In the graph, we can see the gradual progression from the LS-coupling states ${}^3\mathrm{P}_0$, ${}^3\mathrm{P}_1$, ${}^3\mathrm{P}_2$ and ${}^1\mathrm{P}_1$ to the jj-coupling counterparts $[1/2,1/2]_0$, $[1/2,1/2]_1$, $[1/2,3/2]_1$ and $[1/2,3/2]_2$. In the figure, we have also added experimental data for a number of atomic $ns\,n'\mathrm{p}$ states. In order to include the data in the plot, the coefficients $G^{(1)}_{ns:n'\mathrm{p}}$ and $\xi_{n'\mathrm{p}}$ have been least-squares fitted to the data (the procedure for this is explained in chapter 10 of [5]). This shows three examples of trends, all with the level 3p 4s of Si I as reference, where in each case the levels get less and less LS-coupling character. The multiplets indicated in red show increasing values of n for the s-electron; the green ones progressively higher ionisation stages within an isoelectronic series and the blue points, for which we get quite close to good jj-coupling, for incrementally higher Z.

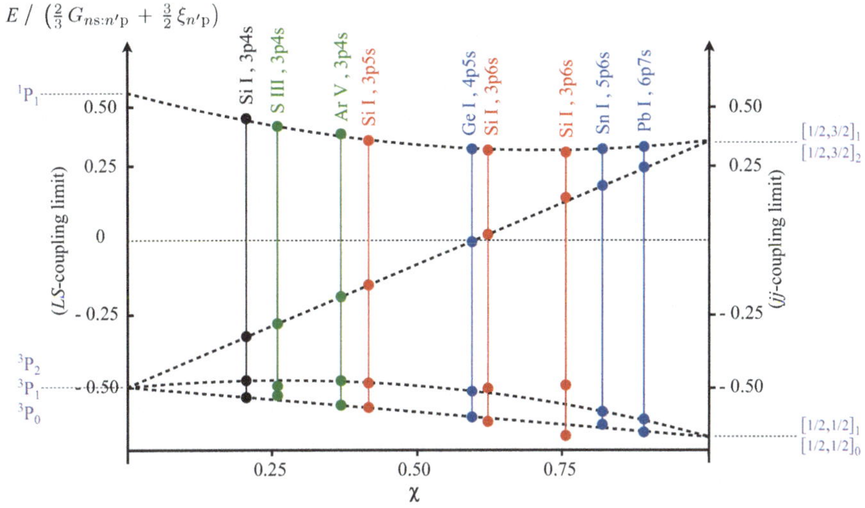

Fig. 9.6 Diagram showing a gradual transition from a pure *LS*-coupling situation to *jj*-coupling for a nsn'p configuration, as a function of the parameter χ — see (9.31) — which indicates the relative importance of the Hamiltonians H_{ee} and H_{SO}. The energies refer to differential ones: $E = E_{level} - [I(ns) + I(n'p) + F^{(0)}_{ns:n'p}]$. In the figure is included experimental values [4] for some multiplets that demonstrate various trends (see the text). In order to include these values in the figure, the values of the G-integral and the radial fine-structure function have for each element been fitted to empirical data. The fact that for some elements — notably Si I, 3p6s — even the best fit deviates from the lines is an effect of configuration interaction (see Sect. 9.5). That multiplet cannot be accurately described by a single configuration. Note: this figure is inspired by a similar one in chapter 10 of [5].

We will give one more example — for the equivalent electron configuration p^2. There are five energy levels to determine in the intermediate range, and in *LS*-coupling and *jj*-coupling they are:

$$LS : \begin{cases} np^2\ {}^1S_0 \\ np^2\ {}^1D_2 \\ np^2\ {}^3P_2 \\ np^2\ {}^3P_1 \\ np^2\ {}^3P_0 \end{cases} , \qquad jj : \begin{cases} [np^2_{3/2}]_0 \\ [np^2_{3/2}]_2 \\ [ns_{3/2}\, np_{1/2}]_2 \\ [ns_{3/2}\, np_{1/2}]_1 \\ [np^2_{1/2}]_0 \end{cases} . \tag{9.35}$$

The energies in the extreme cases have been determined in (7.34), (7.36), (7.39), (8.6) and (8.7). Using the same principles for determining H_{SO} in the $|LSJM_J\rangle$ representation as was applied for the sp-configuration in the previous paragraphs, the full Hamiltonian matrix can be determined to be:

$$H_{\rm to} + H_{\rm SO} = \left[2I(n\mathrm{p}) + F^{(0)}_{n\mathrm{p}:n\mathrm{p}}\right] \mathbb{1} + \frac{F^{(2)}_{n\mathrm{p}:n\mathrm{p}}}{25} \begin{pmatrix} 1 & 0 & 0 & 0 & 0 \\ 0 & -5 & 0 & 0 & 0 \\ 0 & 0 & -5 & 0 & 0 \\ 0 & 0 & 0 & -5 & 0 \\ 0 & 0 & 0 & 0 & 10 \end{pmatrix}$$

$$+ \frac{\xi_{n\mathrm{p}}}{12\sqrt{2}} \begin{pmatrix} 2\sqrt{2} & 10 & 0 & 0 & 0 \\ 10 & 7\sqrt{2} & 0 & 0 & 0 \\ 0 & 0 & -6\sqrt{2} & 0 & 0 \\ 0 & 0 & 0 & -12\sqrt{2} & -24 \\ 0 & 0 & 0 & -24 & 0 \end{pmatrix} , \tag{9.36}$$

with rows and columns ordered in the sequence of: $^1\mathrm{D}_2$, $^3\mathrm{P}_2$, $^3\mathrm{P}_1$, $^3\mathrm{P}_0$ and $^1\mathrm{S}_0$. This leaves us with the task of diagonalising two 2×2 matrices — for $J = 2$ and for $J = 0$. We show the result of a diagonalisation of this matrix in figure 9.7. The energies of the fine-structure levels are also here shown as a function of a transition parameter, which here is:

$$\chi = \frac{3\,\xi_{n\mathrm{p}}}{\frac{3}{5}\,F^{(2)}_{n\mathrm{p}:n\mathrm{p}} + 3\,\xi_{n\mathrm{p}}} . \tag{9.37}$$

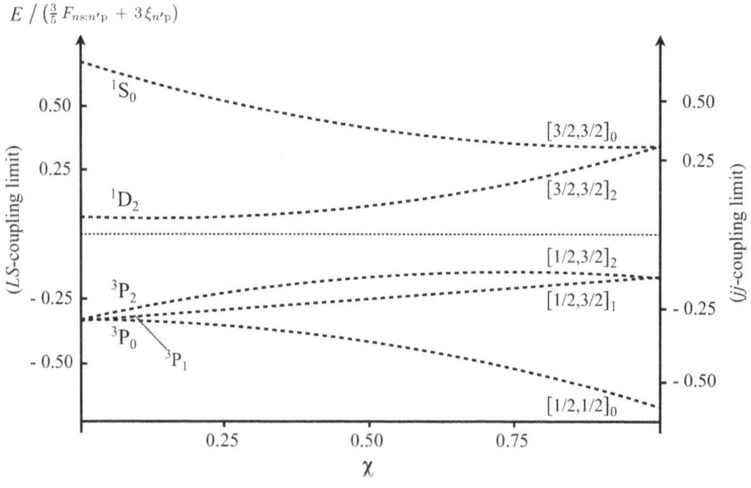

Fig. 9.7 Diagram showing a gradual transition from a pure *LS*-coupling situation to *jj*-coupling for a $nsn'\mathrm{p}$ configuration, as a function of the parameter χ — see (9.37). The energies refer to differential ones: $E = E_{\rm level} - [2I(n\mathrm{p}) + F^{(0)}_{n\mathrm{p}:n\mathrm{p}}]$.

9.5 Configuration Interaction

When we have studied *LS*-coupling and *jj*-coupling in previous parts of the book, and also while discussing schemes such as J_1K-coupling and *LK*-coupling, we have consistently assumed that we can rely on the assumption that the central-field approximation holds. Another way to state this is that all the off-diagonal matrix elements of the perturbation operators H_{ee} and H_{SO} *between* wave functions representing different electronic configurations are taken as zero (as well as those representing interactions with the nucleus and with external fields). In reality, there are many occurrences where this is not the case.

In chapter 3, when we discussed the variational method for finding approximate atomic wave functions and eigenenergies for light atoms, we were already at that point taking an approximation route that does not involve the CFA. For example, if we look at the Hylleraas ansatz of (3.30), the wave function depends explicitly on the interelectronic separation, r_{12}. When this is the case, the CFA cannot accurately describe the state of the atom, and the wave function is not factorisable.

A glance at a table of atomic energy levels will reveal that there are several examples of overlapping electronic configurations. Albeit they may all contain atomic terms that seem well described by *LS*-coupling, energy separations between terms for such atoms are of the same order of, or greater than, that those between the average energies of the configurations — a predicament that seems to invalidate the initial steps in our generally applied perturbative treatment. This does not mean that the CFA and the *LS*-coupling approximation are useless in these instances, but rather that they are incomplete. Describing the state corresponding to an energy level with the standard *LS*-coupling quantum numbers may still be pertinent, and the Landé interval rule may hold quite well, even if the actual term energies are not simple functions of Slater F- and G-integrals, as was assumed in the calculations in section 7.2.

Rather than giving up perturbation theory as the method of choice, we have to adapt it to the situation. As a correction to the CFA, we can treat an interaction between two levels of different configurations as a first level of perturbation. The meaning of the term ‘interaction’ in this context is that the matrix element of the valence electron Hamiltonian H_{val}, defined as in (7.46), has non-vanishing matrix elements of the type:

$$\langle \gamma_{\mathrm{a}} L_{\mathrm{a}} S_{\mathrm{a}} J_{\mathrm{a}} M_{J\mathrm{a}} \,|\, H_{\mathrm{val}} \,|\, \gamma_{\mathrm{b}} L_{\mathrm{b}} S_{\mathrm{b}} J_{\mathrm{b}} M_{J\mathrm{b}} \rangle \equiv U_{\mathrm{ab}} \neq 0\,. \tag{9.38}$$

The integral above is represented in an *LS*-coupling fashion, but any other representation could have been used. The relevant feature is that the configurations are unequal, $\gamma_{\mathrm{a}} \neq \gamma_{\mathrm{b}}$.

Since it is a purely non-diagonal perturbation, we will have to use second-order perturbation theory. For a case with just two interacting energy levels, this means that their energy corrections will be:

$$\Delta E_{\rm a} = \frac{|U_{\rm ab}|^2}{E_{\rm a}^{(0)} - E_{\rm b}^{(0)}} < 0$$
$$\Delta E_{\rm b} = \frac{|U_{\rm ab}|^2}{E_{\rm b}^{(0)} - E_{\rm a}^{(0)}} > 0 \ , \tag{9.39}$$

where the quantities in the denominators are the unperturbed energies (with $E_{\rm a}^{(0)} < E_{\rm b}^{(0)}$). From this we can conclude that the energies of two interacting levels will be pushed apart, and that the perturbed wave functions will be a mixture of the unperturbed ones. In the upcoming sections, we will look closer into how configuration interaction may affect energy levels, and we will give a couple of concrete examples.

9.5.1 Analysing the Non-Diagonal Matrix Elements

It is reasonable to describe an atomic wave function as an expansion of electronic configuration basis functions. The way in which we have used the CFA so far in this book is essentially to truncate such an expansion to just *one* electronic configuration. The meaning of configuration interaction is that at least two configurations must be taken into account in the expansion.

Consider the matrix schematically depicted in figure 9.8. We assume this to be the combined Hamiltonian operator of (7.46):

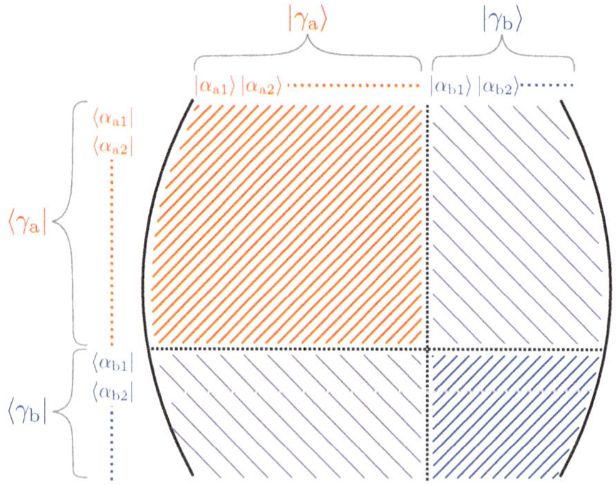

Fig. 9.8 Pictorial description of a configuration interaction matrix for two interacting electronic configurations, $\gamma_{\rm a}$ and $\gamma_{\rm b}$. The operator illustrated is the Hamiltonian $H_{\rm val}$ in (9.40). The red and blue regions are the parts that are diagonal in the respective electronic configurations, and the purple sections contain the off-diagonal elements.

$$H_{\text{val}} = H_{\text{kin}} + H_{\text{nu}} + H_{\text{ee}} + H_{\text{SO}} \; . \tag{9.40}$$

The coupling scheme could be either *LS*-coupling or *jj*-coupling, or something else. In this concrete example, we will consider just two different valence electron configurations: γ_{a} and γ_{b}.

Each element in the matrix may contain contributions such as single-electron integrals $I(nl)$, Slater integrals $R^{(k)}(\text{ab:cd})$ of the two-electron operator $1/r_{12}$ and radial fine-structure functions ξ_{nl} — see (7.16), (7.24) and (7.52). More terms appear if the ensemble of closed orbitals for the two configurations are different. However, in this first example we shall exclude this latter possibility.

The diagonal sub-matrices in figure 9.8, shaded in blue and red, are what we have previously considered, in chapters 7 and 8. What is new is that we will now take into account also the areas shaded in purple — the non-diagonal sub-matrices. As a result, the resulting secular equations will be considerably more complex.

In order to simplify the task at hand, we should consider if there are non-diagonal elements, or terms therein, that can be excluded — either because they are small or because they cancel. To begin with, we can see from (9.39) that we need only include configurations that have eigenenergies that are similar — essentially with overlapping energy spectra. This means that we can limit the size of the secular problems considerably. Additionally, the Hamiltonian H_{val} is invariant with inversion of Cartesian coordinates. This is the same thing as saying that the operator has even parity, and as a consequence, any matrix element between two states of opposite parity will be zero.

As a next step, we look specifically at the electron–electron interaction, H_{ee}. An expansion of matrix elements of this will give terms containing two-electron integrals, but any non-diagonal element between configurations that have three or more electron orbitals that are different will be zero. Furthermore, H_{ee} commutes with $\mathbf{L}^2$ and $\mathbf{S}^2$, which means that in the *LS*-coupling scheme we can also eliminate matrix elements between terms with different values of the combination *LS*.

The integrals of the configuration interaction contribution of H_{ee} will not be as symmetric as for the diagonal terms. Concretely, this means that we cannot take the simplified forms of the Slater integrals — the F- and G-integrals introduced in section 7.2.3. We have to instead take the more general formulation of (7.24), and we have to take into account how to calculate a Slater determinant matrix element for a two-electron operator, that is, we need to follow the rules laid out in (2.37) and (2.38). The matrix elements will be as in (7.23), and the two kinds of radial integrals that will appear in one sum are as in (7.24):

$$R^{(k)}(n_i l_i n_j l_j : n_p l_p n_q l_q) = \int_0^\infty \int_0^\infty R^*_{n_i l_i}(r_1)\, R^*_{n_j l_j}(r_2)\;\; R_{n_p l_p}(r_1)\, R_{n_q l_q}(r_2) \\ \times \frac{(r_<)^k}{(r_>)^{k+1}}\, r_1^2 r_2^2\, \mathrm{d}r_1 \mathrm{d}r_2 \quad , \tag{9.41}$$

and

$$R^{(k)}(n_il_in_jl_j:n_ql_qn_pl_p) = \int_0^\infty\int_0^\infty R^*_{n_il_i}(r_1)\,R^*_{n_jl_j}(r_2)\;R_{n_ql_q}(r_1)\,R_{n_pl_p}(r_2) \times \frac{(r_<)^k}{(r_>)^{k+1}}\,r_1^2r_2^2\,\mathrm{d}r_1\mathrm{d}r_2 \quad . \tag{9.42}$$

In these equations, we have used the indices i and j for the electrons in configuration γ_a, and p and q for γ_b. Note the order of the indices in the two equations.

The other components of H_val, apart from H_ee, are all one-electron operators. This means that matrix elements of these Hamiltonians can only be non-zero if the two involved electron configurations differ with at most one electron. The two contributions H_kin and H_nu will not result in a simple additional term such as (7.16). Rather, finite non-diagonal perturbations from these Hamiltonians must be calculated with (7.14) and (7.15). The spin–orbit Hamiltonian typically contributes very little to configuration interaction, except for very heavy atoms. When it needs to be quantified, (7.52) can still be used, except that the radial fine-structure function must be modified. We need to introduce a *configuration-interaction fine-structure factor*, defined as:

$$\xi_{n_\mathrm{a}l_\mathrm{a}:n_\mathrm{b}l_\mathrm{b}} \equiv \langle\,\gamma_\mathrm{a}\,|\,\xi(r)\,|\,\gamma_\mathrm{b}\,\rangle = \int_0^\infty R^*_{n_\mathrm{a}l_\mathrm{a}}(r)\,\xi(r)\,R_{n_\mathrm{b}l_\mathrm{b}}(r)\,r^2\,\mathrm{d}r\,. \tag{9.43}$$

From the Cauchy-Schwarz inequality, one can find that this factor relates to the individual orbital radial fine-structure function factors as:

$$|\xi_{n_\mathrm{a}l_\mathrm{a}:n_\mathrm{b}l_\mathrm{b}}|^2 \leq |\xi_{n_\mathrm{a}l_\mathrm{a}}|\,|\xi_{n_\mathrm{b}l_\mathrm{b}}| \quad , \tag{9.44}$$

and frequently it is a decent approximation to take the equality as a first-order estimate [5].

9.5.2 An Example of Configuration Interaction

To give a concrete example of how configuration interaction may influence the energy level spectrum of an atom, we will look at some of the lower excited configurations in neutral Mg. In figure 9.9 we show a partial Grotrian diagram, with empirically obtained energies [4].

The first thing to note in the figure is that the energy orders of the configurations, and of the terms within the lowest of them, are the ones we could expect from the aufbau principle and Hund's rules. However, when we reach the states included in the configurations 3s3d and 3s3p, we see that these overlap in energy. At a closer look, another anomaly is that for 3s3d, ^{1}D has a considerably lower energy than ^{3}D,

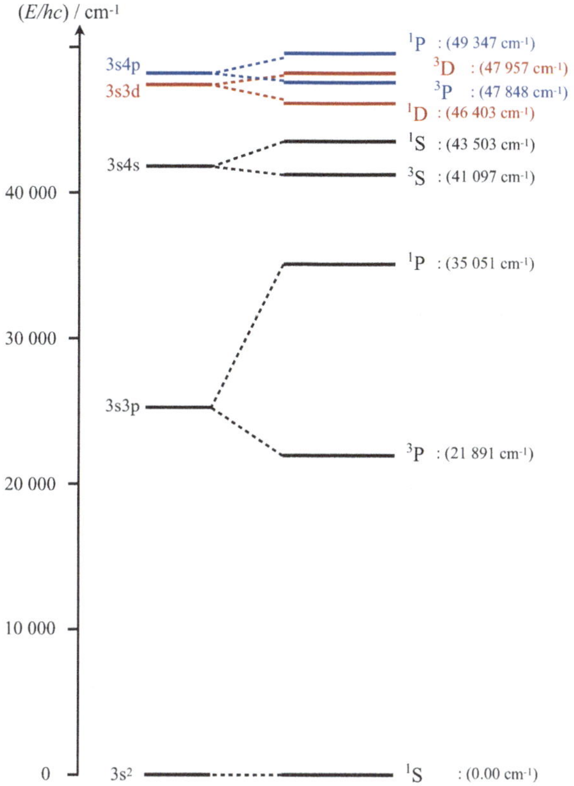

Fig. 9.9 Energy level diagram for the lower states of neutral Mg [4]. The ground state energy has been taken as zero. The three lowest electronic configurations follow an expected progression. However, the respective two terms of the configurations 3s3d (red in the figure) and 3s4p (blue) have overlapping energy ranges. Note also that for 3s3d, the singlet state is lower in energy than the triplet, contrary to Hund's rule.

contrary to the more common energy ordering of triplets and singlets. These deviations from first-order predictions are consequences of configuration interaction.

In order to understand this, we use the methods described in section 7.2 and calculate the energy contributions of H_{ee} for the levels 3s3d ^{1}D and 3s3d ^{3}D, to start with while ignoring configuration interaction. The relevant energies are found to be:

$$\begin{aligned} E^{(LS)}(3\text{s}3\text{d}\ ^1\text{D}) &= I(3\text{s}) + I(3\text{d}) + F^{(0)}_{3\text{s}:3\text{d}} + \frac{1}{5}G^{(2)}_{3\text{s}:3\text{d}} \\ E^{(LS)}(3\text{s}3\text{d}\ ^3\text{D}) &= I(3\text{s}) + I(3\text{d}) + F^{(0)}_{3\text{s}:3\text{d}} - \frac{1}{5}G^{(2)}_{3\text{s}:3\text{d}} \, . \end{aligned} \tag{9.45}$$

That is, the separation between these two terms is predicted to be $\frac{2}{5}\,G^{(2)}_{3\text{s}:3\text{d}}$, and if the G-integral is positive, the term ^{1}D should be higher in energy. We do not know the exact radial wave functions of the shielded electron orbitals. However, with Hartree–

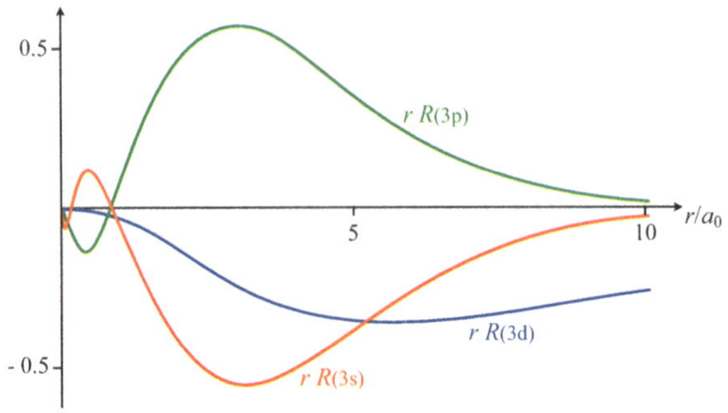

Fig. 9.10 Numerically computed radial wave functions (based on an analysis in [1]) for the outer electron orbitals 3s, 3p and 3d of neutral magnesium. The graph shows the functions $rR(3s)$, $rR(3p)$ and $rR(3d)$ computed in atomic units.

Fock methods (see chapter 14) it is possible to attain approximative wave functions, which in turn enables estimates of the integral. For more details of this calculation, we refer to other works, for example [3, 6], or in particular [7].

We have borrowed the results of numerical calculations of the radial functions from the works cited above, and in figure 9.10, we show a result of such a calculation. This reveals a large overlap for the electron wave functions for the orbitals 3s and 3d for the Mg atom. That has the effect that the G-integral will return a rather large value, and according to (9.45) the energy separation between ^{1}D and ^{3}D will thus be significant. A second lesson from the graph of radial functions is that the two functions $R(3s)$ and $R(3d)$ have the same sign for almost all values of r where they have any significant amplitude. That will give the G-integral a positive value, which in turns means that the triplet should have lower energy than the singlet — in keeping with Hund's rule, but in contradiction to the data.

To identify the cause for these departures from first-order CFA predictions, we should try to trace configurations that may interact with 3s3d. We can exclude 3s3p. It has opposite parity to 3s3d, and thus it cannot interact. Magnesium is a light atom, so we can assume the spin–orbit interaction to be small. The more likely Hamiltonian responsible for the interaction is rather H_{ee}. That means that we can also exclude all configurations that lack the terms ^{1}D and ^{3}D — such as all 3sns and 3sng. Interactions with configurations 3snd, for $n \geq 4$ will not cancel, but should give only a small contribution because of the small overlap with the electronic wave functions with $n \geq 4$.

Remaining as a likely culprit is the doubly excited state 3p^2. This should have a rather high energy, and from (9.39) one could then expect this perturbation to be small since the denominator will be large. The one thing that could offset that, however, is a large numerator. The latter will occur if the 3p electronic orbital has a significant overlap with 3s and/or 3d. We can see in figure 9.10 that this is indeed the case.

A 3p^2 configuration has three possible terms in the LS-coupling scheme: ^{1}S, ^{3}P and ^{1}D. Since H_{ee} only couples like terms, this will result in an interaction with

^{1}D of 3s3d, but not with ^{3}D. According to (9.39), configuration interaction pushes energies apart, and it is indeed this interaction that makes the term energy of 3s3d ^{1}D drop to a value lower than that of 3s3d ^{3}D.

A quantitative estimate of the effect becomes cumbersome since we cannot take advantage of the diagonal-sum rule when specifically calculating non-diagonal matrix elements. The pertinent Slater determinant matrix elements should first be calculated in a $|m_{l1}m_{s1}m_{l2}m_{s2}\rangle$ representation. The resulting matrix must then be transformed to *LS*-coupling representation. This way, one obtains the non-diagonal matrix element between 3s3d ^{1}D and 3p^2 ^{1}D. A solution of the secular equation gives an energy, which is indeed below 3s3d ^{3}D (see for example [1]).

Accounting for the strong interaction with the p^2 configuration, it is clear that the energy level that we call 3s3d ^{1}D is in reality a superposition. The label used in tables only represents the leading term of the expansion. The energy spectrum of neutral Mg is in fact one that has a high degree of configuration mixture. For example, of the levels shown in figure 9.9, the level 3s4p ^{3}P interacts with 3p3d ^{3}P, 3s3d ^{3}D with 3p4f ^{3}D and 3s4p ^{1}P with 3p3d ^{1}P.

9.6 The Difficulties with Assigning Pure State Designations

In a literal sense, there is no such thing as a *pure* angular momentum coupling scheme for a multielectron atom. All the vector arrangements that we have introduced are based on perturbation theory, where in each successive stage of the calculation all (or most) interactions except one are ignored. Added to that is the simplification of typically taking a spherically symmetric time average of the influence of the inner, closed orbital electrons.

When we chose the order in which to couple the angular momentum vectors, we make the approximation that the corresponding representation will leave us with an approximately diagonal Hamiltonian. In reality, there are non-diagonal matrix elements that are not strictly zero in most schemes. If we take the approach that we will simply diagonalise the Hamiltonian, regardless of how complicated it is, then we could in principle formulate the problem in any representation. However, in a grand majority of concrete examples, we will be left with a problem that is unreasonably complicated, and where even an approximate numerical solution might not provide much physical insight.

What we do when we choose a coupling scheme is to decide on the least objectionable option among the various seminal representations, and the true energy levels are then in reality superposition states. When a coupling is 'good' for a certain energy, what this means is that one basis function component in a decomposition is strongly dominant. There are also many cases where no alternative is very good, but *LS*-coupling notation is used nevertheless. This latter choice is made because of the well developed and well-known standards in *LS*-coupling nomenclature, and also because of its compactness. If this is done, one must take great care to note that the true state is strongly mixed in the chosen representation. In some cases, the different angular momentum couplings are so intricate that the only thing that one can do is to label the energy levels with serial numbers.

When the connection is done between a certain quantum number designation and the leading term of a superposition state in some chosen basis, the concept of purity in a designation becomes pertinent. The contribution of each state $|LSJM_J\rangle$ — to take an example with LS-coupling — to a measured energy level is the modulus square of its corresponding coefficient in an expansion over the basis. At this point, it is important not to confuse an energy level assigned to a certain coupling state with the actual basis state. The latter will just be the leading term, albeit in the best cases with an expansion coefficient close to unity.

Making an assignment based only on the leading term can sometimes lead to imprecise designations. An instructive example of this problem is an $np\,n'p'$-configuration, with $p \neq p'$ (see [5] for a more detailed analysis). In LS-coupling notation, the ten possible fine-structure levels are:

$$\begin{array}{lclcl} np\,n'p'\ {}^1S_0 & , & np\,n'p'\ {}^3S_1 & , & np\,n'p'\ {}^1P_1 \\ np\,n'p'\ {}^3P_{2,1,0} & , & np\,n'p'\ {}^1D_2 & , & np\,n'p'\ {}^3D_{3,2,1}\ , \end{array} \tag{9.46}$$

while in jj-coupling, the corresponding states are:

$$\begin{array}{lcl} [np_{1/2}\,n'p'_{1/2}]_{1,0} & , & [np_{1/2}\,n'p'_{3/2}]_{2,1} \\ [np_{3/2}\,n'p'_{1/2}]_{2,1} & , & [np_{3/2}\,n'p'_{3/2}]_{3,2,1,0}\ . \end{array} \tag{9.47}$$

The quantum number that is preserved in a transformation between the two schemes is J since the operator $\mathbf{J}^2$ commutes with all parts of the complete Hamiltonian. There are four states with $J = 1$. For these, we may derive the transformation matrix between LS-coupling and jj-coupling from (9.27). The result is:

$$\begin{array}{c} \\ {}^3S_1 \\ {}^1P_1 \\ {}^3P_1 \\ {}^3D_1 \end{array} \begin{array}{c} \begin{array}{cccc} [1/2\ 1/2]_1 & [1/2\ 3/2]_1 & [3/2\ 1/2]_1 & [3/2\ 3/2]_1 \end{array} \\ \begin{pmatrix} -\frac{1}{3}\sqrt{\frac{1}{3}} & \frac{2}{3}\sqrt{\frac{2}{3}} & -\frac{2}{3}\sqrt{\frac{2}{3}} & \frac{1}{3}\sqrt{\frac{10}{3}} \\ \frac{\sqrt{2}}{3} & -\frac{1}{3} & \frac{1}{3} & \frac{\sqrt{5}}{3} \\ 0 & -\frac{1}{\sqrt{2}} & -\frac{1}{\sqrt{2}} & 0 \\ \frac{2}{3}\sqrt{\frac{5}{3}} & \frac{1}{3}\sqrt{\frac{5}{6}} & -\frac{1}{3}\sqrt{\frac{5}{6}} & -\frac{1}{3}\sqrt{\frac{2}{3}} \end{pmatrix} \end{array} . \tag{9.48}$$

This means that if the configuration is almost 100% jj-coupled, and we describe the energy levels as LS-coupling states according to the leading term, the result will appear as in table 9.1. As can be seen, ${}^3S_1$ is not leading term for any of the jj-coupling states, but is well represented in at least three of them. In this example, an LS-coupling coupling description does not make much sense, and even if the coupling is rather intermediate, it needs to be described with some prudence.

Table 9.1 Leading and second most important terms in an LS-coupling decomposition of the $J = 1$ fine-structure levels of a jj-coupled $np\,n'p'$-configuration (see text).

jj-coupling level	Leading term in a LS-coupling basis	%	next term	%
$[1/2\ 1/2]_1$:	$^3D_1$	74	$^1P_1$	22
$[1/2\ 3/2]_1$:	$^3P_1$	50	$^3S_1$	30
$[3/2\ 1/2]_1$:	$^3P_1$	50	$^3S_1$	30
$[3/2\ 3/2]_1$:	$^1P_1$	56	$^3S_1$	37

The example illustrates the potential pitfalls with energy level designations. Nevertheless, attempts to disentangle parts of the angular momentum couplings for a multielectron atom by vector coupling schemes have been proven to be a useful tool in spectroscopy. It can greatly help in the identification of spectral lines and is also of utility for numerical computations of energies and wave functions.

Further Reading

Quantum theory of atomic structure, by Slater [1]
The theory of atomic spectra, by Condon & Shortley [3]
The theory of atomic structure and spectra, by Cowan [5]
Atomic spectra, by Kuhn [8]
Atomic Spectra and Radiative Transitions, by Sobelman [9]
Atomic Many-Body Theory, by Lindgren & Morrison [10]

References

1. J.C. Slater, *Quantum theory of atomic structure* (McGraw-Hill, New York, 1960)
2. H.A. Bethe, E.E. Salpeter, *Quantum Mechanics of One- and Two-Electron Atoms* (Springer-Verlag, Berlin, 1957)
3. E.U. Condon, G.H. Shortley, *The theory of atomic spectra* (Cambridge University Press, Cambridge, 1935)
4. A. Kramida, Y. Ralchenko, J. Reader, and NIST ASD Team. NIST Atomic Spectra Database (ver. 5.3). [Online]. Available: http://physics.nist.gov/asd (2018). Accessed: 2019-07-14
5. R.D. Cowan, *The theory of atomic structure and spectra* (University of California press, Berkeley, 1981)
6. J.C. Slater, Phys. Rev. **42**, 33 (1932)
7. R.F. Bacher, Phys. Rev. **43**, 264 (1933)
8. H.G. Kuhn, *Atomic spectra* (Longmans, London, 1969)
9. I.I. Sobelman, *Atomic Spectra and Radiative Transitions*, 2nd edn. (Springer, Berlin, 1992)
10. I. Lindgren, J. Morrison, *Atomic Many-Body Theory*, 2nd edn. (Springer Verlag, Berlin, 1986)

Chapter 10
Nuclear Effects

In the first nine chapters of this book, the atomic nucleus has been treated as a point charge with infinite mass. This has sufficed for a decent understanding of atomic structure, but we have now reached a point where we cannot go much further without taking into account some of the real physical qualities of the atomic nuclei.

The atomic nucleus is in itself a composite particle, built up by two different kinds of hadrons — neutrons and protons — that are held together by the nuclear force (nucleon–nucleon interaction). Its mass is finite, as is its volume, and its shape (or charge distribution) is generally not completely spherical.

Many nuclei have a measurable magnetic moment, which we associate with a nuclear spin, in analogy with the electron's magnetic moment and the electron spin. More generally, nuclei have both electric and magnetic moments of various orders.

The finite mass and the finite volume of the nucleus will change the atomic Schrödinger equation, and the consequence is that energy levels are slightly shifted. This is the *isotope shift*, and it will be dealt with in section 10.1. One consequence of the isotope shift is that there are detectable spectroscopic differences between different mass isotopes of the same element.

The *hyperfine structure* is the splitting up of fine-structure levels due to interactions between electrons and electromagnetic moments of the nucleus. It is considerably more complex than the isotope shift, both conceptually and in its effects on atomic structure and atomic spectra. In section 10.2, we will cover the lowest moments of the hyperfine structure, which typically is enough to explain measurable effects.

10.1 Isotope Shift

In this section, we will treat three different effects. The first one is the *reduced mass effect*, which is sometimes taken as synonymous with the isotope shift. However, the fact that the mass is not infinite also gives rise to more subtle perturbation, which

A. Kastberg, *Structure of Multielectron Atoms*, Springer Series on Atomic, Optical, and Plasma Physics 112, https://doi.org/10.1007/978-3-030-36420-5_10

was hinted on already in chapter 2 — the *mass polarisation effect*. Finally, we will look into the *volume effect* — the correction to the energy levels due to the fact that the nucleus is not a point charge. All of these physical phenomena result in differences in the spectra from different isotopes of the same element.

10.1.1 The Reduced Mass Effect

Already in (1.8), in chapter 1, we made the approximation $\mu \approx m_\mathrm{e}$ in the Schrödinger equation for the hydrogen atom. This simplification slightly changes the Hamiltonian, and if we had taken the reduced mass properly into account, the energy levels that we would have found would have been different. However, this is clearly a small correction.

The kinetic energy term in the atomic Hamiltonian is a single-electron operator, and thus it will appear in the same way for a multielectron atom as for hydrogen. All energy levels in the atom will be modified, compared to the model with the infinitely heavy nucleus, as:

$$E_\mu = \frac{\mu}{m_\mathrm{e}} E_\infty \ , \tag{10.1}$$

with:

$$\mu = \frac{m_\mathrm{e} M_\mathrm{n}}{m_\mathrm{e} + M_\mathrm{n}} \ . \tag{10.2}$$

M_n is the mass of the nucleus in question.

To the first order in $m_\mathrm{e}/M_\mathrm{n}$, the corrective factor in (10.1) can be approximated as:

$$\frac{\mu}{m_\mathrm{e}} \approx 1 - \frac{m_\mathrm{e}}{M_\mathrm{n}} \ , \tag{10.3}$$

which means that the correction to the energy can be written as:

$$E_\mu - E_\infty \approx -\frac{m_\mathrm{e}}{M_\mathrm{n}} E_\infty \ . \tag{10.4}$$

The energy is negative for bound states, and therefore (10.4) shows that the reduced mass effect gives a positive contribution to the eigenenergy. Stated in another way, the binding energy is in reality slightly weaker than what one calculates if one assumes an infinitely heavy nucleus. For an atom, this means that different isotopes of the same atom will have slightly different energy levels, and this will be reflected in resolved emission and absorption spectra.

10.1.2 The Mass Polarisation Effect

We encountered the mass polarisation term of the atomic Hamiltonian in (2.6). It is a kinetic energy cross term that emerges in multi-body systems. In CoM coordinates, it corresponds to internal mass re-distributions needed in order to account for conservation of energy. In chapter 2, we immediately neglected the term, due to the fact that it is divided by the nuclear mass, and thus it is always small.

For an N-electron atom, the mass polarisation Hamiltonian is:

$$H_{\text{mpe}} = -\frac{1}{M_{\text{n}}} \sum_{i=1}^{N} \sum_{j>i}^{N} \nabla_{\mathbf{r}_i} \cdot \nabla_{\mathbf{r}_j} \quad , \tag{10.5}$$

where the summation is over all of the electrons of the atom. It is clear that this is irrelevant for one-electron atoms, and with M_{n} in the denominator, it will be sufficient to treat the correction with perturbation theory. Energy levels should be adjusted by the amount:

$$\begin{aligned} E_{\text{mpe}} &= -\frac{1}{M_{\text{n}}} \left\langle \psi^{(0)} \middle| \sum_{i=1}^{N} \sum_{j>i}^{N} \nabla_{\mathbf{r}_i} \cdot \nabla_{\mathbf{r}_j} \middle| \psi^{(0)} \right\rangle \\ &= \frac{1}{M_{\text{n}}} \left\langle \psi^{(0)} \middle| \sum_{i=1}^{N} \sum_{j>i}^{N} \mathbf{p}_i \cdot \mathbf{p}_j \middle| \psi^{(0)} \right\rangle \quad , \end{aligned} \tag{10.6}$$

where $\psi^{(0)}$ is the zero-order wave function.

In the last line of (10.6), we have identified the momentum operators for electrons i and j. Each term of this Hamiltonian can be factorised in electron coordinates, and if $\psi^{(0)}$ is a CFA wave function, it is also separable. The expectation value of the momentum operator for a stationary state is zero, and thus it may appear as if $\Delta E_{\text{mpe}} = 0$. However, even in the CFA, the correct zero-order wave function should be a Slater determinant, and with the operator affecting two electrons for every term in the sum, the matrix elements should be derived with the rule in (2.36), which will produce a different result.

Typically, a reasonably accurate calculation is only available for the smallest atoms, in which case a variational type wave function is used as unperturbed function. For He, and ions isoelectronic with He, the mass polarisation effect has been found to be of the order of 5 meV [1]. For a detailed account of how the mass polarisation effect enters in variational theory calculations, see [2].

10.1.3 The Volume Effect

The nuclear Coulomb potential hitherto used in the book is strictly inversely proportional to r_i. With a nucleus of finite volume, there will be a lower limit for the electronic radius, inside which the $1/r_i$ assumption will no longer be valid. The precise functional behaviour of the potential within the nuclear volume will then depend on the distribution of charges in the nucleus.

The effect of this on the electronic binding energies will be small. It is justified to use a greatly simplified model for the nuclear charge distribution for a rough estimate of the qualitative features of the volume effect. The model we will use is a spherical nucleus of radius r_n, with a sharply defined surface and within which the positive charge is uniformly distributed. For r_n, we take [1]:

$$r_\mathrm{n} = r_0 A^{1/3} \ , \tag{10.7}$$

where $r_0 \approx 1.2\times 10^{-15}\,\mathrm{m} \approx 2.3\times 10^{-5}\, a_0$, and A is the mass number of the isotope in question (the number of nucleons). A non-uniform, and non-spherical, charge distribution would give rise to higher order electric moments. The latter is an important contribution to hyperfine structure, which will be dealt with in section 10.2. The volume isotope shift, which is treated in this section, only concerns the deviation of the $1/r_i$ potential as $r \to 0$.

We model the potential by:

$$V(\mathbf{r}_i) = \begin{cases} \dfrac{Z}{2r_\mathrm{n}}\left(\dfrac{r_i^2}{r_\mathrm{n}^2} - 3\right) & \text{for } r_i \le r_\mathrm{n} \\[2ex] -\dfrac{Z}{r_i} & \text{for } r_i > r_\mathrm{n} \ , \end{cases} \tag{10.8}$$

with r_i and r_n in a.u. This results in a discontinuous first derivative at r_n. A qualitative conclusion that can be drawn immediately is that, at least to the first order, this will especially affect s-states. For $l \ge 1$, the probability amplitude for $r \to 0$ rapidly diminishes. However, even with this simplification, the potential in (10.8) is still one that does not provide easy solutions to the Schrödinger equation. In order to proceed, and to get some more understanding of the effect, we shall make two more simplifications.

To start with, we take the difference between the Coulomb potential for the point-sized nucleus and (10.8) as a perturbation; this is a very benign compromise. More serious is that we will take the hydrogenic wave functions (with $l = 0$) as the zero-order solution.

For $r_i > r_\mathrm{n}$, the wave function is unperturbed. For small r_i, the perturbation Hamiltonian is:

$$H_\mathrm{vol}(r_i \le r_\mathrm{n}) = \frac{Z}{2r_\mathrm{n}}\left(\frac{2r_\mathrm{n}}{r_i} + \frac{r_i^2}{r_\mathrm{n}^2} - 3\right) . \tag{10.9}$$

The energy correction for the hydrogenic function $\psi_{n00}(\mathbf{r}_i) = R_{n0}(r_i)\, Y_{00}(\theta_i, \varphi_i)$ is then:

$$E_\mathrm{vol} = \langle\, \psi_{n00}(\mathbf{r}_i) \,|\, H_\mathrm{vol}(r_i) \,|\, \psi_{n00}(\mathbf{r}_i) \rangle \ . \tag{10.10}$$

With the wave function taken from (1.23), this integral is not difficult to solve. The result is [1]:

$$E_{\text{vol}} = \frac{2r_{\text{n}}^2 Z^4}{5n^3} \left[\frac{M_{\text{n}}}{1+M_{\text{n}}} \right]^3 . \tag{10.11}$$

Since this is in a.u., the nuclear mass, M_{n}, must be entered in units of the electron mass. From this, we can note that a greater nuclear radius gives a more important energy shift (weaker binding energy). This is precisely what could be intuitively expected, since the limit to the range of the $1/r_i$-potential extends further out. We also see a pronounced dependence on Z and n. The volume isotope effect is thus more important for heavy elements, and it diminishes for highly excited states.

10.2 Hyperfine Structure

In the preceding section, we departed from the simplification that the atomic nucleus is a point charge with infinite mass. This must now be taken some steps further. The nucleus is charged, so if we accept that the nucleus has structure, it is unavoidable to consider electric and magnetic moments of order two or larger and their interactions with the atomic electrons. This is what gives rise to *hyperfine structure* (hfs) in atomic spectra, and in atomic structure.

First, we can consider a possible electric dipole moment of the nucleus. With a relatively simple argument, based on the assumption that also the nucleus can be described with quantum mechanics, we can show that this term should be absent, at least if we ignore the possible permanent electric dipole moments of nucleons.

Any finite electric multipole moment must be related to a charge distribution that is not perfectly spherical, and the charge distribution should be proportional to the norm of the wave function. If we take the electric dipole moment as an example, a component along any given direction, say $\hat{e}_x$, must mean a non-zero value of the quantity:

$$\int \phi_{\text{n}}^*(\mathbf{r})\, x\, \phi_{\text{n}}(\mathbf{r})\, d\mathbf{r} , \tag{10.12}$$

where ϕ_{n} is the nuclear wave function. Regardless of the parity of ϕ_{n}, the integrand in (10.12) will be an odd function, and the integral will cancel when taken over all space.

The result will be the same for any direction and for any odd order electric moment. Thus, the largest electric multipole moment that can contribute to the hyperfine structure is the quadrupole one, with terms proportional to x^2, y^2, z^2, xy, yz and zx, and more generally only even order moments will be non-zero. The electric quadrupole moment hyperfine interaction will be treated in section 10.2.5. The second biggest term of this kind is the electric hexadecapole moment — with fourth order spatial contributions. This is a very small effect, and it will be briefly covered in section 10.2.8.

For the magnetic multipole moments, the reasoning has to change somewhat. For that issue, we are not looking at the distribution of charges, but rather at the magnetisation density. The magnetisation vector is $\mathcal{M}(\mathbf{r})$, and the corresponding density is the space integral over $\mathcal{M}^*(\mathbf{r})\mathcal{M}(\mathbf{r})$. According to Maxwell's laws, this will always change sign if we do the transformation $(x,y,z) \rightarrow (-x,-y,-z)$.

A magnetic dipole moment can be described as the integral of the product of the position vector **r** with the magnetisation density. In one direction, this is the entity:

$$\int \mathcal{M}^*(\mathbf{r})\mathcal{M}(\mathbf{r})\, x\, d\mathbf{r}\,, \tag{10.13}$$

and with the reasoning put forward in the preceding paragraph, this will be non-vanishing.

The extrapolation of this argument is that for magnetic hyperfine structure, it is the odd parity moments that will be relevant, and indeed the overall largest contribution to hyperfine structure will typically be the magnetic dipole interaction. We will present this in section 10.2.1, and the sections that follow. The third largest hfs-term (after the electric quadrupole effect) will be that related to the magnetic octupole moment, proportional to third orders of x, y and z. This will be discussed in section 10.2.8.

The qualitative arguments above can be put in a more stringent form, and this will partially be done in the following sections. Our treatment of the hfs will be somewhat qualitative since we will rely on a vector model representation of the interactions.

10.2.1 The Magnetic Dipole Interaction

The fact that atomic nuclei have magnetic dipole moments was initially an hypothesis derived from the observation of small additional structures in atomic spectra. Since these only showed up at high resolution, and they then appeared to play a role as a perturbation to the fine-structure, the term hyperfine structure was coined. In an analogous way to what was done for the electron, the nuclear dipole moment has been associated with a quantum mechanical spin angular momentum.

A crucial difference, however, between an electron and a nucleus is that the latter has internal structure. It is built up by neutrons and protons, and these particles, in turn, are not fundamental either. As far as atomic physics is concerned, it is however quite safe to regard protons and neutrons as particles of spin $\hbar/2$. That is, they are both fermions. As a consequence, a nucleus with an even number of nucleons will be a composite boson, and if it has an overall nuclear spin at all, it has to be integer in $\hbar$. A closer study of nuclear physics (see, for example, [3]) will reveal that if both the number of protons *and* the number of neutrons are even, the individual spins pair up in such a way as to cancel the overall nuclear spin. However, the combination of an odd number of protons and an odd number of neutrons does typically result in a finite, integer, spin. If the total number of nucleons is odd, the total nuclear spin will necessarily be non-zero and half-integer in $\hbar$, and the nucleus in question is then a composite fermion.

This way of referring to a composite nuclear spin relies on some of the same qualitative arguments that we have earlier used when we have applied the vector model to atomic structure. In the present section, we will essentially follow the vector model also for the interaction between the nuclear spin and the electron, and for most examples, the result will be quite satisfactory. A more rigorous treatment of this electromagnetic interaction is presented in appendix F.

If we accept the idea of the nuclear spin as a quantum mechanical angular momentum associated with a magnetic dipole moment, we can introduce quantum numbers for this spin in a way analogous to that used for the electronic part of the atomic structure. As a symbol for the nuclear spin vector operator, we use **I**. Its Cartesian components do not intercommute, so we will describe the spin, relative to the quantisation axis $\hat{e}_z$, with the operators $\mathbf{I}^2$ and I_z, and the corresponding eigenvalue equations are:

$$\begin{aligned} \mathbf{I}^2\,\phi_\mathrm{n} &= I(I+1)\,\phi_\mathrm{n} \\ I_z\,\phi_\mathrm{n} &= M_I\,\phi_\mathrm{n} \quad , \end{aligned} \tag{10.14}$$

with ϕ_n as the nuclear wave function. As usual, M_I can take any value between $-I$ and I, which in turn is positive and either integer, half-integer, or zero.

If the interaction between this spin and the electronic part of the wave function is taken into account, a total atomic wave function can be constructed. From this, all other aspects of the nucleus other than electromagnetic moments can be ignored as far as hyperfine structure is concerned, and since we assume that a perturbative treatment is justified, we may treat the hfs in isolation. Hence, the squared vector operator $\mathbf{I}^2$ should commute with the total Hamiltonian also under a magnetic interaction between **I** and **J**, whereas I_z and J_z will not.

In order to get further, we must also consider the magnitude of the magnetic moment that we take as a consequence of this spin. That is, we need a proportionality factor between the nuclear magnetic moment $\boldsymbol{\mu}_I$ and **I**. To sustain the analogy with the electron, we introduce the relation (in SI-units):

$$\boldsymbol{\mu}_I = \frac{g_I\,\mu_\mathrm{N}}{\hbar}\,\mathbf{I} \ . \tag{10.15}$$

In this equation, we first note that there is a difference in sign compared to (4.2). Two new quantities have also been broached in (10.15), namely g_I and μ_N. The nuclear Landé g-factor, g_I, is specific for each nucleus, and in atomic physics, we typically use empirically determined values for this. It should be noted that g_I can be either positive or negative, depending on the composition of the nucleus, but its approximate order of magnitude is unity.

A consequence of having a g_I of the order of one is that we need another physical constant in (10.15) than the Bohr magneton, to make up for the mismatch in size between a nucleus and an electron. Recall that the definition of μ_B in SI-units is:

$$\mu_\mathrm{B} = \frac{e\hbar}{2m_\mathrm{e}} \ . \tag{10.16}$$

Since this factor is inversely proportional to the mass, it is reasonable to assume that nuclear magnetic moments ought to be of the order of 2000 times smaller than the electronic one. Therefore, the definition of the *nuclear magneton* has been chosen as:

$$\mu_{\mathrm{N}} = \frac{e\hbar}{2m_{\mathrm{p}}} \quad . \tag{10.17}$$

Since this depends on the proton mass, m_{p}, the nuclear magneton is not simply a product of fundamental constants, as the case is with the Bohr magneton. If we want to write μ_{N} in atomic units, we must write:

$$\mu_{\mathrm{N}} = \frac{1}{2m_{\mathrm{p}}} , \tag{10.18}$$

and express the proton mass in units of m_{e}.

10.2.2 The Magnetic Dipole hfs Splitting and Shift

Since the interaction with the nuclear magnetic moment is so much smaller than interelectronic interactions, a perturbative treatment of the magnetic dipole hfs is justified. In what follows, we will also assume that there is no external magnetic field (that aspect will be dealt with in chapter 11). The interaction under study is thus limited to that of the nuclear magnetic moment with the effective magnetic field generated by all the angular momenta of all electrons, at the position of the nucleus.

One aspect of this is that an important difference appears between electrons in an s-orbital and ones with $l_i \neq 0$. Electrons in s-states have non-vanishing probability amplitudes close to the origin, which means that their wave functions will penetrate the nucleus. As we will see, when we assess the magnetic dipole hfs quantitatively, a consequence of this is that we must treat electrons with $l = 0$ separately.

In analogy with the analysis of the spin–orbit interaction (see chapter 4 and appendix F), we take the hfs Hamiltonian as the scalar product between the magnetic moment and the effective magnetic field at the nucleus.

$$H_{\mathrm{hfs-MD}} = -\boldsymbol{\mu}_I \cdot \mathbf{B}_J(\mathbf{r} = \mathbf{0}) \,. \tag{10.19}$$

Here, $\mathbf{B}_J(\mathbf{r} = \mathbf{0})$ is the magnetic field at the origin (the position of the nucleus) produced by the combined effect of all orbiting and spinning electrons. For filled orbitals, we may assume that the different projections of all momenta will combine to zero, and hence we can take $\mathbf{B}_J(\mathbf{r} = \mathbf{0})$ to be proportional to the total electronic angular momentum $\mathbf{J}$ resulting from all the electrons that are in open orbitals. We postpone the analysis of the factor of proportionality somewhat, but take (10.15) into account. The Hamiltonian in (10.19) can then be written as:

$$H_{\mathrm{hfs-MD}} = a_{\mathrm{hfs}}\, \mathbf{I} \cdot \mathbf{J} \quad . \tag{10.20}$$

The factor $a_{\rm hfs}$ in (10.20) is the *magnetic dipole hyperfine structure constant*, or the *hyperfine structure a-constant*. Good values of $a_{\rm hfs}$ typically require empirical data or advanced numerical methods. However, analytical estimates may give qualitative insights, and we will look into this shortly. First, we analyse (10.20) from the vector model approach.

The form of (10.20) is strikingly similar to the Hamiltonian for the spin–orbit interaction in (4.8). This is not surprising since in both cases we deal with the interaction between a magnetic dipole and a magnetic field caused by a finite electronic angular momentum. If we follow the analysis leading up to (4.11), we see that an expectation value of the operator $\mathbf{I}\cdot\mathbf{J}$ is obtainable by taking the sum of $\mathbf{I}$ and $\mathbf{J}$. We introduce the additional angular momentum vector:

$$\mathbf{F} \equiv \mathbf{I} + \mathbf{J} \ , \tag{10.21}$$

and associated quantum numbers defined by:

$$\begin{aligned} \mathbf{F}^2 \chi &= F(F+1)\chi \\ F_z \chi &= M_F \chi \ . \end{aligned} \tag{10.22}$$

In these relations, the wave function χ is that of all the bound electrons plus the electromagnetic moments of the nucleus.

The above definition of the *total atomic angular momentum* facilitates writing the energy perturbation caused by (10.20) as:

$$\begin{aligned} E_{\rm hfs-MD} &= a_{\rm hfs} \, \langle \mathbf{I}\cdot\mathbf{J} \rangle \\ &= \frac{a_{\rm hfs}}{2} \, \langle \mathbf{F}^2 - \mathbf{I}^2 - \mathbf{J}^2 \rangle \\ &= \frac{a_{\rm hfs}}{2} \, [F(F+1) - I(I+1) - J(J+1)] \ . \end{aligned} \tag{10.23}$$

The expression in (10.23) gives the energy correction to fine-structure levels designated by the quantum number J. The variety of possible relative orientations of $\mathbf{J}$ and $\mathbf{I}$ provides a splitting of the fine-structure level. The viable values of the quantum numbers F and M_F are:

$$\begin{aligned} F &= |I-J| \, , \, |I-J|+1 \, , \ldots , I+J \\ M_F &= -F \, , \ldots , F \ . \end{aligned} \tag{10.24}$$

A level J will be split up by the magnetic dipole hyperfine structure into $(2J+1)$ levels if $(I>J)$, or into $(2I+1)$ if $(J>I)$. The orientational degeneracy in M_J, which is present for the fine-structure level, is now instead transferred to the quantum number M_F, with a degeneracy level of $2F+1$.

Equation (10.23) provides us with an interval rule, analogue to the Landé one for fine-structure in the LS-coupling scheme, see (6.37). The difference in energy between two adjacent hfs levels is:

$$E_F - E_{F-1} = a_{\rm hfs} \, F \ . \tag{10.25}$$

If we consider the energy span of an entire hfs-multiplet, that is the difference in energy between the two hfs levels $F_{\max} = I + J$ and $F_{\min} = |I - J|$, this is:

$$E_{F_{\max}} - E_{F_{\min}} = \begin{cases} a_{\text{hfs}}\, I(2J+1) & \text{for } J > I \\ a_{\text{hfs}}\, J(2I+1) & \text{for } I > J \end{cases} \quad . \tag{10.26}$$

10.2.3 The hfs a-Constant

If the atomic system is simple enough (which for most practical purposes means a single valence electron), the constant a_{hfs} can be estimated analytically, and this does provide some general understanding of the phenomenon of magnetic dipole hyperfine structure.

We start by formulating an initial expression for a_{hfs}, based on (10.15) and expectation value expressions formed from (10.19) and (10.20). This gives us a definition of the a-constant as (in SI-units):

$$a_{\text{hfs}} \equiv -\frac{\langle \boldsymbol{\mu}_I \cdot \mathbf{B}_J \rangle}{\langle \mathbf{I} \cdot \mathbf{J} \rangle} = -\frac{g_I\, \mu_N \langle \mathbf{I} \cdot \mathbf{B}_J \rangle}{\hbar \langle \mathbf{I} \cdot \mathbf{J} \rangle} = -\frac{g_I\, \mu_N \left\langle \mathbf{I} \cdot (\mathbf{B}_L + \mathbf{B}_S) \right\rangle}{\hbar \langle \mathbf{I} \cdot \mathbf{J} \rangle} \quad . \tag{10.27}$$

In this equation, we take it as explicit that the magnetic field, $\mathbf{B}_J$, generated by the electrons, is that at the position of the nucleus, and that $\mathbf{J}$ is the total electronic angular momentum. The field has been taken to be the sum of the contributions from the orbital angular momenta and the spins of the electrons.

Handling operator expressions like above is somewhat mathematically frivolous, and we need to proceed with some care. The fields $\mathbf{B}_L$ and $\mathbf{B}_S$ are respectively proportional to $\mathbf{L}$ and $\mathbf{S}$, but with different factors of proportionality. Thus, the total field $\mathbf{B}_J$ is not parallel with $\mathbf{J}$. However, because of the spin–orbit coupling, $\mathbf{L}$ and $\mathbf{S}$ will precess rapidly around $\mathbf{J}$, and if we assume the hyperfine interaction to be a considerably smaller perturbation, we can approximate (10.27) by taking the time-averaged projections of $\mathbf{B}_L$ and $\mathbf{B}_S$ along $\mathbf{J}$.

In a quantum mechanical language, this means that when we calculate a_{hfs} with (10.27), we ignore included matrix elements that are not diagonal in J. With this simplification, the time-averaged expectation value of total field at the nucleus, assumed to be parallel to $\mathbf{J}$, can be written as:

$$\langle \mathbf{B}_J \rangle_{\text{average}} = \langle \mathbf{B}_L \rangle_{\text{average}} + \langle \mathbf{B}_S \rangle_{\text{average}} = \left(\frac{\langle \mathbf{B}_L \cdot \mathbf{J} \rangle}{\langle |\mathbf{J}| \rangle} + \frac{\langle \mathbf{B}_S \cdot \mathbf{J} \rangle}{\langle |\mathbf{J}| \rangle} \right) \left\langle \frac{\mathbf{J}}{|\mathbf{J}|} \right\rangle \quad , \tag{10.28}$$

which yields the following development of (10.27):

$$a_{\text{hfs}} = -\frac{g_I\, \mu_N}{\hbar} \left\langle \frac{(\mathbf{B}_L + \mathbf{B}_S) \cdot \mathbf{J}}{\mathbf{J}^2} \right\rangle \quad . \tag{10.29}$$

In chapter 11, we will treat interactions with an external magnetic field in a similar way (see section 11.1.2 and figure 11.1).

An alternative path for finding the projection of the total field $\mathbf{B}_L+\mathbf{B}_S$ along $\mathbf{J}$ is to use the projection theorem for tensor operators (a special case of the Wigner–Eckart theorem, see appendix C.6). If we settle for basis functions that are eigenvectors to $\mathbf{J}$ and J_z, a direct application of (C.110) results in:

$$\begin{aligned}&\langle LSJM_J \,|\, \mathbf{B}_L+\mathbf{B}_S \,|\, LSJM_J \rangle \\ &\quad = \frac{\langle LSJM_J \,|\, \mathbf{J}\cdot(\mathbf{B}_L+\mathbf{B}_S) \,|\, LSJM_J \rangle}{\hbar^2 J(J+1)} \left| \langle LSJM_J \,|\, \mathbf{J} \,|\, LSJM_J \rangle \right| \quad , \end{aligned} \tag{10.30}$$

analogous to 10.28. Note the theorem must be applied such that the projection is made along $\mathbf{J}$.

In order to get a manageable initial estimate of a_{hfs}, we start by considering a single electron, with the angular momenta $\mathbf{L}_i$ and $\mathbf{S}_i$. For the field $\mathbf{B}_L$, we take the classical expression from the Biot-Savart law:

$$\mathbf{B}_L = -\frac{\mu_0 e}{4\pi} \frac{1}{|r'|^3} \mathbf{v}_{\mathrm{e}} \times \mathbf{r}' \,. \tag{10.31}$$

The vector $\mathbf{r}'$ is here the distance from the charge to the observation point. Using our standard definition of a frame of reference with the origin at the position of the nucleus, we have $\mathbf{r}'=-\mathbf{r}_i$. The vector product with the classical orbital velocity, $\mathbf{v}_{\mathrm{e}}$, gives the orbital angular momentum, and the expression becomes:

$$\mathbf{B}_L = -\frac{\mu_0 e}{4\pi} \frac{1}{r_i^3} \frac{\mathbf{L_i}}{m_{\mathrm{e}}} = -\frac{\mu_0}{4\pi} \frac{2\mu_{\mathrm{B}}}{\hbar r_i^3} \mathbf{L_i} \,. \tag{10.32}$$

The field $\mathbf{B}_S$ can be seen as one generated by an electronic magnetic dipole, $\boldsymbol{\mu}_S$. Classical electromagnetic theory tells us that the magnetic flux density is $\mathbf{B}_S=\nabla\times\mathbf{A}_S$, where the magnetic vector potential is:

$$\mathbf{A}_S = \frac{\mu_0}{4\pi} \frac{1}{|r'|^3} \boldsymbol{\mu}_S \times \mathbf{r}' \,. \tag{10.33}$$

From chapter 4 and appendix E, we learn that $\boldsymbol{\mu}_S=-2\mu_{\mathrm{B}}\mathbf{S}_i/\hbar$. For the curl of the vector product in (10.33), we can use vector calculus:

$$\nabla_{\mathbf{r}'} \times \left(\frac{\boldsymbol{\mu}_S \times \mathbf{r}'}{|r'|^3} \right) = \left(\nabla_{\mathbf{r}'} \cdot \frac{\mathbf{r}'}{r^3} \right) \boldsymbol{\mu}_S - (\nabla_{\mathbf{r}'} \cdot \boldsymbol{\mu}_S) \frac{\mathbf{r}'}{r^3} + \left(\frac{\mathbf{r}'}{r^3} \cdot \nabla_{\mathbf{r}'} \right) \boldsymbol{\mu}_S - (\boldsymbol{\mu}_S \cdot \nabla_{\mathbf{r}'}) \frac{\mathbf{r}'}{r^3} \,. \tag{10.34}$$

The term with the divergence of the position vector will return zero, and if we take the magnetic moment as constant, another two terms go away. We are left with ([19]):

$$\nabla_{\mathbf{r}'} \times \left(\frac{\boldsymbol{\mu}_S \times \mathbf{r}'}{|r'|^3} \right) = -(\boldsymbol{\mu}_S \cdot \nabla_{\mathbf{r}'}) \frac{\mathbf{r}'}{r^3} = -\left[\mu_{Sx} \frac{\partial}{\partial x'} + \mu_{Sy} \frac{\partial}{\partial y'} + \mu_{Sz} \frac{\partial}{\partial z'} \right] \frac{\mathbf{r}'}{r^3}$$
$$= -\frac{\boldsymbol{\mu}_S}{r^3} + \frac{3\,(\mathbf{r}_i \cdot \boldsymbol{\mu}_S)\,\mathbf{r}_i}{r_i^5} \quad . \tag{10.35}$$

This means that the magnetic flux density at the position of the nucleus, resulting from the electron spin, is:

$$\mathbf{B}_S = \frac{\mu_0}{4\pi} \left[-\frac{\boldsymbol{\mu}_S}{r_i^3} + \frac{3\,(\mathbf{r}_i \cdot \boldsymbol{\mu}_S)\,\mathbf{r}_i}{r_i^5} \right] = \frac{\mu_0}{4\pi} \frac{2\,\mu_{\mathrm{B}}}{\hbar\, r_i^3} \left[\mathbf{S}_i - \frac{3\,(\mathbf{r}_i \cdot \mathbf{S}_i)\,\mathbf{r}_i}{r_i^2} \right] \quad . \tag{10.36}$$

The total field is now obtained by adding (10.32) and (10.36). The result is:

$$\mathbf{B}_J = -\frac{\mu_0\, \mu_{\mathrm{B}}}{2\pi} \frac{1}{\hbar\, r_i^3} \left[\mathbf{L}_i - \mathbf{S}_i + \frac{3\,(\mathbf{r}_i \cdot \mathbf{S}_i)\,\mathbf{r}_i}{r_i^2} \right] \quad . \tag{10.37}$$

If we convert this to atomic units, we have:

$$\mathbf{B}_J = -\frac{\alpha^2}{r_i^3} \left[\mathbf{L}_i - \mathbf{S}_i + \frac{3\,(\mathbf{r}_i \cdot \mathbf{S}_i)\,\mathbf{r}_i}{r_i^2} \right] \quad . \tag{10.38}$$

It should be noted that when deriving (10.37), we implicitly assumed that the electron is not in direct contact with the nucleus ($r_i \neq 0$). This means that the applicability of the equation is limited to electronic states with $l \neq 0$. For s-states, a different approach is necessary. Therefore, we will in the following first treat the simpler case, which excludes s-states, and then we estimate the result for electrons with $l = 0$.

For $l \neq 0$, we can proceed with (10.29) and use the magnetic field of (10.37) or (10.38). The a-constant becomes:

$$a_{\mathrm{hfs}} = \frac{g_I\, \mu_0\, \mu_{\mathrm{B}} \mu_N}{2\pi\, \hbar^2} \left\langle \frac{1}{r_i^3} \frac{1}{\mathbf{J}_i^2} \left[\mathbf{L}_i - \mathbf{S}_i + \frac{3\,(\mathbf{r}_i \cdot \mathbf{S}_i)\,\mathbf{r}_i}{r_i^2} \right] \cdot (\mathbf{L}_i + \mathbf{S}_i) \right\rangle \quad , \tag{10.39}$$

or in atomic units:

$$a_{\mathrm{hfs}} = \frac{\alpha^2\, g_I}{2\, m_{\mathrm{p}}} \left\langle \frac{1}{r_i^3} \frac{1}{\mathbf{J}_i^2} \left[\mathbf{L}_i - \mathbf{S}_i + \frac{3\,(\mathbf{r}_i \cdot \mathbf{S}_i)\,\mathbf{r}_i}{r_i^2} \right] \cdot (\mathbf{L}_i + \mathbf{S}_i) \right\rangle \quad . \tag{10.40}$$

When we perform the scalar product at the end of this equation, the term involving $\mathbf{r}_i \cdot \mathbf{L}_i$ will cancel, since by definition $\mathbf{L} = \mathbf{r} \times m\mathbf{v}$. This leaves us with:

$$a_{\mathrm{hfs}} = \frac{\alpha^2\, g_I}{2\, m_{\mathrm{p}}} \left\langle \frac{1}{r_i^3} \frac{1}{\mathbf{J}_i^2} \left[\mathbf{L}_\mathbf{i}^2 - \mathbf{S}_\mathbf{i}^2 + 3 \left(\frac{\mathbf{r}_i \cdot \mathbf{S}_i}{r_i} \right)^2 \right] \right\rangle \quad . \tag{10.41}$$

We can assume LS-coupling and a $|LSJM_J\rangle$-representation and develop the expectation values of the angular momentum operators. The scalar product $(\mathbf{r}_i / r_i) \cdot \mathbf{S}_i$ is the projection of the spin vector along the position vector, and for an electron, this is always plus or minus one half. This means that the square at the end of (10.41) will simply return one quarter. The final result is:

$$a_{\mathrm{hfs}} = \frac{\alpha^2 g_I}{2m_{\mathrm{p}}} \left\langle \frac{1}{r_i^3} \right\rangle \frac{L(L+1)}{J(J+1)}, \tag{10.42}$$

where we have anew used the assumption that we are dealing with a single valence electron, and thus $S = s_i = 1/2$. We have chosen to use the quantum numbers L and J in (10.42), rather than l_i and j_i, in order to have a notation more consistent with what we have previously used for non-hydrogenic atoms with a single valence electron (such as alkali atoms). The expressions (10.42) and (10.23) can also be derived from the operator expression (F.32) in appendix F, with exactly the same outcome.

Equation (10.42) still contains the expectation value of $1/r_i^3$, which we can only calculate analytically if we know the atomic wave function. The equation is nevertheless useful for estimates of relative hfs-intervals, and indeed for comparisons between numerical calculations of wave functions and experimental results. In the case of hydrogenic systems, we do have an explicit expression for the expectation value — see (B.39). Hence, for true one-electron systems, we can write the a-constant as:

$$a_{\mathrm{hfs}} = \frac{\alpha^2 g_I}{2m_{\mathrm{p}}} \frac{Z^3}{n_i^3 (l_i + \frac{1}{2}) j_i(j_i+1)} . \tag{10.43}$$

For a one-electron system with $L = l_i = 0$, the calculation of a_{hfs} must proceed a bit differently from (10.27) and onwards. To start with, we then have $\mathbf{J} = \mathbf{S}$, and the only contribution to $\mathbf{B}_J$ is $\mathbf{B}_S$. In order to estimate the latter, we need to consider the probability of having the electron in direct contact with the nucleus, or in other words the electronic charge density at the origin. We can safely assume that this varies very little in the vicinity of the nucleus, and hence we can take it as the square of the norm of the electronic wave function at the origin, $|\psi(0)|^2$.

From classical electrodynamics, the magnetisation vector is, in SI-units:

$$\mathcal{M} = \mu_S |\psi(0)|^2 = -\frac{g_s \mu_{\mathrm{B}} \mathbf{S}}{\hbar} |\psi(0)|^2 . \tag{10.44}$$

The magnetic flux density in a sphere with homogeneous magnetisation is:

$$\mathbf{B} = \frac{2\mu_0}{3} \mathcal{M} . \tag{10.45}$$

We use $g_s \approx 2$ and convert to atomic units. The relevant field becomes:

$$\mathbf{B}_S = -\frac{8\pi \alpha^2}{3} \mathbf{S} |\psi(0)|^2 . \tag{10.46}$$

When we substitute this into (10.29), and take $\mathbf{B}_L = 0$ and $\mathbf{J} = \mathbf{S}$, the operators will cancel out and we immediately get an expression for the a-constant for s-states, namely:

$$a_{\rm hfs} = \frac{4\pi\,\alpha^2\,g_I}{3\,m_{\rm p}}\,|\psi(0)|^2\,. \tag{10.47}$$

The expression in (10.47) is known as the *Fermi contact term*. To get further than this, the charge density must be calculated. This has been done for one-electron systems, using self-consistent field methods (see chapter 14). We will not cover these calculations here, but instead refer to suggested further reading. We will, however, give the result of the calculation for an s-state in a hydrogenic system. With wave functions from (1.16) and (1.22), it is:

$$a_{\rm hfs} = \frac{8\,\alpha^2\,g_I\,\mu_N}{3}\left(\frac{Z}{n_i}\right)^3\,. \tag{10.48}$$

10.2.4 Magnetic Dipole hfs for Atoms with Many Valence Electrons, and for Heavy Atoms

Atoms with more than one valence electron present a much more formidable case for analytical computations of hyperfine structure constants. The calculation of the a-constant is then severely complicated by the fact that relativistic corrections will depend on j_i, which may be different for the different electrons in the same orbital, and moreover, the interactions between the latter have to be taken into account. However, if one can accept to use empirical data for determinations of $a_{\rm hfs}$, the vector model equation in (10.23) for the energy perturbation still holds as a good approximation.

For very heavy atoms, the expressions in (10.42) and (10.47) agree rather poorly with empirical data, also for atoms with a single valence electron. This is because relativistic corrections — hitherto ignored in this chapter — come into play. Relativistic calculations have been made and one of the conclusions is that it is still practical to use (10.42) and (10.47) as a basis for the hyperfine structure splitting, and to add the relativistic effects as corrections (see, for example, [5] or [6]).

10.2.5 The Electric Quadrupole Interaction

The second most important contribution to the hyperfine structure is the electric quadrupole interaction. That is, the interaction between the quadrupole moment of the nucleus and the angular momenta of the electrons. A prerequisite for the presence of this interaction is that the nuclear and electronic angular momenta involved are not too small. For example, a nucleus with spherical symmetry cannot have a quadrupole moment. The order of magnitude of this perturbation, if it is present, is usually almost the same as that of the magnetic dipole term, but it is nearly always smaller.

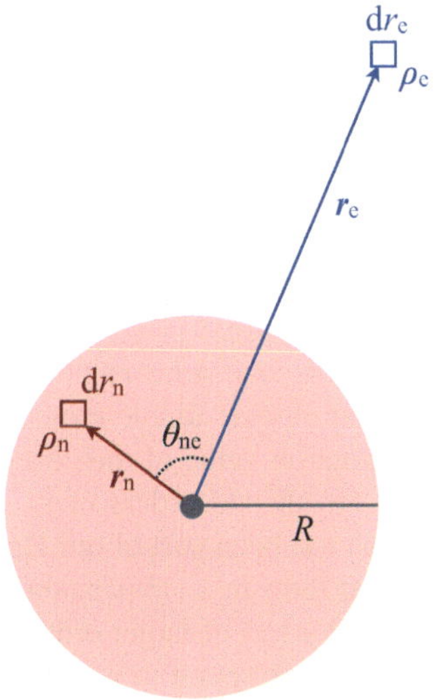

Fig. 10.1 Illustration of the interaction between nuclear positive charge (red) and the electronic charge density. $\mathbf{r}_n$ and $\mathbf{r}_e$ are the position vectors corresponding to volume elements in the nucleus and in the electronic charge cloud. θ_{ne} is the angle between the two vectors above, while ρ_n and ρ_e are, respectively, the position dependent positive and negative charge densities. R is the nuclear charge radius.

In order to unravel this interaction, we start by considering the entire electrostatic interaction between a finite size nucleus and an electronic cloud. An illustration of this is shown in figure 10.1. We consider the interaction between all the positive charge inside a nucleus, of any shape but with approximate maximum extension R, and all the negative charge of the ensemble of electrons.

The potential corresponding to the interaction can be found by integrating the product of both involved charge densities — $\rho_n(\mathbf{r}_n)$ and $\rho_e(\mathbf{r}_e)$ — over all space, for both position vectors (one for the nuclear charge and the one for the electronic). The respective charge densities are defined such that they fulfil the relations:

$$\int \rho_n(\mathbf{r}_n)\,d\mathbf{r}_n = +Ze$$
$$\int \rho_e(\mathbf{r}_e)\,d\mathbf{r}_e = -N_e e \ , \tag{10.49}$$

where N_e is the number of electrons. These quantities are proportional to the moduli squared of the nuclear and electronic wave functions, $|\phi_n(\mathbf{r}_n)|^2$ and $|\psi(\mathbf{r}_e)|^2$. With this in mind, we write the interaction potential as:

$$V_{ne} = \iint \frac{\rho_n(\mathbf{r}_n)\,\rho_e(\mathbf{r}_e)}{r_{ne}}\,d\mathbf{r}_n d\mathbf{r}_e \ , \tag{10.50}$$

where $r_{ne} = |\mathbf{r}_{ne}| \equiv |\mathbf{r}_n - \mathbf{r}_e|$. It is implicit that $\rho_n(\mathbf{r}_n)$ and $\rho_e(\mathbf{r}_e)$ have opposite signs.

The relative position vector can be expressed in terms of the magnitude of the vectors and the angle between them, θ_{ne}, via the cosine law. In turn, $1/r_{ne}$ can be expanded in Legendre polynomials of $\cos\theta_{ne}$, see (C.55), as:

$$\frac{1}{r_{\rm ne}} = \frac{1}{r_{\rm e}} + \frac{r_{\rm n}}{r_{\rm e}^2}\cos\theta_{\rm ne} + \frac{r_{\rm n}^2}{2\,r_{\rm e}^3}(3\cos^2\theta_{\rm ne}-1) + \frac{r_{\rm n}^3}{2\,r_{\rm e}^4}(5\cos^3\theta_{\rm ne}-3\cos\theta_{\rm ne}) + \frac{r_{\rm n}^4}{8\,r_{\rm e}^5}(35\cos^4\theta_{\rm ne}-30\cos^2\theta_{\rm ne}+3)+\ldots\,. \tag{10.51}$$

This will be correct for all combinations of $\mathbf{r}_{\rm n}$ and $\mathbf{r}_{\rm e}$ for which $|\mathbf{r}_{\rm n}| < |\mathbf{r}_{\rm e}|$, which will be true for almost the entire six-dimensional space in question. It is a good first approximation.

The second term in (10.51) is proportional to $r_{\rm n}$. That is, it corresponds to a nuclear electric dipole moment. As explained at the beginning of section 10.2, the expectation value of this term will cancel for parity reasons, and the same will be true for all odd orders of $r_{\rm n}$. The first term in (10.51) is the Coulomb repulsion, which has been part of our Hamiltonians from the beginning of the book. Thus, the largest term in an expansion of the integrand in (10.50), which is related to a nuclear moment of larger order than one, will be that proportional to $r_{\rm n}^2$, and this does indeed correspond to a nuclear quadrupole moment. This part of the total electric interaction is the one that will be the relevant perturbation in this section. The next non-zero term, proportional to $r_{\rm n}^4$, is the electric hexadecapole moment interaction.

There is another restriction in play concerning which multipole moments that may be non-zero. Referring to the additional theorem for spherical harmonics, and to (C.55) and (C.56), the potential in (10.50) can be approximately factorised into one nuclear integral and one electronic. This analysis ignores cross terms in nuclear and electronic parameters, but for the limited study of restrictions in angular momenta, currently undertaken, this is of no consequence. In terms of a series of spherical harmonics, the k'th order of the serial expansion of the potential is:

$$V_{\rm ne}^{(k)} = \frac{4\pi}{2k+1}\sum_{m=-k}^{k}\int \rho_{\rm n}(\mathbf{r}_{\rm n})\, r_n^k\, Y_{km}^*(\theta_n,\varphi_n)\,{\rm d}\mathbf{r}_{\rm n} \times \int \rho_{\rm e}(\mathbf{r}_{\rm e})\,\frac{1}{r_e^{k+1}}\,Y_{km}(\theta_e,\varphi_e)\,{\rm d}\mathbf{r}_{\rm e}\,. \tag{10.52}$$

Here, we have introduced $\theta_{\rm n}$ and $\varphi_{\rm n}$ for the zenith and azimuthal components of the nuclear coordinate $\mathbf{r}_{\rm n}$, and correspondingly $\theta_{\rm e}$ and $\varphi_{\rm e}$ for the electronic part. The charge densities follow from the moduli squared of the nuclear and electronic wave functions. The latter, we write as $\phi_{I,M_I}(\mathbf{r}_{\rm n})$ and $\psi_{J,M_J}(\mathbf{r}_{\rm e})$; that is we specify them with the respective angular momentum quantum numbers. For the present analysis, other quantum numbers do not matter.

With this taken into account the k'th component of the electric multipole interaction will be proportional to:

$$V_{\rm ne}^{(k)} \propto \frac{4\pi}{2k+1}\sum_{m=-k}^{k}\int r_n^k\, Y_{km}^*(\theta_n,\varphi_n)\,\left|\phi_{I,M_I}(\mathbf{r}_{\rm n})\right|^2{\rm d}\mathbf{r}_{\rm n} \times \int \frac{1}{r^{k+1}}\,Y_{km}(\theta_e,\varphi_e)\,\left|\psi_{J,M_J}(\mathbf{r}_{\rm e})\right|^2{\rm d}\mathbf{r}_{\rm e}\,. \tag{10.53}$$

In (10.53), both the nuclear and the electronic integrals will include integrations over a product of three spherical harmonics. This will lead to triangular conditions of the same type that we used when calculating *LS*-coupling term energies in (7.22), and which is further detailed in (D.10). The consequence is that the k'th order electronic multipole moment interactions can only be present if both the following conditions are fulfilled:

$$\begin{aligned} k &\leq 2I \\ k &\leq 2J \ . \end{aligned} \tag{10.54}$$

For the electric quadrupole moment, this means that it will only contribute to the hyperfine structure if the nuclear spin *and* the total electronic angular momenta adhere to the minimum conditions $I \geq 1$ and $J \geq 1$.

10.2.6 The Nuclear Electric Quadrupole Moment

We have pinned down the third term in (10.51) — the one proportional to r_n^2 — as the one responsible for the electric quadrupole hyperfine structure interaction. Since we deal with the effect as a perturbation, it is adequate to write this term as a Hamiltonian. The electric quadrupole part of the hyperfine structure is thus:

$$H_{\mathrm{hfs-EQ}} = \int \rho_\mathrm{n}(\mathbf{r}_\mathrm{n})\, \rho_\mathrm{n}(\mathbf{r}_\mathrm{e})\, \frac{r_\mathrm{n}^2}{2 r_\mathrm{e}^3} \left(3\cos^2\theta_\mathrm{ne} - 1\right) \mathrm{d}\mathbf{r}_\mathrm{n} \mathrm{d}\mathbf{r}_\mathrm{e} \ . \tag{10.55}$$

With the cosine law, this can be expanded in Cartesian coordinates as follows:

$$H_{\mathrm{hfs-EQ}} = \int \rho_\mathrm{n}(\mathbf{r}_\mathrm{n})\, \rho_\mathrm{e}(\mathbf{r}_\mathrm{e})\, \frac{1}{2 r_\mathrm{e}^5} \left[3\left(x_\mathrm{n}x_\mathrm{e} + y_\mathrm{n}y_\mathrm{e} + z_\mathrm{n}z_\mathrm{e}\right)^2 - 1\right] \mathrm{d}\mathbf{r}_\mathrm{n} \mathrm{d}\mathbf{r}_\mathrm{e} \ . \tag{10.56}$$

The integrand in (10.56) is convenient to express as a product between two tensors. We make the following definitions for the tensor components:

$$\begin{aligned} \mathsf{Q}_{xx} &\equiv \int \left(3x_\mathrm{n}^2 - r_\mathrm{n}^2\right) \rho_\mathrm{n}(\mathbf{r}_\mathrm{n})\, \mathrm{d}\mathbf{r}_\mathrm{n} \\ \mathsf{Q}_{xy} &\equiv \int 3 x_\mathrm{n} y_\mathrm{n}\, \rho_\mathrm{n}(\mathbf{r}_\mathrm{n})\, \mathrm{d}\mathbf{r}_\mathrm{n} \\ \mathsf{E}_{xx} &\equiv -\int \left(3x_\mathrm{e}^2 - r_\mathrm{e}^2\right) \frac{\rho_\mathrm{e}(\mathbf{r}_\mathrm{e})}{r_\mathrm{e}^5}\, \mathrm{d}\mathbf{r}_\mathrm{e} \\ \mathsf{E}_{xy} &\equiv -\int 3 x_\mathrm{e} y_\mathrm{e} \frac{\rho_\mathrm{e}(\mathbf{r}_\mathrm{e})}{r_\mathrm{e}^5}\, \mathrm{d}\mathbf{r}_\mathrm{e} \ , \end{aligned} \tag{10.57}$$

with the remaining components — Q_{yy}, Q_{yz}, E_{zz} and so on — defined in analogous fashions. This allows us to write the Hamiltonian in (10.56) in the following compact form:

$$H_{\text{hfs-EQ}} = -\frac{1}{6}\sum_{\alpha}\sum_{\beta} \mathsf{Q}_{\alpha\beta}\,\mathsf{E}_{\alpha\beta} \,, \tag{10.58}$$

where the summation indices α and β run over the coordinates x, y and z.

We define the nuclear electric quadrupole moment as:

$$\begin{aligned} eQ_{\text{n}} &\equiv \int r_{\text{n}}^2\,(3\cos^2\theta_{\text{n}} - 1)\,\rho_{\text{n},M_I=I}(\mathbf{r}_{\text{n}})\,\mathrm{d}\mathbf{r}_{\text{n}} \\ &= \int (3\,z_{\text{n}}^2 - r_{\text{n}}^2)\,\rho_{\text{n},M_I=I}(\mathbf{r}_{\text{n}})\,\mathrm{d}\mathbf{r}_{\text{n}} \,. \end{aligned} \tag{10.59}$$

The angle θ_{n} is the zenith angle for the positive charge coordinate. We take $\hat{e}_z$ as a quantisation axis for the nuclear spin $\mathbf{I}$, and the nuclear charge density in (10.59) is that for the stretched spin projection, that is, $M_I = I$. Physically, this quantity will give us information of the shape of the nucleus. For a homogeneous charge distribution, a positive Q_{n} corresponds to a nucleus that is elongated along the $\hat{e}_z$-axis (prolate), whereas a negative Q_{n} means a flattened nucleus (oblate).

It is always possible to choose a coordinate system such that the off-diagonal elements of Q vanish. If we assume axial symmetry, the elements Q_{xx} and Q_{yy} will be the same and will differ from Q_{zz} by a constant factor. A way to express this mathematically is that the quadrupole moment (10.59) will equal the expectation value of the diagonal tensor element Q_{zz} for $M_I = I$, that is:

$$\langle I,I\,|\,\mathsf{Q}_{zz}\,|\,I,I\rangle = e\,Q_{\text{n}} \,. \tag{10.60}$$

It is appropriate to express the tensor Q in (10.57) in terms of the nuclear spin vector operators. The result is (for an example of one diagonal and one off-diagonal element):

$$\begin{aligned} \mathsf{Q}_{xx} &= \gamma\,(3I_x^2 - \mathbf{I}^2) \\ \mathsf{Q}_{xy} &= \frac{3\gamma}{2}\,(I_xI_y + I_yI_x) \,, \end{aligned} \tag{10.61}$$

where γ is a constant, which should have the dimension of an electric quadrupole moment. The exact expression for γ (in SI-units) can be found from (10.60) through:

$$e\,Q_{\text{n}} = \gamma\hbar^2\left[3I^2 - I(I+1)\right] = \gamma\hbar^2 I\,(2I-1) \,. \tag{10.62}$$

We thus get the following expressions for the tensor elements:

$$\begin{aligned} \mathsf{Q}_{xx} &= \frac{e\,Q_{\text{n}}}{\hbar^2 I\,(2I-1)}\,(3\,I_x^2 - \mathbf{I}^2) \\ \mathsf{Q}_{xy} &= \frac{3\,e\,Q_{\text{n}}}{2\hbar^2 I\,(2I-1)}\,(I_xI_y + I_yI_x) \,. \end{aligned} \tag{10.63}$$

A more stringent proof than (10.61) and (10.63) may be derived from (10.56) and (10.57), but it is somewhat elaborate, and typically requires group theory. We will omit the detailed derivation in this volume, and instead we refer to other works (for example, [7] or [6]).

The value of Q_n can only be calculated if the nuclear charge distribution, and in other words the nuclear wave function, is known. In the special case that a charge consists of one single particle in a symmetric field it is, however, possible to derive the following analytical expression [20]:

$$Q_\mathrm{n} = -\frac{2I-1}{2(I+1)} \left\langle r_\mathrm{n}^2 \right\rangle . \tag{10.64}$$

The electronic part of (10.56) and (10.57) can be treated in exactly the same way, but with the total electronic angular momentum $\mathbf{J}$ being the relevant vector operator. The components of the tensor E are:

$$\begin{aligned} \mathsf{E}_{xx} &= -\frac{e q_J}{J(2J-1)} \left(3J_x^2 - \mathbf{J}^2\right) \\ \mathsf{E}_{xy} &= \frac{3 e q_J}{2J(2J-1)} \left(J_x J_y + J_y J_x\right) , \end{aligned} \tag{10.65}$$

with the other components following the same pattern. In (10.65), the integral q_J is defined by:

$$\begin{aligned} e q_J &\equiv \int \frac{1}{r_\mathrm{e}^3} \left(3\cos^2\theta_\mathrm{e} - 1\right) \rho_{\mathrm{e},M_J=J}(\mathbf{r}_\mathrm{e})\, \mathrm{d}\mathbf{r}_\mathrm{e} \\ &= \int \frac{1}{r_\mathrm{e}^5} \left(3z_\mathrm{e}^2 - r_\mathrm{e}^2\right) \rho_{\mathrm{e},M_J=J}(\mathbf{r}_\mathrm{e})\, \mathrm{d}\mathbf{r}_\mathrm{e} \quad . \end{aligned} \tag{10.66}$$

A physical interpretation of q_J is that it is the second gradient along $\hat{\mathrm{e}}_z$ of the electric potential caused by all charge exterior to the nucleus, for the state $M_J = J$:

$$\frac{\partial^2}{\partial z^2} V(\mathbf{r}_\mathrm{e}) = q_J \quad . \tag{10.67}$$

In the case of a one-electron system, this will equal the analytical expression:

$$q_J = -\frac{2J-1}{2(J+1)} \left\langle r_\mathrm{e}^{-3} \right\rangle , \tag{10.68}$$

in analogy with (10.64) [20].

10.2.7 The Electric Quadrupole Hamiltonian

For the total electric quadrupole interaction Hamiltonian, we need to substitute all components of (10.63) and (10.65) into (10.58). In order to avoid a vast number of terms in the expression, we first write the elements of (10.63) and (10.65) in a more generic form:

$$\begin{aligned} \mathsf{Q}_{\alpha\beta} &= \frac{eQ_{\mathrm{n}}}{\hbar^2 I(2I-1)} \left[\frac{3}{2}(I_\alpha I_\beta + I_\beta I_\alpha) - \delta_{\alpha\beta}\,\mathbf{I}^2\right] \\ \mathsf{E}_{\alpha\beta} &= -\frac{eq_J}{\hbar^2 J(2J-1)} \left[\frac{3}{2}(J_\alpha J_\beta + J_\beta J_\alpha) - \delta_{\alpha\beta}\,\mathbf{J}^2\right] . \end{aligned} \tag{10.69}$$

With this notation, the full Hamiltonian becomes:

$$\begin{aligned} H_{\mathrm{hfs-EQ}} = \frac{1}{6\hbar^4} \frac{e^2 Q_{\mathrm{n}} q_J}{I(2I-1)J(2J-1)} \sum_\alpha \sum_\beta & \left[\frac{3}{2}(I_\alpha I_\beta + I_\beta I_\alpha) - \delta_{\alpha\beta}\,\mathbf{I}^2\right] \\ \times & \left[\frac{3}{2}(J_\alpha J_\beta + J_\beta J_\alpha) - \delta_{\alpha\beta}\,\mathbf{J}^2\right] . \end{aligned} \tag{10.70}$$

In order to calculate energy perturbations from (10.70), the products in the summations have to be expanded. Furthermore, the operators need — if possible — to be restated in solely ones that are diagonal for wave functions represented by the quantum numbers I, M_I, J and M_J.

The multiplications in (10.70) will produce the following kinds of terms:

$$\sum_\alpha \sum_\beta I_\alpha I_\beta \, J_\alpha J_\beta \tag{10.71}$$

$$\sum_\alpha \sum_\beta I_\alpha I_\beta \, J_\beta J_\alpha \tag{10.72}$$

$$\sum_\alpha \sum_\beta I_\alpha I_\beta \, \delta_{\alpha\beta}\,\mathbf{J}^2 \tag{10.73}$$

$$\sum_\alpha \sum_\beta \delta_{\alpha\beta}\,\mathbf{I}^2 J_\alpha J_\beta \tag{10.74}$$

$$\sum_\alpha \sum_\beta \delta_{\alpha\beta}\,\mathbf{I}^2\mathbf{J}^2 \; . \tag{10.75}$$

Starting from the bottom, the last of these can be written as:

$$\sum_\alpha \sum_\beta \delta_{\alpha\beta}\mathbf{I}^2\mathbf{J}^2 = 3\,\mathbf{I}^2\mathbf{J}^2 \, . \tag{10.76}$$

Equations (10.73) and (10.74) will yield the same result, namely:

$$\sum_\alpha \sum_\beta I_\alpha I_\beta \, \delta_{\alpha\beta}\,\mathbf{J}^2 = \sum_{\alpha\beta} \delta_{\alpha\beta}\,\mathbf{I}^2 J_\alpha J_\beta = \mathbf{I}^2 \sum_\alpha J_\alpha^2 = \mathbf{I}^2\mathbf{J}^2 \, . \tag{10.77}$$

The term at the top of the list, (10.71), can be expressed as a scalar product:

$$\sum_\alpha \sum_\beta I_\alpha I_\beta J_\alpha J_\beta = \left(\sum_\alpha I_\alpha J_\alpha\right)\left(\sum_\beta I_\beta J_\beta\right) = (\mathbf{I}\cdot\mathbf{J})^2 \,. \qquad (10.78)$$

For the remaining term, (10.72), we use the commutations rules of (C.3) to find:

$$\begin{aligned}\sum_\alpha \sum_\beta I_\alpha I_\beta J_\beta J_\alpha &= \sum_\alpha I_\alpha^2 J_\alpha^2 + I_x I_y (J_x J_y - \hbar J_z) + I_y I_x (J_y J_x + \hbar J_z) \\ &\quad + I_y I_z (J_y J_z - \hbar J_x) + I_z I_y (J_z J_y + \hbar J_x) + I_z I_x (J_z J_x - \hbar J_y) + I_x I_z (J_x J_z + \hbar J_y) \\ &= \sum_\alpha \sum_\beta I_\alpha I_\beta J_\alpha J_\beta - \hbar^2 \sum_\alpha I_\alpha J_\alpha = (\mathbf{I}\cdot\mathbf{J})^2 - \hbar^2\,\mathbf{I}\cdot\mathbf{J} \qquad (10.79)\end{aligned}$$

When we finally put all this together, the Hamiltonian in (10.70) turns into:

$$\begin{aligned}H_{\text{hfs-EQ}} &= \frac{1}{6\hbar^4}\,\frac{e^2 Q_\text{n} q_J}{I(2I-1)J(2J-1)} \\ &\quad \times \left\{\frac{9}{2}(\mathbf{I}\cdot\mathbf{J})^2 + \frac{9}{2}\left[(\mathbf{I}\cdot\mathbf{J})^2 - \hbar^2\,\mathbf{I}\cdot\mathbf{J}\right] - 3\mathbf{I}^2\mathbf{J}^2 - 3\mathbf{I}^2\mathbf{J}^2 + 3\mathbf{I}^2\mathbf{J}^2\right\} \\ &= \frac{e^2 Q_\text{n} q_J}{2\hbar^4 I(2I-1)J(2J-1)}\left[3(\mathbf{I}\cdot\mathbf{J})^2 - \frac{3\hbar^2}{2}\mathbf{I}\cdot\mathbf{J} - \mathbf{I}^2\mathbf{J}^2\right] \,. \qquad (10.80)\end{aligned}$$

Since we typically do not have access to analytical values for the quantities Q_n and q_J, it makes sense to introduce an empirically obtainable *electric quadrupole hyperfine structure constant*, in a way analogous to what was done in (10.20) for the magnetic dipole effect. This constant is also called the *hyperfine structure b-constant*, and we define it as:

$$b_\text{hfs} \equiv e^2 Q_\text{n} q_J \,. \qquad (10.81)$$

Accommodating for this, we rewrite the Hamiltonian (10.80) as:

$$H_{\text{hfs-EQ}} = b_\text{hfs}\,\frac{3(\mathbf{I}\cdot\mathbf{J})^2 - \frac{3}{2}\mathbf{I}\cdot\mathbf{J} - \mathbf{I}^2\mathbf{J}^2}{2I(2I-1)J(2J-1)} \,, \qquad (10.82)$$

where powers of $\hbar$ have been removed by converting to atomic units.

From (10.82), we can compute the magnitude of the energy shift. The scalar product $\mathbf{I}\cdot\mathbf{J}$ can be written by help of the total atomic angular momentum $\mathbf{F}$ as in (10.23). To disencumber the notation, we introduce the definition:

$$K \equiv F(F+1) - I(I+1) - J(J+1) \,. \qquad (10.83)$$

With this, the energy shift due to the perturbation (10.82) becomes:

$$E_{\mathrm{hfs-EQ}} = \langle H_{\mathrm{hfs-EQ}} \rangle = \frac{1}{8} \langle b_{\mathrm{hfs}} \rangle \frac{3K(K+1) - 4I(I+1)J(J+1)}{I(2I-1)J(2J-1)} . \quad (10.84)$$

As for the a-constant, the expectation value $\langle b_{\mathrm{hfs}} \rangle$ is commonly written without the mean brackets. Its value is usually either calculated numerically or derived from spectra.

If we assume that higher orders can still be ignored, we can write the total hyperfine structure energy, from (10.23) and (10.84) as:

$$E_{\mathrm{hfs}} = \frac{a_{\mathrm{hfs}}}{2} K + \frac{b_{\mathrm{hfs}}}{8} \frac{3K(K+1) - 4I(I+1)J(J+1)}{I(2I-1)J(2J-1)} . \quad (10.85)$$

For states with $I \geq 1$ and $J \geq 1$, and when b_{hfs} is not completely negligible compared to a_{hfs}, this will lead to a deviation from the interval rule stated in (10.25).

10.2.8 Higher Order Multipole Moments

If we refer back to (10.50) and (10.51), it is clear that we have ignored an infinite number of higher order terms of electrostatic interactions. Using the same symmetry arguments as the one in section 10.2.5 — and around (10.12) — we may exclude all interactions with odd electric multipole moments. Thus, after the quadrupole term, the next electric hfs-term is the interaction with the hexadecapole moment, proportional to $r_{\mathrm{n}}^4 r_{\mathrm{e}}^{-5}$.

Also for magnetic interactions, a thorough analysis must include higher order moments than the nuclear magnetic dipole. Following the arguments connected to (10.13), we learn that for the magnetic interactions between the nucleus and the angular momenta of the electrons, it is the odd order moments that need to be retained. This means that the largest magnetic effect, following the magnetic dipole hfs, will be the interaction with the nuclear magnetic octupole moment.

As can be expected, these higher order hfs effects are small, but with modern spectroscopic techniques, there are several examples for which they have been detected. For many precision experiments, these small effects are important. This includes the influence of the magnetic octupole hfs in alkali atoms, where a precise understanding of the structure is needed, for example, for measurements of the elec tric dipole moment of the electron and for atomic frequency standards.

In the following, we will present Hamiltonians and expectation values for the magnetic octupole hfs and the electric hexadecapole moment, in terms of element and state dependent parameters called the hyperfine structure *c and d constants*. For derivations, we refer to more specialised literature (for example, [8, 9] or [6]).

The expression for the magnetic octupole hfs-interaction energy shift is:

$$E_{\text{hfs-MO}} = \frac{c_{\text{hfs}}}{4} \frac{1}{I(I-1)(2I-1)J(J-1)(2J-1)} \times \Big\{ 5K^3 + 20K^2 + 4K\left[3 + I(I+1) + J(J+1) - 3I(I+1)J(J+1)\right] - 20I(I+1)J(J+1) \Big\} , \tag{10.86}$$

and for the electric hexadecapole moment it is:

$$E_{\text{hfs-EH}} = \frac{d_{\text{hfs}}}{8} \frac{1}{I(I-1)(2I-1)(2I-3)J(J-1)(2J-1)(2J-3)} \times \Big\{ 35K^4 + 350K^3 + 20K^2\left[39 + 5I(I+1) + 5J(J+1) - 6I(I+1)J(J+1)\right] + 20K\left[18 + 12I(I+1) + 12J(J+1) - 34I(I+1)J(J+1)\right] - 24I(I+1)J(J+1)\left[27 + 4I(I+1) + 4J(J+1) - 2I(I+1)J(J+1)\right] \Big\} . \tag{10.87}$$

In these equations, K is still defined as in (10.83), and the factors c_{hfs} and d_{hfs} are respectively the hyperfine structure constants for the nuclear octupole and hexadecapole moments. As for the magnetic dipole and electric quadrupole effects, these 'constants' are really derived from expectation values of corresponding operators.

The conditions in (10.54) still apply, and thus the magnetic octupole hfs will cancel unless $I \geq 3/2$ and $I \geq 3/2$. For the electric hexadecapole moment, the conditions for a non-vanishing effect are $I \geq 2$ and $I \geq 2$.

10.2.9 Hyperfine Structure in Multielectron Systems

The treatments in sections 10.2.1 and 10.2.5 have mostly ignored multielectron aspects. The operators in (10.20) and (10.82) have essentially been derived as if they were single-electron operators. Extending this to a multielectron atom, in a first-order approximation, is not an overwhelming problem since the task can be dealt with as a sum one one-electron operators.

For an atom with all valence electrons in the same orbital, we can typically use the CFA, ignore the contribution from filled orbitals, and then just use the total electronic angular momentum quantum number J in (10.20) and (10.82), without further modification. This does mean an extra level of approximation since it ignores interactions between configurations (see chapter 9), which may be important at the level of small hyperfine structure. Another important difference in the multielectron case is that it becomes exceedingly difficult to analytically calculate the hfs constants.

The situation changes a bit if we have an atom with more than one open orbital. We will illustrate this with the simplest possible example. Imagine an excited two-electron atom, such as a neutral alkaline earth, where one electron is in the s-orbital, but the other is excited to a less bound orbital. If the outermost electron is relatively highly excited, it will contribute very little to the magnetic dipole hfs, as is evident in (10.43) and (10.48). In this situation, the operator mainly responsible for the magnetic dipole hfs interaction will be $\mathbf{I}\cdot\mathbf{S}_1$ rather than $\mathbf{I}\cdot\mathbf{J}$.

On the other hand, for the electric quadrupole interaction, we have the requirement of $J \geq 1$, and thus the inner s-electron will not contribute at all. The excited electron does, however, provided that it is in a p-orbital, or any other with $l_2 > 0$.

10.2.10 Some Concrete Examples of Atomic Hyperfine Structure

Within the approximations used in preceding sections, and with knowledge of the nuclear spin, it is relatively straightforward to identify the hyperfine structure for a fine-structure level, using the vector model. The only quantum numbers needed are J, I, F and M_F, and provided that there is no external magnetic field, an F-level will be $(2F+1)$-fold degenerate in M_F. A quantitative estimate of the hfs-splittings is, however, more difficult. In the following, we will exemplify the usage of the vector model for low lying levels in one alkali atom and in one two-electron system.

As a first example, we take the caesium atom. There is a unique interest in the hyperfine structure of Cs since the hfs-splitting in the ground state of Cs currently (2019) constitutes the definition of a second. Thus, a stabilisation of an oscillator on a resonance between the involved hfs levels is a realisation of a frequency standard.

There is only one stable isotope of Cs, namely ^{133}Cs, which has the nuclear spin $I = 7/2$. For the lowest electronic configurations, 6s and 6p, this will give a hyperfine structure as illustrated in figure 10.2. The successive application of perturbative terms eventually sees the fine-structure levels $^2S_{1/2}$, $^2P_{1/2}$ and $^2P_{3/2}$ split up into respectively two, two and four hfs states.

The hyperfine structure Hamiltonian that gives rise to the splitting shown in the right part of figure 10.2 includes all terms in the expansion of the magnetic and electrostatic interactions. However, the restrictions in orders given in (10.54) must be taken into account. Thus, for $^2S_{1/2}$ and $^2P_{1/2}$, the magnetic dipole interaction will be the only hfs interaction in play. For the fine-structure level $^2P_{3/2}$, J and I are both high enough to cause additional contributions to the hyperfine structure by both the electric quadrupole and the magnetic octupole interactions.

As a second example, we take the ground state and the two lowest excited configurations of neutral barium, but instead of showing a Grotrian diagram, we demonstrate the hyperfine structure in table form. In table 10.1, we illustrate the successive steps in the perturbation theory, ending with all relevant terms of the Hamiltonian H_{hfs}.

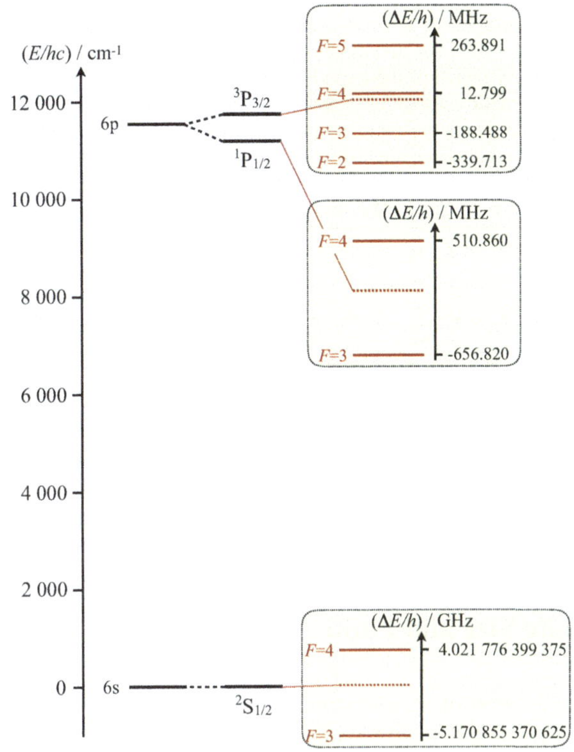

Fig. 10.2 Diagram showing the ground state and lowest excited states of neutral caesium, including hyperfine structure. The empirical data are from [10] and [11], and the ground state has been taken as zero energy. The hyperfine structures of the three fine-structure levels are shown in the yellow blow-ups, and the splittings are there given in frequency units. The hyperfine structure of the ground state constitutes the definition of the second, and the numbers given for that in the figure are thus exact.

Barium has many stable isotopes, with the most abundant one being ^{138}Ba. This is a nucleus with an even number of neutrons and an even number of protons. As a result, the nuclear spin is zero, and this isotope will not have any magnetic dipole hyperfine structure. Table 10.1 instead shows the energy levels for the most abundant isotope with an odd number of nuclei, ^{137}Ba, which has nuclear spin $I = 3/2$. This nuclear spin will prohibit hyperfine interactions of higher orders than the magnetic octupole one. However, of the nine fine-structure levels included in the table, $^3P_2$, $^3D_2$, $^3D_3$ and $^1D_2$ will have non-zero interaction terms for all orders up to, and including, the magnetic octupole. For the three levels $^3P_1$, $^1P_1$ and $^3D_1$, the magnetic dipole moment and the electric quadrupole moment will contribute. The remaining two levels, $^1S_0$ and $^3P_0$, have a vanishing total electronic angular momentum, and thus all hyperfine structure cancels — within the level of approximation that we are currently adhering to.

Table 10.1 Tabular illustration of a successive perturbative treatment of some low lying energy levels in neutral ^{137}Ba ($I = 3/2$). The first three columns represent a standard LS-coupling situation. The following three show which of the hyperfine structure Hamiltonian terms (magnetic dipole, electric quadrupole, magnetic octupole) that are non-zero for the nine different fine-structure levels. The last column shows the possible values of the quantum number F. Empirical quantitative data for the various energies can be found in [10, 12–14] and [15].

H_{CF}	H_{LS}	H_{SO}	H_{hfs-MD}	H_{hfs-EQ}	H_{hfs-MO}	F
6s6p	^{1}P	$J=1$	✓	✓	–	1/2, 3/2, 5/2
	^{3}P	$J=2$	✓	✓	✓	1/2, 3/2, 5/2, 7/2
		$J=1$	✓	✓	–	1/2, 3/2, 5/2
		$J=0$	–	–	–	3/2
6s5d	^{1}D	$J=2$	✓	✓	✓	1/2, 3/2, 5/2, 7/2
	^{3}D	$J=3$	✓	✓	✓	3/2, 5/2, 7/2, 9/2
		$J=2$	✓	–	–	1/2, 3/2, 5/2, 7/2
		$J=1$	✓	✓	–	1/2, 3/2, 5/2
6s^2	^{1}S	$J=0$	–	–	–	3/2

10.3 The Finite Size and Finite Mass Nucleus

As a first approximation of atomic structure, it is quite reasonable to think of the nucleus as being point-sized and having infinite mass. In this chapter, we have expanded the study of atomic structure to include the effects of a nucleus with more complicated, and more realistic, properties.

In terms of experimental observations, and the history of nuclear effects on atoms, the situation is similar to that for other aspects of atomic structure. The experimental signatures were detected in spectroscopy before the theory was developed. Hyperfine splittings in emission spectra were observed already towards the end of the 19th century, well before quantum mechanics was introduced and before basic atomic structure was understood. The connection between the small detected structures in the spectra and the atomic nucleus was suggested in 1924, by Pauli.

The nucleus is in itself a composite object and nuclear structure is a field of research in its own right. However, this does not come within the scope of the current book. Instead of detailing why a specific nucleus has the properties that it has, we settle with taking some simplified structural properties of different compound nuclei as they are, building on empirical data, and then we analyse how the interaction with the electronic cloud changes the energy eigenfunctions of the compound atom. For a study of nuclear structure, there exist many volumes specialised on the subject, see, for example, [3] or [16].

The nuclear masses that we consider will typically be close to the number of nucleons multiplied by 1800 times the electron mass (the masses of the neutron and the proton are very similar, and the binding energies are small in comparison). More exact data can readily be obtained from tabulated data.

Typical nuclear radii are of the order of 10 fm, and as a consequence of the properties of the strong nuclear interaction, it increases only a little across the periodic table [17]. The spatial extension enters our analyses when we consider the volume isotope shift. For hyperfine structure, the deviations of the nuclear geometry from exactly spherical will result in orders of multipole moments. Further to that, another important property that we have considered is the compound nuclear spin, which results in a magnetic dipole moment. The latter is typically the largest hyperfine structure perturbation and the most important nuclear effect on atomic structure.

In terms of spectroscopy, the nuclear effects on atomic energy levels are of the order of 1 cm^{-1} (around 100 μeV or 5 millionth of an atomic energy unit), or smaller. In frequency, this corresponds to radiation in the microwave region. The splitting of energy levels will be so small that in most traditional spectroscopic observations, it will be orders of magnitude smaller than the Doppler width of the spectral line in question, and thus it may not be resolved.

There are many examples of applications of nuclear effects of atomic structure. One is the so-called 21 cm line, used in radio-astronomy. This is a transition between the two hyperfine structure states $F=1$ and $F=0$ of the $1s\,^2S_{1/2}$ ground state in hydrogen. Another one is the current (2019) SI definition of the second, which is the 9.2 GHz hyperfine splitting between the levels $F=4$ and $F=3$ of the $6s\,^2S_{1/2}$ ground state of Cs. Other important applications, where the nuclear properties are of great importance, are magnetic resonance imaging (for example, in medicine) and the Mössbauer effect.

Further Reading

Physics of Atoms and Molecules, by Bransden & Joachain [1]
Atomic Many-Body Theory, by Lindgren & Morrison [4]
Atomic spectra, by Kuhn [5]
Nuclear Moments, by Ramsey [7]
The theory of atomic spectra, by Condon & Shortley [18]
Atomfysik, by Lindgren [19]
Atomic Spectra and Radiative Transitions, by Sobelman [20]
Springer Handbook of Atomic, Molecular, and Optical Physics by Drake [21]

References

1. B.H. Bransden, C.J. Joachain, *Physics of Atoms and Molecules*, 2nd edn. (Prentice Hall, Harlow, England, 2003)
2. G.W.F. Drake, in *Springer Handbook of Atomic, Molecular, and Optical Physics*, ed. by G.W.F. Drake (Springer-Verlag, New York, 2006), p. 199
3. E.M. Henley, A. Garcia, *Subatomic Physics*, 3rd edn. (World Scientific, 2007)

4. I. Lindgren, J. Morrison, *Atomic Many-Body Theory*, 2nd edn. (Springer Verlag, Berlin, 1986)
5. H.G. Kuhn, *Atomic spectra* (Longmans, London, 1969)
6. G.T. Emery, in *Springer Handbook of Atomic, Molecular, and Optical Physics*, ed. by G.W.F. Drake (Springer-Verlag, New York, 2006), p. 253
7. N.F. Ramsey, *Nuclear Moments* (Wiley, New York, 1953)
8. C. Schwartz, Phys. Rev. **97**, 380 (1955)
9. L. Armstrong, *Theory of the Hyperfine Structure of Free Atoms* (Wiley, New York, 1971)
10. A. Kramida, Y. Ralchenko, J. Reader, and NIST ASD Team. NIST Atomic Spectra Database (ver. 5.3). [Online]. Available: http://physics.nist.gov/asd (2018). Accessed: 2019-07-14
11. D.A. Steck. Cesium D line data. [Online]. Available: http://steck.us/alkalidata (2010). Accessed: 2019-05-20
12. M. Gustavsson, G. Olsson, A. Rosén, Zeitschrift für Physik A Atoms and Nuclei **290**, 231 (1979)
13. S.G. Schmelling, Phys. Rev. A **9**, 1097 (1974)
14. G. zu Putlitz, Annalen der Physik **466**, 248 (1963)
15. H.J. Kluge, H. Sauter, Zeitschrift für Physik **270**, 295 (1974)
16. K.P. Lieb, in *Constituents of Matter*, ed. by W. Raith (Walter de Gruyter, Berlin, 1997), p. 625
17. E.D. Commins, *Quantum mechanics: an experimentalists approach* (Cambridge University Press, New York, 2014)
18. E.U. Condon, G.H. Shortley, *The theory of atomic spectra* (Cambridge University Press, Cambridge, 1935)
19. I. Lindgren, S. Svanberg, *Atomfysik* (Universitetsförlaget, Uppsala, 1974)
20. I.I. Sobelman, *Atomic Spectra and Radiative Transitions*, 2nd edn. (Springer, Berlin, 1992)
21. G.W.F. Drake (ed.), *Springer Handbook of Atomic, Molecular, and Optical Physics* (Springer-Verlag, New York, 2006)

Chapter 11
The Zeeman Effect

Heretofore in the book, we have exclusively dealt with individual atoms isolated from the environment. In this chapter, we will introduce the *Zeeman effect* — the interaction between an atom and an external magnetic field. It is an interaction which is both prevalent and that has important consequences. Atomic energy levels will shift, and degeneracies are lifted.

The external field imposes a preferred direction in space and thus breaks spherical symmetry. Therefore, the specific degeneracies broken are those of the projection quantum numbers M_J and M_F, and the choice of quantisation axis will now matter for the interaction Hamiltonian. The magnetic field gives rise to forces when it interacts with magnetic moments. The latter are consequences of electronic and nuclear angular momenta, and thus, different projections of the momenta along the field direction result in different potential energies. This is why groups of otherwise degenerate fine-structure levels (or hfs levels) split up in non-degenerate states, characterised by quantum numbers M_J (or M_F).

Once again we resort to perturbation theory, and as far as possible, we will stick to the CFA. Given this, it is reasonable to treat the external field classically, in a first approximation. We also limit the analysis to static fields that are homogeneous over the volume taken up by the atomic charge distribution.

In many ways, the treatment of the Zeeman effect is similar to that of the spin–orbit effect, which is natural since the latter is also a magnetic interaction. Indeed, with the most accurate formalism, these two effects should be treated together (see appendix F). However, in a first heuristic approach, the Zeeman effect Hamiltonian might be taken as:

$$H_Z = -\mathbf{B} \cdot \boldsymbol{\mu} \quad . \tag{11.1}$$

$\mathbf{B}$ is the magnetic flux density and $\boldsymbol{\mu}$ is the total magnetic moment of the compound atom. That is, we treat the interaction as a classical magnetic moment in a field. With the origin as usual taken at the centre-of-mass, the contributions to $\boldsymbol{\mu}$ are the orbital angular momenta of all electrons, the spins of all electrons, and the nuclear spin.

A. Kastberg, *Structure of Multielectron Atoms*, Springer Series on Atomic, Optical, and Plasma Physics 112, https://doi.org/10.1007/978-3-030-36420-5_11

In fact, (11.1) is far too simplistic. We will not get a good handle of the Zeeman effect unless we study the different contributions to $\boldsymbol{\mu}$ in some more detail. Not least important is the fact that the gyromagnetic ratios (the g-factors) are different for spin and orbital angular momenta.

We have seen in previous chapters that handling atomic structure with perturbation theory constitutes an art of placing the various interaction Hamiltonians in diminishing energy order. What is new here, compared to effects hitherto covered, is that the Hamiltonian in (11.1) depends on a variable parameter. This variation and the relative magnitudes of the other energy contributions will provide a rough outline for the sections that follow.

We will commence the chapter by considering magnetic fields sufficiently feeble to render its ensuing energy contribution the smallest perturbation. This, in turn, splits up into two cases, namely atoms that respectively do (section 11.2) and do not (section 11.1) experience hyperfine structure. Thereafter, we study how the physics changes when the magnetic field is so strong that (11.1) dominates over either the spin–orbit, or the hyperfine structure Hamiltonian, and we also cover the intermediate case. We will end the chapter by looking at higher order effects.

There are some inconsistencies in common terminology that should be mentioned. In parts of traditional atomic physics literature, the interaction with an external magnetic field is only called the Zeeman effect if its Hamiltonian is the smallest energy in the system, whereas the behaviour of atoms in stronger fields is referred to as the Paschen-Back effect. Other sources refer to the term Zeeman effect as a general identification for any quantum system interacting with a magnetic field.

A more troublesome ambiguity is the term 'quadratic Zeeman effect'. To the lowest order, the interaction Hamiltonian is linear in $|\mathbf{B}|$. As will be seen in (11.2), there is also a term quadratic in the field. The energy contribution from the latter is referred to as either the quadratic or the diamagnetic Zeeman effect, and it is small. The magnetic fields needed to make it significant are usually those found in the interior of stars. However, for a range of high precision experiments, the term quadratic Zeeman effect is often used with a different meaning. When the Zeeman effect is taken as a final perturbation, matrix elements of the Hamiltonian that are non-diagonal in J (or F) are ignored. As the field increases and the Paschen-Back regime is approached, there will be a quadratic correction to the linear energy shift, which is highly important is many metrological applications. We address this in section 11.4 and we will refer to it as the 'intermediate regime'.

11.1 Weak Magnetic Field — The Case with No Hyperfine Structure

We begin the treatment of the Zeeman effect with an external field small enough that it can be seen as a perturbation to all internal interactions. We will simplify matters by assuming an atom that has no nuclear spin, that is an atom without a magnetic dipole hyperfine structure. The inclusion of the latter is a relatively straightforward generalisation, and it will be dealt with in section 11.2.

11.1.1 The Zeeman Effect Hamiltonian

To derive a more elaborate Zeeman effect Hamiltonian than (11.1), we will use both geometric arguments within the vector model method, and the Wigner-Eckart theorem (see appendix C.6). To begin with, we formulate a Hamiltonian, and we use the standard expression from electrodynamics for a charge in a magnetic field (see for example [1], and also appendix E). For a particle of mass m_e in an external field, we have (in a.u.):

$$H = \frac{1}{2}\left[\mathbf{p}+\mathbf{A}(\mathbf{r})\right]^2 + V(\mathbf{r}) = -\frac{1}{2}\nabla^2 - \mathrm{i}\,\mathbf{A}(\mathbf{r})\cdot\nabla + \frac{1}{2}\left[\mathbf{A}(\mathbf{r})\right]^2 + V(\mathbf{r}) \; . \qquad (11.2)$$

$\mathbf{A}(\mathbf{r})$ is the electromagnetic vector potential describing the field, and $V(\mathbf{r})$ represents all other potential energies acting on the particle. For an electron bound to an atomic nucleus, the first and last terms in (11.2) are, respectively, the usual kinetic energy and the sum of all internal potential energies. Besides these familiar terms, we now also have two that depend on $\mathbf{A}(\mathbf{r})$ — one linearly and one quadratically. The sum of these two will be referred to as H_{mag}.

In the Coulomb gauge, the vector potential is related to the magnetic flux density $\mathbf{B}(\mathbf{r})$ through [1]:

$$\mathbf{A}(\mathbf{r}) = \frac{1}{2}\left[\mathbf{B}(\mathbf{r})\times\mathbf{r}\right] . \qquad (11.3)$$

If we substitute this in (11.2) and isolate the two terms associated with magnetic interactions, we get:

$$H_{\text{mag}} = \frac{1}{2}\mathbf{B}\cdot\mathbf{L} + \frac{1}{8}\left[\mathbf{B}(\mathbf{r})\times\mathbf{r}\right]^2 = H_{\text{Z}} + H_{\text{dZ}} \; . \qquad (11.4)$$

For the first term in (11.4), we have used the definition of the orbital angular momentum, ($\mathbf{L}=\mathbf{r}\times\mathbf{p}$), and the properties of a vector triple product. After the last equality, we have introduced the symbols H_{Z} and H_{dZ} for the respective Hamiltonians for the *paramagnetic Zeeman effect* and the *diamagnetic Zeeman effect*.

The term linear in $\mathbf{B}(\mathbf{r})$ in (11.4) is very similar to (11.1), assuming that the magnetic moment is proportional to the angular momentum, with the factor of proportionality being the Bohr magneton, $\mu_{\text{B}} = e\hbar/(2m_{\text{e}})$, and the fact that the latter is one half in atomic units. However, the classical particle in the analysis above ignores the existence of spin.

With a relativistic analysis, the electron spin becomes a natural contribution to the total magnetic moment. Here, we will for the moment settle with the empirical knowledge of the spin contribution to the term H_{Z} in (11.4), and with its approximate magnitude. What we are aiming for is to find an expression for the total magnetic moment of the atom, taking into account all electron spins and the orbital angular momenta of all electrons. In the course of this analysis, we will also see that we cannot ignore the spin–orbit coupling.

In the limit where the external field is weak, it is reasonable to begin with studying the paramagnetic, or 'linear', Zeeman effect for a one-electron atom, and to then form a Hamiltonian for the multielectron case by summing over all electrons.

Seen separately, the contributions to the total electronic magnetic moment from $\mathbf{L}_i$ and $\mathbf{S}_i$ are (in SI-units):

$$\begin{aligned} \boldsymbol{\mu}_l &= -\frac{g_l\,\mu_\mathrm{B}}{\hbar}\,\mathbf{L}_i \\ \boldsymbol{\mu}_s &= -\frac{g_s\,\mu_\mathrm{B}}{\hbar}\,\mathbf{S}_i \ . \end{aligned} \tag{11.5}$$

The factors of proportionality, the *g-factors* or the *gyromagnetic ratios*, are $g_l \approx 1$ and $g_s \approx 2$. The orbital angular momentum g-factor, g_l, is strictly equal to one only for an infinitely massive nucleus. For a more precise value, the ratio of the electron mass to the nuclear mass should be subtracted from unity [2]. The value of g_s cannot be explained by classical physics; a relativistic analysis is necessary. An introduction to the theory of the electron spin and to relativistic quantum mechanics is included in appendix E.

Using the approximate g-factors in (11.5), we can write the linear Zeeman Hamiltonian as:

$$H_\mathrm{Z} = -\left[\boldsymbol{\mu}_l + \boldsymbol{\mu}_s\right]\cdot\mathbf{B}(\mathbf{r}_i) \approx \frac{\mu_\mathrm{B}}{\hbar}\left[\mathbf{L}_i + 2\mathbf{S}_i\right]\cdot\mathbf{B}(\mathbf{r}_i) \ . \tag{11.6}$$

We have here chosen to express (11.6) in SI-units, in order to explicitly show the role of the Bohr magneton. With atomic units, the prefactor $\mu_\mathrm{B}/\hbar$ is replaced by $1/2$. This equation is also derived in (F.20), in which the spin contribution is included in a more rigorous way.

We choose the $\hat{e}_z$-axis to be everywhere parallel with the external field:

$$\mathbf{B}(\mathbf{r}_i) = B(\mathbf{r}_i)\,\hat{e}_z \ . \tag{11.7}$$

This simplifies the vector product in H_dZ, and it means that for $\mathbf{L}_i$ and $\mathbf{S}_i$ we can take the projection operators:

$$H_\mathrm{Z} = \frac{\mu_\mathrm{B}}{\hbar}\left[L_{zi} + 2S_{zi}\right]B(\mathbf{r}_i) \ . \tag{11.8}$$

The diamagnetic Hamiltonian with such a field, in SI-units, is — from (11.4):

$$H_\mathrm{dZ} = \frac{e^2}{8m_\mathrm{e}}B^2(\mathbf{r}_i)\left(x^2+y^2\right) = \frac{e^2}{8m_\mathrm{e}}B^2(\mathbf{r}_i)\,r_i^2\sin^2\theta_i \ , \tag{11.9}$$

where θ_i is the angle between the $\hat{e}_z$-axis and the electron position vector $\mathbf{r}_i$ (the standard definition of the zenith angle), and x and y are the respective Cartesian coordinates of the latter. The total magnetic interaction Hamiltonian, from (11.8) and (11.9), is:

$$H_\mathrm{mag} = H_\mathrm{Z} + H_\mathrm{dZ} = \frac{\mu_\mathrm{B}}{\hbar}\left[L_{zi} + 2S_{zi}\right]B(\mathbf{r}_i) + \frac{e^2}{8m_\mathrm{e}}B^2(\mathbf{r}_i)\,r_i^2\sin^2\theta_i \ , \tag{11.10}$$

In atomic units, this is:

$$H_{\text{mag}} = H_{\text{Z}} + H_{\text{dZ}} = \frac{1}{2}\,[L_{zi} + 2S_{zi}]\,B(\mathbf{r}_i) + \frac{1}{8}\,B^2(\mathbf{r}_i)\,r_i^2 \sin^2\theta_i\,. \tag{11.11}$$

Measured in atomic units, the relative proportion of the diamagnetic term to the paramagnetic one is of the order of:

$$\frac{\langle H_{\text{dZ}}\rangle}{\langle H_{\text{Z}}\rangle} \propto \frac{n^4\,B(\mathbf{r}_i)}{Z^2}\quad. \tag{11.12}$$

Here we have assumed that the angular momenta projections are close to unity, and for the expectation value of r_i^2, we have taken the hydrogenic solution from (B.43). In (11.12), $B(\mathbf{r})$ has to be taken in a.u. From appendix A, we learn that the atomic unit for magnetic flux density is of the order of 10^5 T. As a conclusion, for all laboratory magnetic fields, the second-order term will be very small. We will defer the treatment of this diamagnetic Zeeman effect to section 11.5, and here we will continue with the assumption that we only need to retain the linear term.

For an atom with more than one electron, and as long as the interaction with the magnetic field is weak, we may take the linear (paramagnetic) Zeeman Hamiltonian as a sum of all single-electron terms:

$$H_{\text{Z}} = \sum_{i=1}^{N} \frac{1}{2}\,[L_{zi} + 2S_{zi}]\,B(\mathbf{r}_i) = \frac{1}{2}\,[L_z + 2S_z]\,B\quad. \tag{11.13}$$

Here, it suffices to take the summation over the valence electrons — closed orbitals do not contribute to the overall **L** or **S** — and in the last step we have taken the field as homogeneous. We have also assumed *LS*-coupling. The total electronic orbital angular momentum and spin, and their projections, in (11.13) have the usual definitions — see section 6.1. Equation (11.13) can now be used to calculate the first-order Zeeman shift.

11.1.2 The Paramagnetic Zeeman Shift

An assumption we have made in the preceding paragraphs is that we can treat the interaction with the external field as a perturbation when all internal effects have been taken into account. Since the field can be arbitrarily weak, there will always exist a domain in which this will be true.

To the first order, this means that the energy shift to be computed from (11.13) is:

$$\begin{aligned} E_{\text{Z}} &= \frac{B}{2}\,\langle\, LSJM_J\,|\,(L_z + 2S_z)\,|\,LSJM_J\,\rangle \\ &= \frac{B}{2}\langle LSJM_J\,|\,(J_z + S_z)\,|\,LSJM_J\,\rangle = \frac{B}{2}\big(M_J + \langle LSJM_J\,|S_z|\,LSJM_J\rangle\,\big)\,. \end{aligned} \tag{11.14}$$

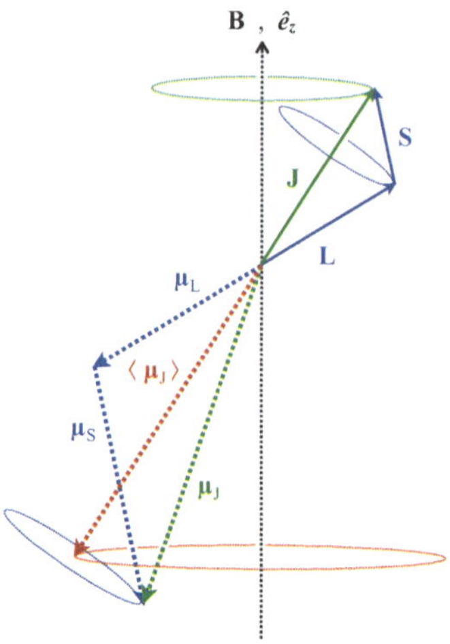

Fig. 11.1 Schematic illustration, in a vector model picture, of the influence of the Zeeman effect on an *LS*-coupled state. The dominating magnetic effect is the spin–orbit interaction, resulting in a fast precession of **L** and **S**. The fact that $g_s \neq g_l$ will render $\boldsymbol{\mu}_J$ non-parallel to **J**. A way to compute the overall coupling constant, g_J, is to consider the time average of $\boldsymbol{\mu}_J$, which can be arrived at by taking the projections of $\boldsymbol{\mu}_l$ and $\boldsymbol{\mu}_s$ along **J** (see the text).

A couple of assumptions have gone into (11.14). One is that we have again taken the magnetic field to be homogeneous over the entire volume of the atom, and thus we have dropped the spatial dependence from B. The second one is that we have assumed *LS*-coupling (and a $|JM_J\rangle$-representation) in the formulation of the matrix elements. The latter is strictly speaking not a restriction. The Hamiltonian is valid regardless of representation, but it is practical to choose basis functions that render the zero-order Hamiltonian as close to diagonal as possible.

We will attack the second term in (11.14) in different ways. The first one is by a classical vector model reasoning, where we are essentially taking a single-electron view of the issue at hand.

We illustrate the task in figure 11.1. Due to the spin–orbit interaction, the vectors **L** and **S** intercouple, and their vector sum is **J**. The effect of the external magnetic field is that the vectors **L** and **S** couple also to this, with a strength given by their respective g-factors. The combination of the two predicaments above means that the instantaneous torques on **L** and **S** will vary periodically. Since the two g-factors are different ($g_s \approx 2$ and $g_l \approx 1$), the total instantaneous electronic magnetic moment, for which we introduce the notation $\boldsymbol{\mu}_J$, will not be parallel with **J**.

At any given moment, $\boldsymbol{\mu}_J$ will interact with the magnetic field resulting in a torque around $\hat{e}_z$. However, if the field is weak, this will be a small effect, dwarfed by the fast precessions of **L** and **S**. The latter means that also $\boldsymbol{\mu}_J$ will precess, and it makes sense to define its time average. We call this the effective total magnetic moment, $\boldsymbol{\mu}_{J,\text{eff}}$, and we can think of this as representing the average overall interaction with the external field.

Mathematically, this amounts to forming $\boldsymbol{\mu}_{J,\text{eff}}$ by taking the projections of $\boldsymbol{\mu}_l$ and $\boldsymbol{\mu}_s$ along **J** and ignoring (or rather averaging out) the components of the individual moments that are orthogonal to their vector sum. The goal of the exercise is to find a factor of proportionality between $\boldsymbol{\mu}_{J,\text{eff}}$ and **J** — a g_J-factor, analogous to g_l

and g_s. For this, we will have to accept the caveat that g_J will not be a constant, but a function of **J**, **L** and **S**, or in practical terms of the corresponding quantum numbers J, L and S.

Calculating the vector projections, in a.u., we first get:

$$\left|\boldsymbol{\mu}_{J,\text{eff}}\right| = \left|\frac{\boldsymbol{\mu}_l \cdot \mathbf{J}}{|\mathbf{J}|} + \frac{\boldsymbol{\mu}_s \cdot \mathbf{J}}{|\mathbf{J}|}\right| = \frac{1}{2\,|\mathbf{J}|}\left|\mathbf{L}\cdot\mathbf{J} + 2\mathbf{S}\cdot\mathbf{J}\right| \ . \tag{11.15}$$

From the identity $\mathbf{J} \equiv \mathbf{L} + \mathbf{S}$, we can derive the equalities:

$$\begin{aligned} \mathbf{L}\cdot\mathbf{J} &= \frac{1}{2}\left(\mathbf{J}^2 + \mathbf{L}^2 - \mathbf{S}^2\right) \\ 2\mathbf{S}\cdot\mathbf{J} &= \left(\mathbf{J}^2 - \mathbf{L}^2 + \mathbf{S}^2\right) \ . \end{aligned} \tag{11.16}$$

We substitute this in (11.15) and obtain:

$$\left|\boldsymbol{\mu}_{J,\text{eff}}\right| = \frac{1}{4\,|\mathbf{J}|}\left|3\mathbf{J}^2 - \mathbf{L}^2 + \mathbf{S}^2\right| \ . \tag{11.17}$$

The result in (11.17) means that we can express g_J as:

$$\boldsymbol{\mu}_{J,\text{eff}} \equiv -\frac{1}{2}\, g_J\, \mathbf{J} = -\frac{1}{4}\left(\frac{3\mathbf{J}^2 - \mathbf{L}^2 + \mathbf{S}^2}{\mathbf{J}^2}\right)\mathbf{J} \, . \tag{11.18}$$

Here we have again used the fact that the Bohr magneton has the value one half in a.u. From (11.18), we find an identity for g_J. The time-averaged projections made above amount to neglecting matrix elements that are non-diagonal in the $|LSJM_J\rangle$-representation and if we take expectation values of the vector operators in (11.18), we find the following expression for g_J:

$$\begin{aligned} g_J &= \frac{1}{2}\left\langle\frac{3\mathbf{J}^2 - \mathbf{L}^2 + \mathbf{S}^2}{\mathbf{J}^2}\right\rangle = \frac{1}{2}\left(2 + \left\langle\frac{\mathbf{J}^2 - \mathbf{L}^2 + \mathbf{S}^2}{\mathbf{J}^2}\right\rangle\right) \\ &= 1 + \frac{J(J+1) - L(L+1) + S(S+1)}{2J(J+1)} \ , \end{aligned} \tag{11.19}$$

where in the last line, we have taken the quantum mechanical expressions for the expectation values of $\mathbf{L}^2$, $\mathbf{S}^2$ and $\mathbf{J}^2$ in LS-coupling. Equation (11.19) is known as *Landé's formula*, and g_J is referred to as *Landé's g-factor*. Having established this, the linear Zeeman shift for a state specified by the angular momentum quantum number M_J, for a homogeneous magnetic field parallel with the $\hat{e}_z$-axis, may be written as:

$$E_Z = \frac{g_J}{2} M_J B \ , \tag{11.20}$$

or in SI-units:

$$E_Z = \mu_B\, g_J\, M_J B \ . \tag{11.21}$$

When going from the absolute values in (11.17) to the definitions of g_J in (11.18) and (11.19), we choose a sign convention such that a positive field and a positive M_J give a positive Zeeman shift.

Albeit the classical derivation above is crude, the resulting final expressions for E_Z and g_J are quite accurate for *LS*-coupled atomic states. In the terminology of quantum mechanics, what we have done above is to neglect all basis states that are mixtures of different J. This is justified by assuming that the precession of **J** is so slow, compared to the spin–orbit interaction, that we may regard it as a constant of the motion at an initial level of approximation.

A shorter and more mathematical derivation of g_J gives the same result as above. For this, we commence by *postulating*:

$$\boldsymbol{\mu}_{J,\mathrm{eff}} = -\frac{1}{2} g_J \mathbf{J} \ , \tag{11.22}$$

and

$$E_Z = \frac{g_J}{2} M_J B = -\left\langle LSJM_J \left| \mu_{Jz} \right| LSJM_J \right\rangle B \ , \tag{11.23}$$

before we attempt to infer an expression for g_J.

We are looking for a projection of $\boldsymbol{\mu}_J$ along $\hat{e}_z$. We have chosen *LS*-coupling, but the same methodology would work also in other coupling schemes. According to the projection theorem — a special case of the Wigner-Eckart theorem, see (C.110) — we can project the matrix element in (11.23) as:

$$\begin{aligned} \left\langle LSJM_J \left| \mu_{Jz} \right| LSJM_J \right\rangle &= \frac{\langle LSJM_J \,|\mathbf{J}\cdot\boldsymbol{\mu}_J|\, LSJM_J \rangle}{J(J+1)} \langle LSJM_J \,|J_z|\, LSJM_J \rangle \\ &= \frac{M_J}{J(J+1)} \langle LSJM_J \,|\mathbf{J}\cdot\boldsymbol{\mu}_J|\, LSJM_J \rangle \ . \end{aligned} \tag{11.24}$$

We expand the operator $\mathbf{J}\cdot\boldsymbol{\mu}_J$ as:

$$\begin{aligned} \mathbf{J}\cdot\boldsymbol{\mu}_J = (\mathbf{L}+\mathbf{S})\cdot\left(-\frac{g_l}{2}\mathbf{L} - \frac{g_s}{2}\mathbf{S}\right) &= -\frac{1}{2}\left[g_l\left(\mathbf{L}^2+\mathbf{S}\cdot\mathbf{L}\right) + g_s\left(\mathbf{S}\cdot\mathbf{L}+\mathbf{S}^2\right)\right] \\ &= -\frac{1}{4}\left[g_l\left(\mathbf{J}^2+\mathbf{L}^2-\mathbf{S}^2\right) + g_s\left(\mathbf{J}^2-\mathbf{L}^2+\mathbf{S}^2\right)\right] \ . \end{aligned} \tag{11.25}$$

We take the expectation value of this and substitute it in (11.24) to get:

$$\begin{aligned} \left\langle LSJM_J \left| \mu_{Jz} \right| LSJM_J \right\rangle = -\frac{M_J}{4J(J+1)} \Big\{ & g_l\left[J(J+1)+L(L+1)-S(S+1)\right] \\ & + g_s\left[J(J+1)-L(L+1)+S(S+1)\right]\Big\} \ . \end{aligned} \tag{11.26}$$

Combining this with (11.23), and using again the approximate values of $g_s \approx 2$ and $g_l \approx 1$, we end up with the same expression for g_J as in (11.19). The fact that we get the same result is not surprising, since we use the same projection assumption, and then merely an alternative formalism. For a more accurate value of g_J, experimental values can be used, or more sophisticated methods from quantum chemistry.

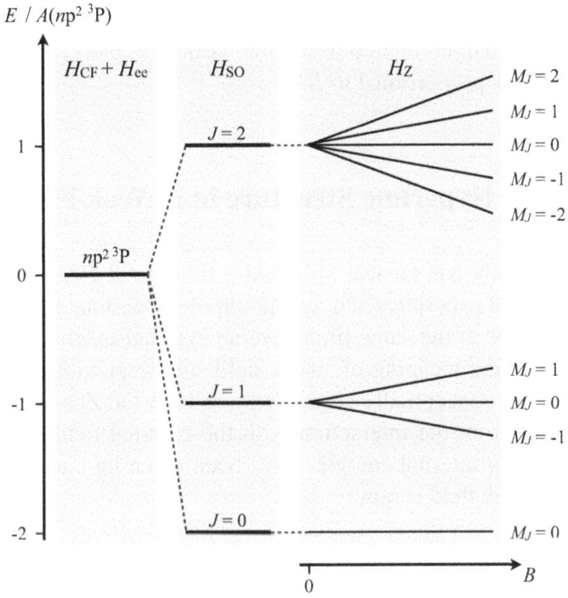

Fig. 11.2 Illustration of the Zeeman splitting for an idealised triplet term of an $n\mathrm{p}^3$ atom. The term is first divided into three fine-structure levels. The degenerate M_J levels are then split up proportionally to the magnetic field. The Zeeman splitting will be linear in B as long as it is small compared to the fine-structure. A magnetic substate with $M_J=0$ experiences no shift at this level of approximation.

We see from (11.20) that a consequence of the interaction with the external magnetic field is that the degeneracy in M_J is broken. A fine-structure level J will experience a *Zeeman splitting*, which gives $2J+1$ different energy eigenstates. The energy difference between adjacent states M_J and M_J+1 is:

$$\Delta E_\mathrm{Z} = \frac{g_J B}{2}\,, \tag{11.27}$$

or in SI-units:

$$\Delta E_\mathrm{Z} = g_J\, \mu_\mathrm{B}\, B\,. \tag{11.28}$$

As long as the interaction Hamiltonian can be treated as a perturbation to all internal interactions, this Zeeman splitting will increase linearly with the magnetic flux density B to a good level of approximation. The Zeeman splitting in (11.28) is often given in frequency units, as:

$$\nu_\mathrm{L} \equiv \frac{\Delta E_\mathrm{Z}}{h}\,, \tag{11.29}$$

where ν_L is known as the *Larmor frequency*.

In figure 11.2, we show an illustration of the Zeeman effect, on an idealised LS-coupled state belonging to the two-electron term $n\mathrm{p}^{2\ 3}\mathrm{P}$. The fine-structure level $J=2$ is five-fold degenerate in the absence of a magnetic field, and $J=1$ three-fold.

The inclusion of the Zeeman Hamiltonian breaks the degeneracy, and as long as the Zeeman splitting is small compared to the fine-structure, the Zeeman effect contribution to the energy is proportional to B.

11.2 Atoms with Hyperfine Structure in a Weak Field

If the atom under study has nuclear spin and a finite total electronic angular momentum ($J > 0$), it will experience a magnetic dipole hyperfine structure interaction (see chapter 10). If we at the same time have an external magnetic field present at the site of the atom, the meaning of 'weak field' in the previous section changes. As in section 11.1, a conceptually simple way to treat the Zeeman effect is if the energy contribution from the interaction with the external field can be applied as a perturbation after *all* internal energies have been taken into account. That is, the requirement for a weak field is now:

$$H_{\mathrm{Z}} < H_{\mathrm{hfs}} < \{H_{\mathrm{SO}}, H_{\mathrm{ee}}\} < H_{\mathrm{CFA}} \,. \tag{11.30}$$

Equation (11.30) should be understood as a comparison of the energy contributions from the respective Hamiltonians, and we have left it open if the atom obeys *LS*-coupling, *jj*-coupling or some other more complicated coupling configuration. The important point is that the hyperfine structure should be added as a perturbation before the Zeeman effect.

Since the hyperfine structure is small, the region within which this approximation is valid will be much more limited than the corresponding interval was in section 11.1. Even with a relatively modest magnetic field, from a laboratory point of view, the field may be beyond what can safely be regarded as weak, and we then have to turn to methods that will be described in forthcoming sections. However, in the weak field limit, the practical treatment of the Zeeman effect, in the presence of hyperfine structure, can be pursued in a fashion very similar to that presented in the previous section.

11.2.1 The Zeeman Effect Hamiltonian in the Presence of Hyperfine Structure

We have to take into account the magnetic moments both of the electronic ensemble, and of the nucleus. To that end, we reformulate the Zeeman Hamiltonian in 11.6 as:

$$H_{\mathrm{Z}} = -(\boldsymbol{\mu}_{J,\mathrm{eff}} + \boldsymbol{\mu}_I) \cdot \mathbf{B}(\mathbf{r}) = \frac{1}{\hbar}(g_J\, \mu_{\mathrm{B}}\, \mathbf{J} - g_I\, \mu_{\mathrm{N}}\, \mathbf{I}) \cdot \mathbf{B}(\mathbf{r}) \,. \tag{11.31}$$

This is in SI-units. The Landé g-factor, g_J, is the one defined in (11.19) — a function of J, L and S — and accordingly $\boldsymbol{\mu}_{J,\mathrm{eff}}$ is the time-averaged overall electronic magnetic moment calculated in (11.15) and following equations. The nuclear g-factor, g_I, is the one introduced in chapter 10, in (10.15), where we also defined the nuclear magneton μ_N. In the analyses that follows, we are also going to need the total atomic angular momentum vector operator $\mathbf{F}$, as introduced in (10.21) and (10.22).

The operator H_Z in (11.31) commutes with J_z, I_z and F_z, but not with $\mathbf{F}^2$. In contrast, the hfs Hamiltonian (10.20) commutes with $\mathbf{F}^2$ and F_z, but not with J_z and I_z. Thus, the only rigorously good quantum number is now M_F. However, for very weak fields we can ignore the matrix elements that are non-diagonal in F, and we can calculate the energy perturbation due to (11.31) in the representation $|JIFM_F\rangle$. We can follow a path analogous to that used in section 11.1.2 for calculating $\mu_{J,\text{eff}}$ — as illustrated in figure 11.1 — with the important difference that g_J and g_I are not constants of the motion. Another difference is that this time the two g-factors are very different in magnitude. Other than that, we can use the same reasoning and formalism that was used in conjunction with figure 11.1, exchanging $\mathbf{J}$, $\mathbf{I}$ and $\mathbf{F}$ for $\mathbf{L}$, $\mathbf{S}$ and $\mathbf{J}$.

Accordingly, we start by projecting the moments in (11.31) along $\mathbf{F}$. This amounts to formulating a new total magnetic moment, for an atom with magnetic dipole hfs, defined as:

$$\boldsymbol{\mu}_{F,\text{eff}} \equiv \boldsymbol{\mu}_{J,\text{eff}} + \boldsymbol{\mu}_I \ , \tag{11.32}$$

and a corresponding Hamiltonian as:

$$H_Z = -\boldsymbol{\mu}_{F,\text{eff}} \cdot \mathbf{B}(\mathbf{r}) = \frac{\mu_B}{\hbar}\, g_F\, \mathbf{F} \cdot \mathbf{B}(\mathbf{r}) \ . \tag{11.33}$$

Here we have introduced a new g-factor, g_F, and we note that this should include a small term, proportional to μ_N/μ_B. The inclusion of μ_N is the reason for why, at this point, it is convenient to use to SI-units.

Comparing (11.31) and (11.33), the absolute value of the magnetic moment $\mu_{F,\text{eff}}$ may be found with the same strategy as that used in (11.15):

$$\begin{aligned} |\mu_{F,\text{eff}}| &= \left| -\frac{\mu_B\, g_J\, \mathbf{J} \cdot \mathbf{F}}{\hbar\, |\mathbf{F}|} + \frac{\mu_N\, g_I\, \mathbf{I} \cdot \mathbf{F}}{\hbar\, |\mathbf{F}|} \right| \\ &= \left| -\frac{\mu_B\, g_J}{2\hbar\, |\mathbf{F}|} \left(\mathbf{F}^2 + \mathbf{J}^2 - \mathbf{I}^2\right) + \frac{\mu_N\, g_I}{2\hbar\, |\mathbf{F}|} \left(\mathbf{F}^2 + \mathbf{I}^2 - \mathbf{J}^2\right) \right| \ . \end{aligned} \tag{11.34}$$

Next step is to take the expectation value of the Hamiltonian in (11.33). From this, we arrive at the following expression for g_F:

$$\begin{aligned} g_F &= \frac{g_J}{2} \left\langle \frac{\mathbf{F}^2 + \mathbf{J}^2 - \mathbf{I}^2}{|\mathbf{F}^2|} \right\rangle - \frac{\mu_N}{\mu_B} \frac{g_I}{2} \left\langle \frac{\mathbf{F}^2 + \mathbf{I}^2 - \mathbf{J}^2}{|\mathbf{F}^2|} \right\rangle \\ &= g_J\, \frac{F(F+1) + J(J+1) - I(I+1)}{2F(F+1)} \\ &\quad - \frac{\mu_N}{\mu_B}\, g_I\, \frac{F(F+1) + I(I+1) - J(J+1)}{2F(F+1)} \ . \end{aligned} \tag{11.35}$$

11.2.2 Energy Levels

The ratio between the nuclear magneton and the Bohr magneton is of the same order as that between the electron and nuclear masses. Thus, in a first-order calculation, the second term in (11.35) can be ignored, and with that in mind we can formulate a simplified expression for g_F:

$$g_F \approx g_J \frac{F(F+1)+J(J+1)-I(I+1)}{2F(F+1)} \ . \tag{11.36}$$

Important to note here is that even though the magnetic moment of the nucleus has been neglected, the nuclear spin still influences the retained term. This is because we have calculated a time-averaged magnetic moment, whose instantaneous value is affected by the hfs magnetic dipole induced precession of **I** and **J**.

As before, we chose the magnetic field direction as the quantisation axis $\hat{e}_z$, and we assume the field to be homogeneous. The Hamiltonian (11.33) can then be simplified as:

$$H_Z = \frac{\mu_B}{\hbar} g_F F_z B \ . \tag{11.37}$$

This is in SI-units. Since we no longer have any term that includes the nuclear magneton, it is straightforward to return to atomic units and write:

$$H_Z = \frac{1}{2} g_F F_z B \ , \tag{11.38}$$

where yet again the magnetic flux density must be given in a.u.

Zeeman shifted energy levels are found from an expression analogous to (11.20):

$$E_Z = \frac{g_F}{2} M_F B \ , \tag{11.39}$$

and the splitting between two adjacent levels is:

$$\Delta E_Z = \frac{g_F B}{2} \ . \tag{11.40}$$

This time, it is the degeneracy in M_F which is broken. Equation (11.39) will be valid as long as the field is so weak that the Zeeman splitting does not become comparable to the hyperfine splitting.

11.3 The Paschen-Back Effect — Strong Fields

In the two preceding sections, we have been preoccupied with *weak fields*. The meaning of 'weak' in this context is that the magnetic field is so small that the B-field Hamiltonian can be treated as a perturbation with respect to all internal degrees of freedom of the atom. In this section, we will diverge from this. We will assume that the field is strong enough to be taken as perturbation *before* the smallest of all the internal Hamiltonians. Many times, the actual field is intermediate between 'weak' and 'strong'.

When we say that the perturbation due to the magnetic field should be stronger than the smallest of the internal energies, we have not addressed which one that is, and the answer to that question will depend on what the most pertinent coupling scheme is for the atom under study, and whether or not the atom has hyperfine structure.

The least complex situation would be a one-electron atom with a light nucleus without spin. In that case, the Paschen-Back regime is that for which H_Z is considerably more important than the one-electron spin–orbit interaction. From there, going to a multielectron atom with almost pure *LS*-coupling, and no hyperfine structure, is a generalisation that can be made in the same way as previously, that is by adopting the CFA and *LS*-coupling formalism. We will begin by studying this circumstance in section 11.3.1.

In section 11.3.2, we will turn to atoms that are *LS*-coupled, but that do have hyperfine structure. For this regime, the Paschen-Back effect becomes important for a much smaller magnetic field, since the interaction with the field only needs to be greater than H_{hfs}.

For *jj*-coupling, the Paschen-Back regime is rarely relevant. Atoms that obey approximate *jj*-coupling have large Z, and accordingly pronounced spin–orbit interactions. The magnetic field must for such state be *very* strong if it should overcome internal Hamiltonians. This could mean that the diamagnetic Zeeman effect — see (11.9) — starts to be relevant, and it would also bring with it a substantial mixing of electron configurations. The same is true for the *complete Paschen-Back effect*, which appears when the magnetic field is so strong that its interaction with the electrons swamps *all* angular, electronic interaction. We will briefly comment on this in section 11.5.

In section 11.4, we will study the general contingency, when neither the weak field nor the strong field limit can be assumed with fidelity. To deal with that, one typically has to work with a representation in which the Hamiltonian is non-diagonal. This is a transitional situation between the weak field Zeeman effect and the Paschen-Back effect.

11.3.1 Perturbation Approach in the Absence of Hyperfine Structure

The situation addressed in this section is that for which the order of energy contributions is the following:

$$H_{\text{CFA}} > H_{\text{ee}} > H_Z > H_{\text{SO}} \quad , \tag{11.41}$$

and the perturbation theory will be applied in that order. This is pertinent for a strong field, but also for an atom with very weak total spin–orbit coupling. For an *LS*-coupled atom, this means that the total electronic orbital angular momenta, **L** and **S**, will couple more strongly to the external magnetic field than they do to each other. This can also be stated as that **L** and **S** become decoupled in the presence of the *B*-field. The spin–orbit interaction will be added only as the very last perturbation to the Paschen-Back state and, when it is, it will be on an individual electron basis.

A consequence of this is that the representation $|\gamma LSJM_J\rangle$, which we typically use for fine-structure states in *LS*-coupling, will no longer yield even an approximately diagonal total Hamiltonian. The coupling of **L** and **S** to **B** will make the vector **J** more or less meaningless. The non-diagonal terms in the $|\gamma LSJM_J\rangle$-representation, which could be ignored for weak fields, will now instead be dominant.

The logical representation when **L** and **S** are decoupled is instead $|\gamma LSM_LM_S\rangle$. For the order in which we take the Hamiltonians in (11.41), this is not a problem. The electron–electron Hamiltonian, H_{ee} is diagonal in any representation where first **L** and **S** are formed by the individual electronic momenta $\mathbf{L}_i$ and $\mathbf{S}_i$ (this is essentially the definition of *LS*-coupling). We do, however, need the field not to be too strong, to assure that **L** and **S** will not also be broken up.

We should note that even if J is not a good quantum number in the Paschen-Back regime, M_J is still a useful parameter since it must be the sum of the projections M_L and M_S. This is of importance for the intermediate field regime because it means that a large matrix can be divided up in several smaller ones for different values of M_J (see section 11.4).

The Hamiltonian is the same as in (11.13), namely

$$H_Z = \frac{1}{2}(L_z + 2S_z)B \ , \tag{11.42}$$

where we, as in previous sections, take the quantisation axis $\hat{e}_z$ as parallel with the homogeneous magnetic field **B**, and use the approximate values for g_l and g_s. The difference is now that we need to take the expectation value of this in the representation $|LSM_LM_S\rangle$. This means that the field-dependent energy perturbation of a term ^{2S+1}L will be:

$$E_Z = \frac{1}{2}(M_L + 2M_S)B\,. \tag{11.43}$$

Note that, in both (11.43) and (11.42), we have used atomic units and we have retained the subscript 'Z' also for the Paschen-Back regime (it is the same Hamiltonian). The *LS*-coupling term will here be divided up in $(2L+1)(2S+1)$ states. Some of these may be energy degenerate, and the separation between different energies will be in integer multiples of $B/2$ (or $\mu_B B$ in SI-units).

If the atom has a weak spin–orbit interaction, this must now be added as a perturbation to the Paschen-Back states. As when we previously treated the spin–orbit interaction perturbatively, the first-order correction is obtained by taking the expectation value of the operator H_{SO}. The difference is that this time, we must do it in another representation than customary. What we have are two angular momentum vectors, **L** and **S**, precessing around the magnetic field direction (chosen as $\hat{e}_z$), and now these precessions will be slightly perturbed by the spin–orbit interaction — rather than the other way around. Along the $\hat{e}_z$ direction, this interaction will slightly change the projections of **L** and **S**.

We can use the same solution to the problem as the one we showed in section 7.3.2. That is, we take (7.53) as the Hamiltonian and the energy contribution will be exactly as in (7.55):

$$\begin{aligned} E_{\mathrm{SO}} &= \langle LSM_LM_S \,|H_{\mathrm{SO}}|\, LSM_LM_S \rangle \\ &= A(\gamma LS) \left\langle LSM_LM_S \,\middle|\, L_xS_x + L_yS_y + L_zS_z \,\middle|\, LSM_LM_S \right\rangle \\ &= A(\gamma LS)\, M_LM_S \quad , \end{aligned} \tag{11.44}$$

with $A(\gamma LS)$ being the fine-structure factor for the LS-coupling term. One way to interpret this, alternative to that given in section 7.3.2, is that for specific values of M_L and M_S, the time averages of the operators for the projections along $\hat{e}_x$ and $\hat{e}_y$ are zero, due to the fast precession of **L** and **S** around $\hat{e}_z$.

The spin–orbit interaction perturbs the Paschen-Back states, and if some of the latter have the same energies, a degeneracy may be broken. We illustrate this in figure 11.3, for a specific example of a fictional $n\mathrm{p}^2\ {}^3\mathrm{P}$ atomic state. The strong field is shown in the right part of the figure, and to the left we have included, as a reference, the situation for a weak field, analogous to figure 11.2. Before the spin–orbit interaction has been taken into account, the term splits up in nine Paschen-Back states, with seven different energies. The spin–orbit perturbation removes the remaining degeneracy. Since M_J must be the sum of M_L and M_S, the states in the left part of the diagram can be directly related to the ones to the right. This is useful when studying the intermediate regime (see section 11.4).

11.3.2 The Case with Hyperfine Structure and a Strong Field

The situation treated in the previous section requires a fairly strong field in order to be relevant. For an atom with hyperfine structure, however, the Paschen-Back regime comes into play already for relatively modest laboratory magnetic fields. For this eventuality, it is sufficient that the field-dependent Zeeman Hamiltonian will provide a larger energy contribution than H_{hfs}.

The reasoning will be very similar to that in the previous section. For simplicity, we assume LS-coupling, and that after the fine-structure has been taken into account, the strongest of the remaining perturbations is the interaction with the magnetic field. We take the latter as being homogeneous and directed along $\hat{e}_z$. We use the Hamiltonian from (11.31), which in atomic units is:

$$H_{\mathrm{Z}} = \frac{1}{2}\left(g_J J_z - \frac{g_I}{m_{\mathrm{p}}} I_z\right) B \quad , \tag{11.45}$$

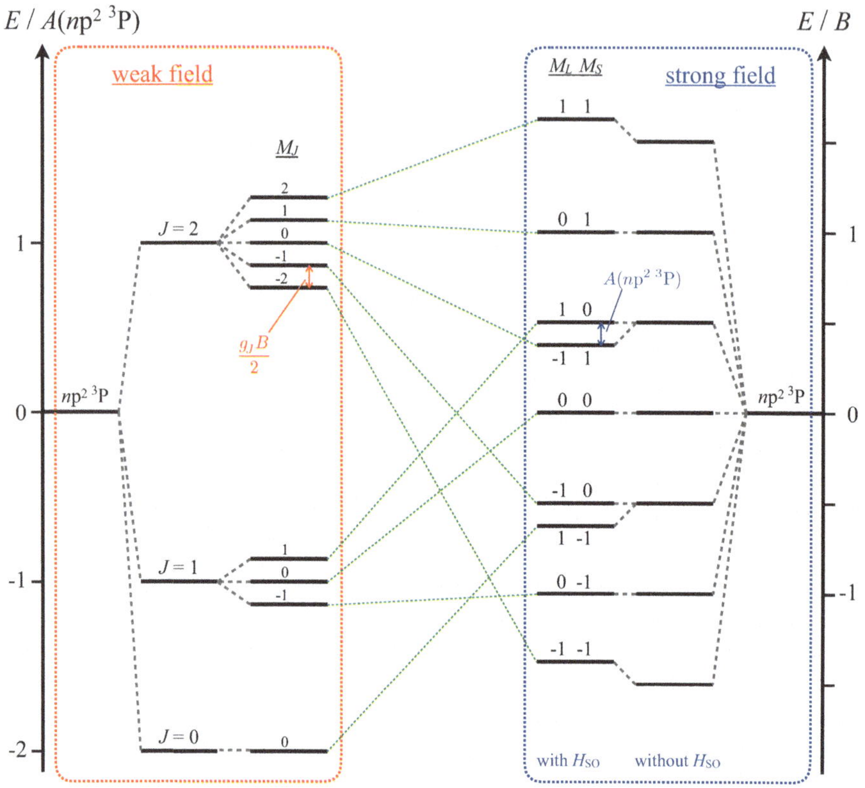

Fig. 11.3 The right part of the figure illustrates the Paschen-Back effect for an idealised triplet term of an np^3 atom. For comparison, the weak field version (from figure 11.2) is shown in the left part. Energies and the field are in atomic units, and the two energy scales are very different. In each domain, some constant magnetic flux density (small or great) has been assumed. The dashed green lines combine states with definite M_J between the two regimes. Note that states with the same values of M_J do not cross

where B and the proton mass, m_{p}, must be given in a.u. The perturbation energy is:

$$E_Z = \frac{1}{2}\left(g_J M_J - \frac{g_I}{m_{\mathrm{p}}} M_I\right) B \ . \tag{11.46}$$

A difference from the situation in section 11.3.1 is the g-factors. The Landé factor g_J is the one in (11.19), and the factor for the nucleus, g_I, is dependent on the isotope under study (see section 10.2.1). The second term in (11.46) will typically be negligible. In analogy with the Paschen-Back effect for the fine-structure, we here have the angular momenta **J** and **I** being decoupled, due to their strong individual coupling to the magnetic field. This makes F irrelevant, and the representation we must use is $|IJM_IM_J\rangle$. This has already been made implicit in (11.46).

The final step is to take the hyperfine structure Hamiltonian as the last perturbation. We here restrict the study to the magnetic dipole hfs and take the interaction from (10.20). In the chosen representation, the hfs perturbation to the Paschen-Back levels is :

$$\begin{aligned} E_{\text{hfs-MD}} &= \langle IJM_IM_J \,|a_{\text{hfs}}\,\mathbf{I}\cdot\mathbf{J}|\, IJM_IM_J \rangle \\ &= a_{\text{hfs}} \left\langle IJM_IM_J \,\middle|\, J_xI_x + J_yI_y + J_zI_z \,\middle|\, IJM_IM_J \right\rangle \\ &= a_{\text{hfs}}\, M_IM_J \ . \end{aligned} \tag{11.47}$$

We illustrate also this with an example, and we will use empiric energies for the ground state of Na. The diagram is shown in figure 11.4. In the right part of the

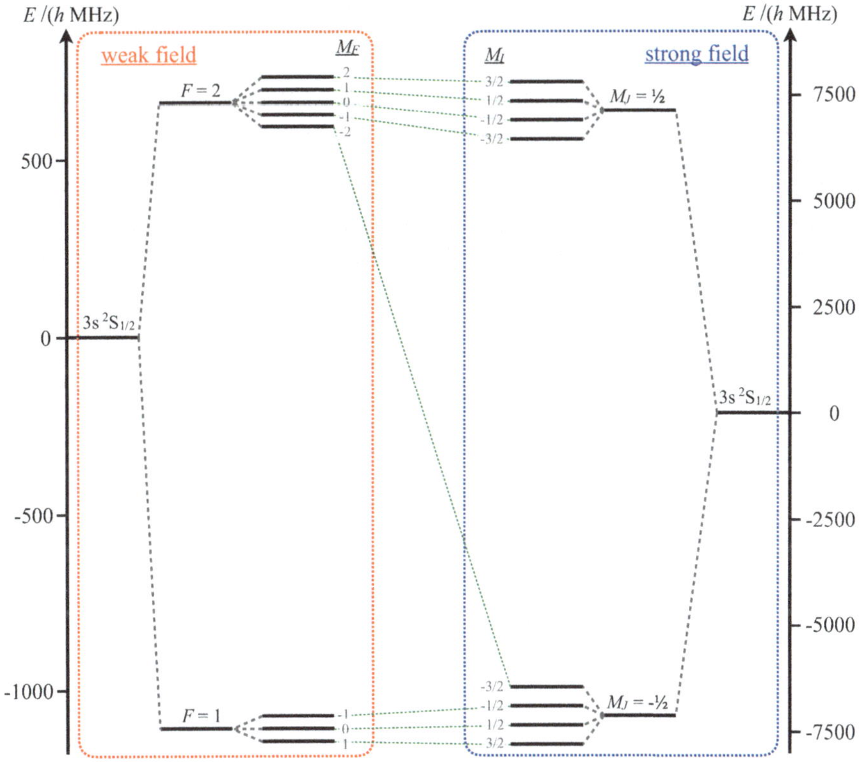

Fig. 11.4 Energy diagram showing the ground state of Na, $3s\,^2S_{1/2}$, in the presence of an external magnetic field. The data for the magnetic dipole hyperfine structure constant and the g-factors have been taken from [3]. The left section of the figure shows a weak field case, and for the Zeeman split levels, we have assumed a magnetic field of 5 mT. The right part of the figure is for a strong magnetic field, taken as 0.5 T. For a strong field, the most important perturbation to the fine-structure level is the interaction with the field, resulting in the Paschen-Back levels. These states are subsequently perturbed by the magnetic dipole hfs interaction. The dashed green lines correspond to constant values of $M_F = M_I + M_J$, and connects the weak and strong field domains. Note that the energy scales differ with about one order of magnitude.

figure, it appears as if, before the hfs has been taken into account, there are just two Paschen-Back levels — both four-fold degenerate. This is not strictly true, and the application of the hfs perturbation is not a break of degeneracy per se, since all eight states are non-degenerate already at the Paschen-Back level. This is due to the small nuclear magnetic moment term in (11.46).

11.4 Intermediate Fields

After having studied interactions between atoms and magnetic fields, both weak and strong, we have to examine the situation for which the field is rather intermediate. This is when the magnetic field is too strong to make the Zeeman approximations of (11.20) or (11.39) appropriate, but still not strong enough to fulfil the Paschen-Back approximations of (11.43) or (11.46). When two Hamiltonians are of the same order of magnitude, neither one nor the other can be given precedence in a prudent perturbative calculation. The matrix representing the sum of Hamiltonians must be formulated. It may not be diagonal in any simple representation, and in that case, it must instead be diagonalised.

Before we go down this road, we will first attempt to avoid this by applying second-order perturbation theory, and still take Hamiltonians one by one. This is not really simpler than the full solution, and it will only give quantitatively reasonable results if the field is such that the deviation from either of the limiting cases is relatively small. It nevertheless has a value to consider this approach, because of some qualitative insights that might be gained.

The second-order correction to (11.14) for an energy level E_p — taking as example an isotope without hyperfine structure — is:

$$E_{Zp}^{(2)} = \frac{B^2}{4} \sum_{q \neq p} \frac{\left| \left\langle L_p S_p J_p M_{Jp} \mid (L_z + 2S_z) \mid L_q S_q J_q M_{Jq} \right\rangle \right|^2}{E_p^{(0)} - E_q^{(0)}} . \tag{11.48}$$

This sum will rarely contain many terms of appreciable magnitude. First, it should be enough to restrict the summation to within one configuration. Secondly, if H_{SO} and H_{Z} have comparable contributions, the sum of these two operators will not commute with either of $\mathbf{J}^2$, L_z or S_z. However, it will commute with J_z, and

$$M_J = M_L + M_S \tag{11.49}$$

will be a good quantum number. This means that the only terms that need to be retained in (11.48) are those for which $M_{Jq} = M_{Jp}$. If the configuration has a Zeeman level with a unique M_J, this will not have a second-order perturbative term and will remain linear all the way to the high-field domain. For M_J values that are represented twice, these energies will bend quadratically for increasing field, and due to the denominator in (11.48), two such Zeeman levels will bend away from each other. A most important consequence of that is that in the transition region between low and high fields, two levels with the same M_J will not cross each other.

11.4.1 Diagonalising the Intermediate Field Hamiltonian

The spin–orbit Hamiltonian H_{SO} is diagonal in the quantum numbers $|LSJM_J\rangle$, but not in $|LSM_LM_S\rangle$. For H_Z, the opposite is true. Since we will have to solve a secular equation either way, this means that we can choose any of the two representations above. In the example that follows, we will opt for $|LSM_LM_S\rangle$ and then search for the matrix elements of H_{SO}. In our initial example, we will work under the assumption that there is no hyperfine structure. If there is, the formalism will still follow the same essential lines, and we will deal with the particularities that arise for atoms with hfs in the next section.

The spin–orbit interaction Hamiltonian can, according to (7.54), be reformulated as:

$$H_{SO} = A(\gamma LS)\,\mathbf{L}\cdot\mathbf{S} = A(\gamma LS)\left[L_zS_z + \frac{1}{2}(L_+S_- + L_-S_+)\right] . \tag{11.50}$$

The diagonal matrix elements will turn out as in (7.55) and, with the use of (C.20) and (C.22), we can derive expressions for the non-diagonal ones:

$$\begin{aligned}
&\langle L,S,M_L+1,M_S-1\,|H_{SO}|\,LSM_LM_S\rangle \\
&\quad = \frac{A(\gamma LS)}{2}\sqrt{S(S+1)-M_S(M_S-1)}\,\sqrt{L(L+1)-M_L(M_L+1)} \\
&\langle L,S,M_L-1,M_S+1\,|H_{SO}|\,LSM_LM_S\rangle \\
&\quad = \frac{A(\gamma LS)}{2}\sqrt{S(S+1)-M_S(M_S+1)}\,\sqrt{L(L+1)-M_L(M_L-1)} .
\end{aligned} \tag{11.51}$$

There is no need to attempt to diagonalise the entire matrix. If the latter is organised after M_J, there will be blocks along the diagonal whose smaller secular equations can be dealt with individually. Among these, none will be of greater dimension than $2S+1$ if $S<L$, or $2L+1$ if $L<S$. For the special case of a single valence electron, and $M_J = M_L \pm 1/2$, (11.51) simplifies to:

$$\begin{aligned}
&\langle L,\,S,\,M_L\pm 1,\,\mp 1/2\,|H_{SO}|\,L,\,S,\,M_L,\,\pm 1/2\rangle \\
&\quad = \frac{A(\gamma LS)}{2}\sqrt{L(L+1)-M_L(M_L\pm 1)} \\
&\quad = \frac{A(\gamma LS)}{2}\sqrt{L(L+1)-M_J^2+1/4} ,
\end{aligned} \tag{11.52}$$

and the largest matrices left to diagonalise will be of dimension two.

11.4.2 Intermediate Field and Hyperfine Structure

When the smallest internal Hamiltonian is that of the hyperfine structure, already a moderate magnetic field will qualify as being of intermediate strength. The proper form of the Zeeman Hamiltonian is in this situation that of (11.45). For the hyperfine structure, the complexity of the Hamiltonian will depend on the magnitudes of the nuclear spin **I** and the total electronic angular momentum **J** — see section 10.2 — and also on the coupling constants for the various orders of the hfs.

If either of the limitations $I \leq 1/2$ or $J \leq 1/2$ are in play, the only hfs component will be the magnetic dipole interaction between the nucleus and the electrons. For higher electronic angular momentum and greater nuclear spin, more terms will be present, but as a first-order approximation, we may neglect these with the rationale that they are typically smaller than the magnetic dipole effect. This simplification carries the advantage that the remaining hfs Hamiltonian in (10.20) will be simple and formally similar to the spin–orbit interaction in fine-structure. A straightforward adaptation of the analysis from the preceding section can then be applied, merely by exchanging a few angular momenta and g-factors.

We write the intermediate field Hamiltonian as:

$$\begin{aligned} H_{\text{interm}} &= \frac{1}{2}\left(g_J J_z - \frac{g_I}{m_{\text{p}}} I_z\right) B + a_{\text{hfs}}\, \mathbf{I}\cdot\mathbf{J} \\ &= \frac{1}{2}\left(g_J J_z - \frac{g_I}{m_{\text{p}}} I_z\right) B + a_{\text{hfs}} \left[I_z J_z + \frac{1}{2}\left(I_+ J_- + I_- J_+\right)\right] . \end{aligned} \tag{11.53}$$

We use the representation $|IJM_IM_J\rangle$, and in analogy with (11.46), (11.47) and (11.51), the matrix elements will be:

$$\begin{aligned} &\langle IJM_IM_J \,|H_{\text{interm}}|\, IJM_IM_J\rangle = \frac{1}{2}\left(g_J M_J - \frac{g_I}{m_{\text{p}}} M_I\right) B + a_{\text{hfs}}\, M_I M_J \\ &\langle I,J,M_I{+}1,M_J{-}1 \,|H_{\text{interm}}|\, IJM_IM_J\rangle \\ &\quad = \frac{a_{\text{hfs}}}{2}\sqrt{J(J{+}1)-M_J(M_J{-}1)}\,\sqrt{I(I{+}1)-M_I(M_I{+}1)} \\ &\langle I,J,M_I{-}1,M_J{+}1 \,|H_{\text{interm}}|\, IJM_IM_J\rangle \\ &\quad = \frac{a_{\text{hfs}}}{2}\sqrt{J(J{+}1)-M_J(M_J{+}1)}\,\sqrt{I(I{+}1)-M_I(M_I{-}1)} . \end{aligned} \tag{11.54}$$

This time, it is the total projection quantum number M_F which remains unchanged for a varying magnetic field. The total matrix in (11.54) can thus be divided into sub-matrices, one for each M_F, and they will have the maximum dimension of whatever is the smallest value of $2I{+}1$ or $2J{+}1$.

11.4.3 The Breit–Rabi Formula

Special attention has been given to states for which either $I = 1/2$ or $J = 1/2$. Those constraints reduce the maximum size of the secular equations to dimension two, and it becomes easier to set up explicit expressions for the energies as functions of the field strength. One reason for the extra emphasis on this situation is its relevance for hydrogen and alkalis — frequently used in metrology applications.

If we take $J = 1/2$ (valid for the ground state of all alkalis), and thus $M_F = M_I \pm 1/2$, the diagonal and non-diagonal matrix elements are, from (11.54):

$$\langle M_I, \pm 1/2 \,|\, H_{\text{interm}} \,|\, M_I, \pm 1/2 \rangle$$
$$= \left[-\frac{g_I B M_F}{2 m_{\text{p}}} - \frac{a_{\text{hfs}}}{4} \right] \pm \left[\frac{g_J B}{4} + \frac{g_I B}{4 m_{\text{p}}} + \frac{a_{\text{hfs}} M_F}{2} \right] \equiv d_1 \pm d_2 \,, \quad (11.55)$$

and

$$\langle M_I \pm 1, \mp 1/2 \,|\, H_{\text{interm}} \,|\, M_I, \pm 1/2 \rangle = \frac{a_{\text{hfs}}}{2} \sqrt{I(I+1) - M_F^2 + \frac{1}{4}} \equiv n \,. \quad (11.56)$$

To shorten the notation, we have written the state vectors without I and J, and we have introduced d_1, d_2 and n. There will be two unambiguous states corresponding to the aligned spins with $M_F = I + 1/2$ and $M_F = -I - 1/2$. The energies for these are given directly by (11.55) and they will be linear for all values of B.

Other values of M_F are represented by 2×2 non-diagonal matrices. A general solution to the secular equation for a matrix of the form:

$$\begin{pmatrix} d_1 + d_2 & n \\ n & d_1 - d_2 \end{pmatrix} \quad (11.57)$$

gives the eigenvalues as:

$$\lambda = d_1 \pm \sqrt{d_2^2 + n^2} \,. \quad (11.58)$$

This means that the energies of a pair of levels with the same values of M_F are:

$$E_{\text{interm}} = -\frac{a_{\text{hfs}}}{4} - \frac{g_I B M_F}{2 m_{\text{p}}}$$
$$\pm \frac{a_{\text{hfs}} \left(I + \frac{1}{2}\right)}{2} \left[1 + \frac{\left(g_J + \frac{g_I}{m_{\text{p}}}\right) B M_F}{a_{\text{hfs}} \left(I + \frac{1}{2}\right)^2} + \frac{\left(g_J + \frac{g_I}{m_{\text{p}}}\right)^2 B^2}{4 a_{\text{hfs}}^2 \left(I + \frac{1}{2}\right)^2} \right]^{1/2} . \quad (11.59)$$

This is known as the *Breit–Rabi formula*. The plus sign is for $M_F = M_I + 1/2$ and minus for $M_F = M_I - 1/2$.

In many works, the Breit–Rabi formula is formulated together with the following two substitutions:

$$\Delta W \equiv a_{\text{hfs}}\left(I+\frac{1}{2}\right)$$
$$x \equiv \left(g_J - \frac{g_I}{m_{\text{p}}}\right)\frac{B}{2\Delta W} \quad . \tag{11.60}$$

ΔW is the separation between hfs energies in the absence of a field. With the definitions above, (11.59) can be rewritten as:

$$E_{\text{interm}} = -\frac{\Delta W}{2(2I+1)} - \frac{g_I B M_F}{2m_{\text{p}}} \pm \frac{\Delta W}{2}\sqrt{1+\frac{4M_F x}{(2I+1)}+x^2} \quad . \tag{11.61}$$

This is in keeping with more standard notations, except for the caveat that we have written (11.59) and (11.61) in atomic units. For quantitative results on an absolute scale, B and m_{p} must be in its respective a.u., and a_{hfs}, ΔW and E_{interm} in E_{h}. For this expression, not so much simplicity is gained by the use of atomic units, and it is more common in the literature that the Breit–Rabi formula is given in SI-units:

$$E_{\text{interm}} = -\frac{\Delta W}{2(2I+1)} - g_I \mu_{\text{N}} B M_F \pm \frac{\Delta W}{2}\sqrt{1+\frac{4M_F x}{2I+1}+x^2} \quad , \tag{11.62}$$

wherein now:

$$\Delta W \equiv a_{\text{hfs}}\hbar^2\left(I+\frac{1}{2}\right)$$
$$x \equiv (g_J\mu_{\text{B}} + g_I\mu_{\text{N}})\frac{B}{\Delta W} \quad . \tag{11.63}$$

We will exemplify the Breit–Rabi formula by calculating energies for the intermediate region in figure 11.4. That is, the range of magnetic field flux densities that renders both the Zeeman and the Paschen-back approximations inadequate. This is shown in figure 11.5.

The small deviation from a linear Zeeman shift for a weak field, included in the last term of (11.61) and (11.62), is sometimes referred to as the second-order, or quadratic, Zeeman effect — see also (11.48). There are some justifications for this terminology since it will indeed lead to a quadratic deviation in the transition between the Zeeman and Paschen-Back regions. Unfortunately, it also leads to an ambiguity since the diamagnetic Zeeman effect — the last term in (11.11) — is also dubbed 'quadratic'. We will aim to avoid some of the confusion by strictly using the term 'intermediate field' for the situation described in this section, and 'diamagnetic' for the very strong field regime dealt with in the following section.

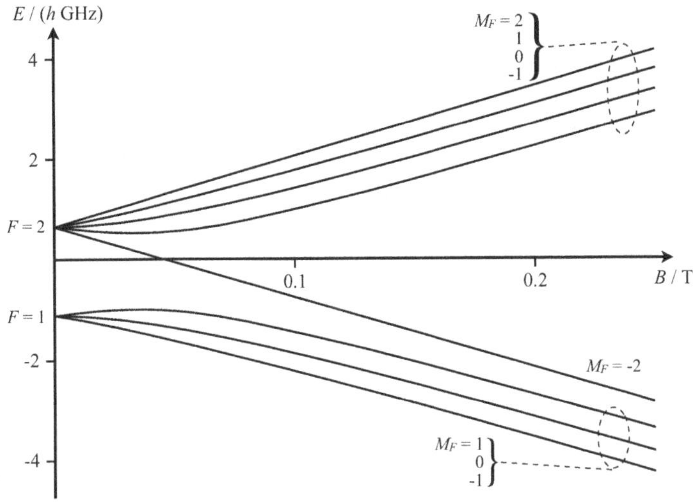

Fig. 11.5 Energy levels of the ground state of Na, $3s\,^2S_{1/2}$, as functions of a magnetic field B. The two hyperfine structure states $F = 2$ and $F = 1$ are split by $h \times 1.77$ GHz for $B = 0$. In the region intermediate between the Zeeman and Paschen-Back effects, energies have been computed with the Breit–Rabi formula (11.59).

11.5 The Diamagnetic Zeeman Effect and Very Strong Fields

The diamagnetic Zeeman effect is in this book used as the nomenclature for the term quadratic in B in (11.9), which we have hitherto ignored. In SI-units, and with a homogeneous field, it is:

$$H_{\mathrm{dZ}} = \frac{e^2}{8m_\mathrm{e}}\left[\mathbf{B}(\mathbf{r}_i) \times \mathbf{r}_i\right]^2 = \frac{e^2 B^2}{8m_\mathrm{e}} r_i^2 \sin^2\theta_i \ . \tag{11.64}$$

This is the Hamiltonian for a single electron. After the second equal sign in (11.64), we have assumed that the field is uniform and directed along the $\hat{e}_z$-axis, and θ_i refers to the angle between the field axis and the electron position vector $\mathbf{r}_i$. A classical point of view of the diamagnetic Zeeman effect is that when we take this quadratic term into account, we do not only consider the magnetic moment emanating from the unperturbed electronic orbits, but we include in the analysis also that the orbits themselves will shift under the interaction with the external field. The phenomenon is equivalent to magnetic induction, and it will scale with the enclosed area of an orbiting electron. Thus, the quadratic scaling with the orbital radius is qualitatively explained.

From (11.12), and the thereafter following discussion, we learned that the diamagnetic Zeeman shift is very small compared to the linear one for any typical laboratory magnetic field. Exceptions from this rule are Rydberg states (electrons excited to orbits with high n, see section 13.1), for which the quartic scaling with the princi-

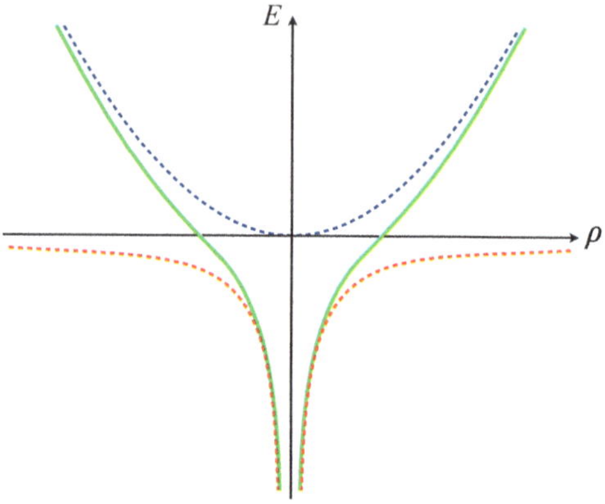

Fig. 11.6 Illustration of the combined potential of a single electron bound to a nucleus, in the presence of a very strong magnetic field. The field is taken as homogeneous and directed along $\hat{e}_z$, and the potential shown is that in the xy-plane ($\rho \equiv \sqrt{x^2+y^2}$). The dotted red line is the $1/r$ Coulomb potential, the blue one the harmonic oscillator cyclotron potential (11.65), and the full green line is the sum of the two.

pal quantum number may render the effect important. However, for the diamagnetic Zeeman effect to become a dominating perturbation, we typically need magnetic fields of the order of the ones present in neutron stars or white dwarfs. There, the energy contribution can get so large that it even dominates over the Coulomb attraction to the nucleus.

For such strong fields, it is practical to rewrite (11.64) in Cartesian coordinates

$$H_{\mathrm{dZ}} = \frac{e^2 B^2}{8 m_{\mathrm{e}}} \left(x^2 + y^2\right) \equiv \frac{m_{\mathrm{e}}\, \omega_{\mathrm{dZ}}^2}{8} \left(x^2 + y^2\right) . \tag{11.65}$$

In the last line, $\omega_{\mathrm{dZ}} = eB/m_{\mathrm{e}}$ corresponds to the classical cyclotron frequency of an electron in a magnetic field, and the Hamiltonian above is that of a two-dimensional harmonic oscillator. If (11.65) is added to the nuclear attraction Coulomb potential, the overall potential will, in the xy-plane, be of a form schematically illustrated in figure 11.6.

Mathematically, this is cumbersome because if both contributions are of the same order, the problem will be separable neither in cylindrical nor in spherical coordinates. The situation needs to be dealt with piecewise for different ranges of the magnetic field.

11.5.1 The Landau Region

When the field is so strong that (11.65) is by far the most important energy contribution, the system will be a harmonic oscillator in a plane orthogonal to the field direction, and will essentially experience free motion along $\hat{e}_z$ (assuming that this has been chosen parallel to $\mathbf{B}$). The zero-order energies of an electron will be:

$$E_{\mathrm{dZ}} = \hbar\omega_{\mathrm{dZ}} \left(n_{\mathrm{dZ}} + \frac{1}{2}\right) + \frac{\hbar^2 k_z^2}{2 m_{\mathrm{e}}} , \tag{11.66}$$

with $\omega_{\rm dZ}$ being the cyclotron frequency as defined in (11.65), k_z the wave vector along $\hat{e}_z$, and $n_{\rm dZ} \geq 0$ an integer quantum number. For a constant kinetic energy along $\hat{e}_z$, this corresponds to discrete cyclotron orbits, with energies separated by $\hbar\omega_{\rm dZ}$. These are called *Landau levels* and the domain is the *Landau region* or *the complete Paschen-Back effect*.

11.5.2 The Quasi-Landau and Mixing Regions

In a transitional region, the magnetic field may be not quite strong enough to supersede the nuclear attraction, but still important enough that the total Zeeman interaction strongly influences the atomic state. When the nuclear Coulomb interaction and $H_{\rm Z}+H_{\rm dz}$ contributes approximately in equal parts to the total energy, the situation will indeed be akin to that illustrated in figure 11.6, and analytical treatments will be very difficult.

Perturbation theory can be attempted, but will only give reasonable results if the two energy contributions differ sufficiently. In the general case, more complex routes have to be taken. This *quasi-Landau region* is of great general interest, because with a smooth variation of the magnetic field, the system will undergo a transition to classical chaos.

For a magnetic field a bit weaker still than that of the quasi-Landau situation, the Zeeman effect may still cause a pronounced mixing in l, or even in n. More extensive descriptions of the Landau, quasi-Landau and mixed regions can be found in, for example, [4] or [5].

11.5.3 The Diamagnetic Zeeman Effect as a Weak Perturbation

For sufficiently weak fields, the treatment in sections 11.1–11.4 holds and, for further precision, the Hamiltonian in (11.64) can be added as a perturbation. In a.u. the first-order energy correction for a one-electron atom is:

$$E_{\rm dZ} = \frac{B^2}{8} \left\langle r_i^2 \sin^2\theta_i \right\rangle \ . \tag{11.67}$$

The analytical treatment of the expectation value, for a general atom, is hard and good quantitative calculations typically requires numerical methods for the solution of the wave function.

Something we can infer from (11.67) is that non-diagonal matrix elements will be non-zero for angular momentum states that differ with $\Delta l = 2$, or more generally between states with equal parity. When applied to hydrogenic systems, which are degenerate in l, this means that degenerate perturbation theory must be used. For lower states of an alkali, this is not an issue, since the quantum defect removes the degeneracy. For Rydberg states, (11.67) is of limited value. The scaling of the Hamiltonian with n^4 often results in a pronounced mixing of states with different l, as discussed in section 11.5.2.

11.6 Magnetic Interactions

Matter is built up by charges and, accordingly, magnetism in various forms must necessarily be of core importance for the understanding of the structure of matter. Also an isolated atom will experience magnetic interactions, and for a complete understanding of the effect of a field external to the atom, the two effects can never be entirely decoupled. Some of the starting points for more complete treaties of magnetic interactions in atoms are touched upon in appendix F.

One of the most significant things that happen when we place an atom in a magnetic field — even a very weak one — is that spherical symmetry is broken, and with it, the orientational degeneracy must be abandoned. Another way of stating this is that the choice of quantisation axis no longer is arbitrary and that the degeneracy in the projection quantum number M_J is removed. It is rational that M_J is often referred to as the *magnetic quantum number*.

The Zeeman effect was discovered in spectroscopy some decades before quantum mechanics was developed, and before atomic structure was well understood. Nevertheless, the connection to moving charges was made very early on and a classical theory was developed, which is remarkably accurate, albeit for some limited cases. The continued studies of the effect of magnetic fields on atomic spectra were subsequently a crucial ingredient in what led to the establishment of the presence of an electron spin, and of atomic structure as a whole. The history of this development is described in many sources, for example in [6].

Careful studies of the Zeeman effect, and its impact on spectroscopy, are still highly relevant. For example, studies of stellar spectra, combined with an understanding of the magnetic properties of atoms, are used to get a measure of the magnetic fields inside stars. One of the most stringent experimental tests of QED that has been done, at the time of writing this, relates to measurements of the electron's magnetic dipole moment and the g_s-factor [7]. Descriptions of the theory underlying QED corrections to the interaction between atoms and magnetic fields can be found, for example, in [8].

Further Reading

Quantum theory of atomic structure, by Slater [9]
Atomic spectra, by Kuhn [6]
Atomfysik, by Lindgren & Svanberg [10]

Atomic Many-Body Theory, by Lindgren & Morrison [11]
Atoms, by Kleinpoppen [5]
Physics of Atoms and Molecules, by Bransden & Joachain [4]

References

1. L.D. Landau, L.E. M., *The Classical Theory of Fields — Course of Theoretical Physics, volume 2*, 4th edn. (Butterworth-Heinemann, Waltham, 1987)
2. W.E. Lamb, Phys. Rev. **85**, 259 (1952)
3. D.A. Steck. Sodium D line data. [Online]. Available: http://steck.us/alkalidata (2010). Accessed: 2019-05-20
4. B.H. Bransden, C.J. Joachain, *Physics of Atoms and Molecules*, 2nd edn. (Prentice Hall, Harlow, England, 2003)
5. H. Kleinpoppen, in *Constituents of Matter*, ed. by W. Raith (Walter de Gruyter, Berlin, 1997), p. 1
6. H.G. Kuhn, *Atomic spectra* (Longmans, London, 1969)
7. J. Baron, W.C. Campbell, D. DeMille, J.M. Doyle, G. Gabrielse, Y.V. Gurevich, P.W. Hess, N.R. Hutzler, E. Kirilov, I. Kozyryev, B.R. O'Leary, C.D. Panda, M.F. Parsons, E.S. Petrik, B. Spaun, A.C. Vutha, A.D. West, Science **343**, 269 (2014)
8. I. Lindgren, *Relativistic Many-Body Theory, A New Field-Theoretical Approach*, 2nd edn. (Springer-Verlag, Switzerland, 2016)
9. J.C. Slater, *Quantum theory of atomic structure* (McGraw-Hill, New York, 1960)
10. I. Lindgren, S. Svanberg, *Atomfysik* (Universitetsförlaget, Uppsala, 1974)
11. I. Lindgren, J. Morrison, *Atomic Many-Body Theory*, 2nd edn. (Springer Verlag, Berlin, 1986)

Chapter 12
The Stark Effect

In the preceding chapter, we covered the concept of an atom interacting with a static, external magnetic field. In the present one, we instead consider a static electric field. The influence of such a field is more straightforward to understand, at least in a simplified picture. An atom is a congruence of positive and negative charges. In a uniform and static electric field, one can reasonably assume that the positive charge will be pulled in the positive field direction, and the opposite for the negatively charged electrons. If this happens, a time-averaged electric dipole moment is induced. Another way to phrase this is that for an atom in an electric field, it will be energetically favourable to have its inherent charges displaced as above. Thus, also in a quantum mechanical picture, one may expect that the energies of bound electron states will change.

This interaction is called the *Stark effect*, and it is actually considerably more complex than as indicated in the previous paragraph — even in a classical picture. The Stark effect is generally seen as less important than the Zeeman effect. This is partly due to the fact that finite magnetic fields are prevalent in nature, but net static electric fields less so. Moreover, the effect is typically smaller than the Zeeman one, albeit that obviously depends on the respective field strengths.

As hinted at above, it is possible to give a classical description of the Stark effect. The first-order effect thus found agrees quite well with the more proper quantum version. However, the second and third orders do not. Another thing to note in this context is that the Stark effect works quite differently for hydrogen than it does for all other atoms. This is because the first-order effect tends to cancel for all atoms with more than a single electron.

The first and second orders are more commonly referred to as the *linear Stark effect* and the *quadratic Stark effect*. Both terminologies are pertinent since quantum mechanically they are derived by first- and second-order perturbation theory, respectively; and while the former has an energy shift proportional to the electric field, that of the latter scales with the square of the field.

A. Kastberg, *Structure of Multielectron Atoms*, Springer Series on Atomic, Optical, and Plasma Physics 112, https://doi.org/10.1007/978-3-030-36420-5_12

The principal reason for the difference between hydrogen and other atoms occurs because the Stark effect Hamiltonian has odd parity. As a consequence, the first-order perturbation term cancels. The only exception from this is when two states of different parities happen to be degenerate, and this is the case for hydrogen.

12.1 The Linear Stark Effect

The Hamiltonian describing the first-order interaction between an external electric field and an atom is:

$$H_{\mathrm{St}} = -\mathbf{d} \cdot \mathcal{E}(\mathbf{r}) \ , \tag{12.1}$$

where **d** is the electric dipole moment of the atom. If we define the *positive* $\hat{e}_z$-axis to be parallel with the (homogeneous) electric field, and consider an atom with a single electron, this Hamiltonian becomes:

$$H_{\mathrm{St}} = z\mathscr{E} \ . \tag{12.2}$$

This is here expressed in atomic units (see appendix A for the a.u. for the electric field amplitude $\mathscr{E}$), and z is the position coordinate along the quantisation axis of the sole electron (with the origin at the centre-of-mass). The sign of this Hamiltonian term appears differently in different books, because of different sign conventions for the field. With our choice of sign, the energy increases for an electron located far on the positive side of the nucleus.

This operator only acts on spatial coordinates. This means that H_{St} commutes with all spin operators, and thus we can leave spin out of the analysis of the Stark effect. A second observation is that the Hamiltonian in (12.2) has odd parity. A consequence of this is that if the interaction is taken as a perturbation, its expectation value for any zero-order state should cancel. This is why the first-order perturbation term of the Stark effect should always be zero. The exception to this rule can only occur if two states of opposite parity are energy degenerate. In the then ensuing degenerate perturbation analyses, there will be contributions of the kind:

$$\langle \psi_1 \,|\, H_{\mathrm{St}} \,|\, \psi_2 \rangle \ , \tag{12.3}$$

which will be finite if the states ψ_1 and ψ_2 have opposite parity.

One-electron systems are degenerate in l. For example, the hydrogen states 2s and 2p have the same energy, if fine-structure is ignored. This last point is important. In order to get away with a first-order perturbative calculation — as the one which will follow — the Stark effect must be larger than the fine-structure, while still being small enough to be treated as a perturbation (small compared to the quantisation in n). A first-order Stark effect may occur also for very weak electric fields, but this requires a different treatment and we will briefly discuss this in section 12.3.

A moderately excited alkali atom will typically not have a first-order Stark shift. Albeit it has a single valence electron, the l-dependent quantum defect will be large enough to break the degeneracy present in hydrogen for all low lying states (see section 5.4). For a highly excited alkali, or for any atom where one electron is excited to a high principal quantum number, the one-electron approximation will often be good enough to take states of differing l as degenerate. Thus, a first-order Stark effect analysis along the same lines as for hydrogen is then pertinent.

As was the case with the Zeeman effect, the Stark effect will break the degeneracy in M_J. This is to be expected, since again a field will impose a preferred direction, and will break spherical symmetry. However, as we will see, this time the lifting of the degeneracy will be incomplete. There will be levels with different M_J that will have the same energy.

12.1.1 The Stark Effect Hamiltonian

If we take the non-relativistic Hamiltonian for a one-electron atom, and add the Stark effect potential energy from (12.2), we get the following Schrödinger equation:

$$\left(-\frac{1}{2}\nabla^2 - \frac{Z}{r} + \mathscr{E}z\right)\psi(\mathbf{r}) = E\psi(\mathbf{r})\,. \tag{12.4}$$

An effect of the last term in the Hamiltonian is that the potential energy will go to minus infinity for large negative values of z. This has a couple of consequences. One is that there will no longer be any strictly bound states. There will always be a finite possibility for the electron to tunnel out of the nuclear bond. However, for week fields and low excitation, a quasi-discrete spectrum will ensue nevertheless, whence the Stark term can be taken as a perturbation. On the other hand, for highly excited levels, near the so-called series limit, the states may be dissociative. A short discussion about this will be presented in section 12.3.

A second consequence of the term linear in z in the Hamiltonian in (12.4) is that even in the absence of couplings between electrons, the Schrödinger equation will no longer be separable in spherical coordinates. Separation of variables is, however, possible in parabolic coordinates. This latter predicament gives us two possible routes for treating the linear Stark effect Hamiltonian for a one-electron atom. Within the restriction of having a truly hydrogenic system, and while ignoring spin–orbit coupling, we can solve (12.4) in parabolic coordinates. The other option is to begin with the known solutions for an unperturbed hydrogen atom and to then add the term $\mathscr{E}z$ as a perturbation to these. The two methods give the same answer, and in the following, we will demonstrate them both.

12.1.2 Perturbative Treatment of the Linear Stark Effect for a One-Electron Atom

To exemplify a perturbation calculation of the linear Stark effect for a hydrogenic atom, we take the simplest example possible, that is, the excited states for which $n=2$. What we have to do is to apply the perturbation $H_{\mathrm{St}}=\mathscr{E}z$ to the hydrogenic wave functions ψ_{200}, ψ_{21-1}, ψ_{210} and ψ_{211}, using degenerate perturbation theory (with the zero-order eigenfunctions written in the format ψ_{nlm_l}). We can take the functional form of these functions from (1.16) and (1.22). From (1.24) we find that the zero-order energies are $E_2^{(0)}=-Z^2/8$.

In spherical coordinates, we can rewrite the perturbation Hamiltonian as:

$$H_{\mathrm{St}}=\mathscr{E}r\cos\theta \tag{12.5}$$

(a zenith angle $\theta>\pi/2$ gives a negative energy contribution). The secular equation for the degenerate perturbative calculation will include non-diagonal matrix elements of this operator, whereas all diagonal elements will vanish because of the parity of the operator. However, there is no dependence on the coordinate angle φ in (12.5), and therefore the orthogonality of the azimuthal functions will make all elements non-diagonal in m_l zero. Thus, the present problem is reduced to a 2×2-matrix, and the states ψ_{21-1} and ψ_{211} will not be shifted by the linear Stark effect. With a corresponding analysis of the zenith function, one can show that the matrix elements of (12.5) will cancel also for non-diagonal elements that do not fulfil $l'=l\pm 1$.

This leaves us with the following eigenvalue equation to solve:

$$\begin{pmatrix} 0 & \langle\,\psi_{200}\,|H_{\mathrm{St}}|\,\psi_{210}\,\rangle \\ \langle\,\psi_{210}\,|H_{\mathrm{St}}|\,\psi_{200}\,\rangle & 0 \end{pmatrix}\begin{pmatrix} c_1 \\ c_2 \end{pmatrix}=E_{\mathrm{St}}\begin{pmatrix} c_1 \\ c_2 \end{pmatrix}. \tag{12.6}$$

The coefficients c_1 and c_2, for the two solutions, will determine the new eigenstates:

$$\begin{aligned} \psi_{\mathrm{a}} &= c_{1\mathrm{a}}\,\psi_{200}+c_{2\mathrm{a}}\,\psi_{210} \\ \psi_{\mathrm{b}} &= c_{1\mathrm{b}}\,\psi_{200}+c_{2\mathrm{b}}\,\psi_{210}\ , \end{aligned} \tag{12.7}$$

and the two resulting values of E_{St} will be the corresponding Stark shift energies.

The non-diagonal elements in (12.6) can be calculated as:

$$\begin{aligned} \langle\,\psi_{200}\,|H_{\mathrm{St}}|\,\psi_{210}\,\rangle &= \langle\,\psi_{210}\,|H_{\mathrm{St}}|\,\psi_{200}\,\rangle \\ &= \mathscr{E}\frac{Z^4}{16\pi}\int_0^{\infty} r^4\mathrm{e}^{-Zr}\left(1-\frac{Zr}{2}\right)\mathrm{d}r\int_0^{\pi}\cos^2\theta\sin\theta\,\mathrm{d}\theta\int_0^{2\pi}\mathrm{d}\varphi \\ &= -\frac{3}{Z}\mathscr{E}\,. \end{aligned} \tag{12.8}$$

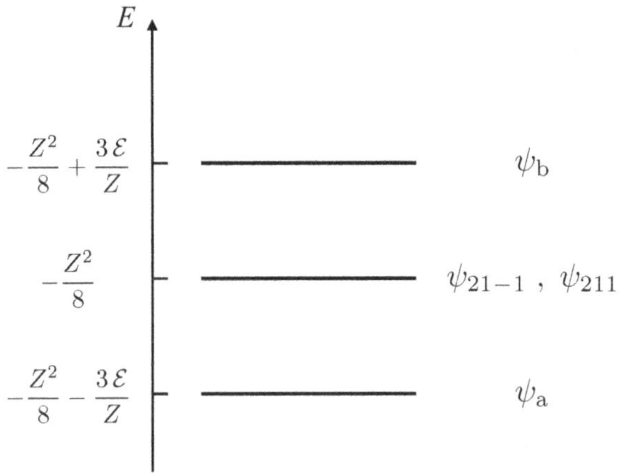

Fig. 12.1 Stark shifted energies of the four hydrogenic states with $n=2$ (degenerate in absence of a perturbation). The interaction with the external electric field mixes the states ψ_{200} and ψ_{210} and partially breaks the degeneracy.

From this, we can find the two normalised Stark shifted states as:

$$\begin{aligned}\psi_a &= \frac{1}{\sqrt{2}}\left(\psi_{200}+\psi_{210}\right)\\ \psi_b &= \frac{1}{\sqrt{2}}\left(\psi_{200}-\psi_{210}\right) \ ,\end{aligned} \tag{12.9}$$

with the energies:

$$\begin{aligned}E_a &= -\frac{Z^2}{8}-\frac{3\mathcal{E}}{Z}\\ E_b &= -\frac{Z^2}{8}+\frac{3\mathcal{E}}{Z} \ .\end{aligned} \tag{12.10}$$

This is shown in figure 12.1. We have two degenerate, unshifted, states and the two mixed states, ψ_a and ψ_b. For the latter two, l is not a good quantum number, but m_l is. Mathematically, this is a consequence of the fact that H_{St} commutes with L_z, but not with $\mathbf{L}^2$.

For higher lying states, the calculation is done in the same way. We will give one more example — for $n = 3$. The zero-order energy is:

$$E_3^{(0)} = -\frac{Z^2}{18} \ , \tag{12.11}$$

and there are nine degenerate states. Two of these, ψ_{32-2} and ψ_{322}, will be unshifted since they have unique values of m_l. For the two states with $m_l = -1$ and $m_l - 1$, we get the matrices:

$$\begin{pmatrix} 0 & \langle \psi_{31-1} | H_{St} | \psi_{32-1} \rangle \\ \langle \psi_{32-1} | H_{St} | \psi_{31-1} \rangle & 0 \end{pmatrix}, \tag{12.12}$$

and

$$\begin{pmatrix} 0 & \langle \psi_{311} | H_{St} | \psi_{321} \rangle \\ \langle \psi_{321} | H_{St} | \psi_{311} \rangle & 0 \end{pmatrix}, \tag{12.13}$$

for which eigenvalues and eigenvectors have to be sought. For $m_l = 0$, the matrix will be of dimension 3×3:

$$\begin{pmatrix} 0 & \langle \psi_{310} | H_{St} | \psi_{300} \rangle & 0 \\ \langle \psi_{300} | H_{St} | \psi_{310} \rangle & 0 & \langle \psi_{320} | H_{St} | \psi_{310} \rangle \\ 0 & \langle \psi_{310} | H_{St} | \psi_{320} \rangle & 0 \end{pmatrix}. \tag{12.14}$$

The outcome is that the nine states will have the following five different field-dependent energies:

$$\begin{aligned} E_a &= -\frac{Z^2}{18} - \frac{9\mathscr{E}}{Z} \\ E_b &= -\frac{Z^2}{18} - \frac{9\mathscr{E}}{2Z} \\ E_c &= -\frac{Z^2}{18} \\ E_d &= -\frac{Z^2}{18} + \frac{9\mathscr{E}}{2Z} \\ E_e &= -\frac{Z^2}{18} + \frac{9\mathscr{E}}{Z}. \end{aligned} \tag{12.15}$$

In figure 12.2, we show the corresponding energy level diagram, with the state mixing indicated.

12.1.3 The Linear Stark Effect, for a One-Electron Atom, with Parabolic Coordinates

The perturbation calculations emanating from the standard zero-order solutions to the one-electron Schrödinger equation (in spherical coordinates), works fine for relatively low principal quantum numbers. For larger n, another approach is more suitable. This is one in which coordinates and the corresponding quantisations are chosen such that the Stark effect perturbation Hamiltonian becomes diagonal. That can be accomplished with parabolic coordinates. This kind of solution brings with it the extra advantage that more physical trends are revealed than what is the case for the solution shown in section 12.1.2.

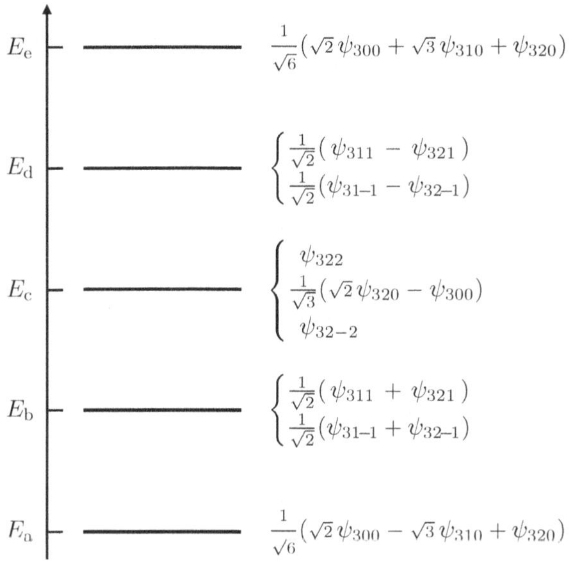

Fig. 12.2 Stark shifted energies of the nine hydrogenic states with $n=3$. The two states ψ_{322} and ψ_{32-2} are unshifted and degenerate also with the superposition state ψ_c, which has a zero net Stark shift. The energies E_d and E_b are two-fold degenerate.

In appendix B.2, we demonstrate how the one-electron Schrödinger equation can be separated, and solved, in parabolic coordinates [1]. The definition of parabolic coordinates that we will use is:

$$\begin{aligned} \xi &\equiv r+z \\ \eta &\equiv r-z \\ \varphi &\equiv \arctan\left(\frac{y}{x}\right) . \end{aligned} \tag{12.16}$$

That is, the azimuthal angle is the same as for spherical coordinates, and the coordinates ξ and η have the dimensions of length.

The solutions to the field-free Schrödinger equation in these coordinates is derived in (B.33):

$$\begin{aligned} \psi_{n_1 n_2 m}(\xi,\eta,\varphi) = {} & \frac{Z^{3/2}}{\sqrt{\pi}\,n^2}\sqrt{\frac{n_1!\,n_2!}{[(n_1+|m|)!\,(n_2+|m|)!]^3}}\,(\rho_1\,\rho_2)^{|m|/2} \\ & \times \mathrm{e}^{-(\rho_1+\rho_2)/2}\,\mathrm{L}_{n_1+|m|}^{(|m|)}(\rho_1)\,\mathrm{L}_{n_2+|m|}^{(|m|)}(\rho_2)\,\mathrm{e}^{im\varphi} . \end{aligned} \tag{12.17}$$

The wave function in (12.17) is discretised in the quantum numbers n_1, n_2 and m, corresponding to the respective coordinates ξ, η and φ. The parameter and quantum number definitions are:

$$\begin{aligned} n &\equiv n_1 + n_2 + |m| + 1 \\ \rho_1 &\equiv \frac{Z\xi}{n} \\ \rho_2 &\equiv \frac{Z\eta}{n} \quad . \end{aligned} \tag{12.18}$$

The functions $\mathrm{L}_{n_i+|m|}^{|m|}(\rho_i)$ are associated Laguerre polynomials, as defined in (B.14) and (B.15). The definition of the principal quantum number n in (12.18) is commensurate with the one used in the standard solution of the hydrogenic problem. The zero-order energies are obviously also the same as when we solved the Schrödinger equation in spherical coordinates. However, an important difference is that an eigenstate expressed in the quantum numbers n_1, n_2 and m is not necessarily symmetric with respect to the plane $z=0$. Thus, states written as in (12.17) may not have definite parities.

In the presence of a homogeneous electric field, the full Schrödinger equation for an hydrogenic atom — see (B.22) — is:

$$\left(-\frac{1}{2}\nabla^2 - \frac{2Z}{\xi+\eta} + H_{\mathrm{St}} \right) \psi_{n_1 n_2 m}(\xi,\eta,\varphi) = E\, \psi_{n_1 n_2 m}(\xi,\eta,\varphi) \quad , \tag{12.19}$$

and the Stark Hamiltonian (12.2) in parabolic coordinates is:

$$H_{\mathrm{St}} = \frac{\mathscr{E}}{2}(\xi - \eta) \quad . \tag{12.20}$$

As in the previous section, we have assumed a homogeneous field directed along the positive $\hat{e}_z$-axis. This equation is cylindrically symmetric, which makes it diagonal in m. Due to the orthogonality of associated Laguerre polynomials, it will also be diagonal in n_1 and n_2. This means that the Stark shift energy can be calculated as [2]:

$$\begin{aligned} E_{\mathrm{St}} &= \langle\, \psi_{n_1 n_2 m}\, | H_{\mathrm{St}} |\, \psi_{n_1 n_2 m} \rangle \\ &= \frac{\mathscr{E}}{8} \int_0^\infty \int_0^\infty \int_0^{2\pi} |\, \psi_{n_1 n_2 m}(\xi,\eta,\varphi)\,|^2\, (\xi^2 - \eta^2)\, \mathrm{d}\xi \mathrm{d}\eta \mathrm{d}\varphi \,. \end{aligned} \tag{12.21}$$

where we have used the parabolic coordinates volume element:

$$\mathrm{d}V = \frac{1}{4}(\xi+\eta)\,\mathrm{d}\xi \mathrm{d}\eta \mathrm{d}\varphi \quad . \tag{12.22}$$

The integrals in (12.21) can be separated, and an analytical solution can be obtained. For the details of the solution, we refer to [2] or [3], and here we give the result. The energy shift is found to be:

$$E_{\mathrm{St}} = \frac{3\mathscr{E}}{2Z}\, n\,(n_1 - n_2) \quad , \tag{12.23}$$

and the total energy is:

$$E_{n_1 n_2 m} = -\frac{Z^2}{2n^2} + \frac{3\mathscr{E}}{2Z}\, n(n_1 - n_2) \ . \tag{12.24}$$

It can easily be confirmed that this agrees with (12.10) and (12.15).

The extreme values of the Stark shift in (12.23), for a given n, will appear for the quantum number combinations $(n_1 = n-1\,, n_2 = 0)$ and $(n_1 = 0\,, n_2 = n-1)$. This makes the energy span of a Stark shifted level n:

$$\frac{3\mathscr{E}}{Z}\, n(n-1)\,. \tag{12.25}$$

Thus, we see that this energy width, and the linear Stark shift, scale with n^2. The increase of the splitting with n is expected, since a higher n means a larger radial distance, and thus a larger dipole.

We can also identify qualitative differences between n_1 and n_2. If $n_1 > n_2$, it will be more probable to find the electron at $z > 0$ than on the negative side of $\hat{e}_z$. That means a dipole moment directed oppositely to the external field, resulting in higher energy.

12.2 Quadratic Stark Effect

The quadratic Stark effect is what emerges from a second-order perturbation treatment of the Hamiltonian in (12.1). As the name suggests, the energy displacement will be proportional to the square of the absolute value of the electric field, and it will also scale with the total electronic angular momentum projection quantum number as M_J^2. Even though a second-order perturbation is typically small compared to the first-order one, this term is the most important one for all atomic states where there is no energy degeneracy between states of opposite parity (as explained in section 12.1). This is the case for essentially all atoms, expect for excited states of hydrogen-like systems and Rydberg states.

The quadratic Stark effect typically requires quite strong fields to be important. However, the effect can be substantial in alkali atoms, due to their large polarisability. For other atoms, the general trends are that heavier atoms have a smaller Stark effect than lighter ones and that highly excited states experience larger shifts than low lying energy levels. In the following, we will first treat the non-degenerate hydrogen atom ground state, which is not covered in section 12.1. After that, we will study the general situation, with multielectron atoms.

12.2.1 The Stark Shift in the Hydrogen Ground State

As in section 12.1, we assume the electric field to be static and homogeneous, and we chose the quantisation axis $\hat{e}_z$ to be parallel with the field direction. The Hamiltonian then takes the form of (12.2). The second-order energy perturbation term of this, for a state ψ_a, is:

$$E_{\mathrm{St}}^{(2)}(\psi_a) = \mathscr{E}^2 \sum_{b \neq a} \frac{|\langle \psi_b | z | \psi_a \rangle|^2}{E_a - E_b} \quad . \tag{12.26}$$

The energies E_a and E_b are the zero-order ones, and they need to be given in atomic units (as do z and $\mathscr{E}$). The sum should run over all states other than the one for which we seek the perturbation. However, the energy interval in the denominator means that the levels close in energy to E_a will provide the dominant terms. We also see that the second-order perturbation will result in a mixing of states. For the specific case of a one-electron atom, the spatial part of the wave function should be described by a wave function ψ_{nlm_l} and, as seen in the previous section, the spin plays no part in the Stark effect.

Excited states of one-electron atoms typically have linear Stark shifts, and the quadratic shift will enter as a correction to that. In order to calculate that correction, (12.26) will not be sufficient since the denominator will become zero for degenerate states. For that eventuality, degenerate perturbation theory must be applied. In contrast, for the non-degenerate ground state, the second-order term will be the largest one and we can use (12.26) to calculate the perturbation. For the ground state, all terms in the sum must be negative, which means that it will be repelled in energy from other states by the quadratic Stark shift.

The sum in (12.26) can be greatly simplified by some symmetry considerations. This analysis is closely related to that of selection rules for radiative transitions. This is quite natural since the leading term of a light-matter interaction Hamiltonian will also be an electric dipole interaction, albeit not a static one.

We have already established that we can exclude all states ψ_b that have the same parity as ψ_a. As we shall see, we will get more restrictions by scrutinising the action of H_{St} on the orbital angular momentum.

The operator for the projection of $\mathbf{L}$ is, see (C.1):

$$L_z = x p_y - y p_x \quad . \tag{12.27}$$

Thus, L_z commutes with z. We write the state vectors in (12.26) in the form $|n, l, m_l\rangle$, and take the matrix element of the commutator between L_z and z. The result is:

$$\langle n_b, l_b, m_{lb} | [L_z, z] | n_a, l_a, m_{la} \rangle = (m_{lb} - m_{la}) \langle n_b, l_b, m_{lb} | z | n_a, l_a, m_{la} \rangle = 0 \, . \tag{12.28}$$

Hence, we can exclude from the sum in (12.26) all states ψ_b for which $m_{lb} \neq m_{la}$.

The constraint for l is a bit more complicated. We begin with the commutation relation (C.35). A consequence of that is that the following matrix element must cancel:

$$\langle n_b, l_b, m_{lb} \,|\, \left[\mathbf{L}^2, [\mathbf{L}^2, z]\right] - 2(\mathbf{L}^2 z + z\mathbf{L}^2) \,|\, n_a, l_a, m_{la} \rangle = 0 \ . \tag{12.29}$$

If we develop the commutators, and take the eigenvalues of $\mathbf{L}^2$, we find:

$$(l_b + l_a + 2)(l_b + l_a)(l_b - l_a + 1)(l_b - l_a - 1)\,\langle n_b, l_b, m_{lb} \,|z|\, n_a, l_a, m_{la} \rangle = 0 \, . \tag{12.30}$$

The quantum numbers l_a and l_b can never be negative, and the criterion that the parities have to be opposite for ψ_a and ψ_b excludes $l_a = l_b = 0$. The conclusion is that the only terms that we need to retain in (12.26) are those for which $l_b = l_a + 1$ or $l_b = l_a - 1$.

The ground state vector of a one-electron atom is $|\,1,0,0\rangle$. The above conditions mean that its corresponding quadratic Stark shift is:

$$E_{\mathrm{St}}^{(2)}(\psi_{100}) = \mathscr{E}^2 \sum_{n_b > 1} \frac{|\langle n_b, 1, 0 \,|z|\, 1,0,0\rangle|^2}{E_{100} - E_{n_b 10}} \, . \tag{12.31}$$

The leading terms in (12.31) can be calculated analytically, and since all terms are negative, a lower bound for the shift can be estimated by including the first few excited levels. For a full derivation of the Stark shift of the ground states of hydrogenic atoms we refer to [4], and here we settle with presenting the result:

$$E_{\mathrm{St}}^{(2)}(\psi_{100}) = -\frac{2.25\,\mathscr{E}^2}{Z^4} \, . \tag{12.32}$$

Taking into account that the atomic unit for electric field strength is as high as about 5×10^{11} V/m, it is evident that very strong fields are needed in order to produce a substantial shift.

We shall use the expressions derived in this section to introduce the concept of polarisability. The interaction with the electric field will create an induced dipole moment in the direction of the field (here set to be parallel with $\hat{\mathrm{e}}_z$). This will be:

$$d_z = -\frac{\partial E_{\mathrm{St}}}{\partial \mathscr{E}} \equiv \alpha_{\mathrm{sp}}\, \mathscr{E} \ . \tag{12.33}$$

Here we have defined the *static polarisability* α_{sp}. By differentiating (12.26), we find the following expression for α_{sp} for the one-electron atom ground state $|\,1,0,0\rangle$:

$$\alpha_{\mathrm{sp}}(\psi_{100}) = 2 \sum_{n_b > 1} \frac{|\langle n_b, 1, 0 \,|z|\, 1,0,0\rangle|^2}{E_{n_b 10} - E_{100}} \, . \tag{12.34}$$

The atomic unit for polarisability is $(e^2 a_0^2 E_{\mathrm{h}}^{-1})$ — see table A.1.

12.2.2 Stark Shifts for Multielectron Atoms

The Stark effect Hamiltonian for a multielectron atom is obtained by summing (12.2) over all electrons, as in:

$$H_{\mathrm{St}} = \mathscr{E} \sum_{i=1}^{N} z_i = -\mathscr{E} d_z \ . \tag{12.35}$$

Here we have anew assumed a homogeneous field parallel with $\hat{e}_z$, and the shorthand notation for the projection of the compound atomic electric dipole moment is identical with that used in 12.33.

This time we need to find matrix elements of the operator d_z, in an appropriate basis of zero-order atomic states. As in previous sections, diagonal matrix elements will cancel for symmetry reasons — d_z is an odd operator. A more esoteric way to phrase this is that a permanent atomic electric dipole moment is forbidden due to space-inversion symmetry (an atomic wave function must have definite parity — odd or even). The same argument could have been applied in section 10.2 in order to justify why the hyperfine structure interaction only exists for odd nuclear multipole magnetic moments, and even for electric ones.

This predicament excludes the first-order perturbative correction. Again, we have to go to the second-order term to find the Stark shifts for a multielectron atom. This term will also be responsible for a mixing of levels. The energy displacement of atomic level ψ_a will be:

$$E_{\mathrm{St}}^{(2)}(\psi_a) = \mathscr{E}^2 \sum_{b \neq a} \frac{|\langle \psi_b \,|d_z|\, \psi_a \rangle|^2}{E_a - E_b} = \mathscr{E}^2 \sum_{b \neq a} \frac{\langle \psi_a \,|d_z|\, \psi_b \rangle \langle \psi_b \,|d_z|\, \psi_a \rangle}{E_a - E_b} \ , \tag{12.36}$$

where in the last line, the mixing of states of opposite parity is shown explicitly.

To proceed, we have to contemplate the choice of basis states and possible selection rules (symmetry considerations making some matrix element vanish). As long as the field is weak, and the Stark effect can be taken as the most feeble perturbation, the choice of coupling scheme can be put on hold, and we can limit the discussion to the dependence on the shift on the compound quantum numbers J and M_J. We can then write the shift as:

$$E_{\mathrm{St}}^{(2)}(\psi_a) = \mathscr{E}^2 \sum_{b \neq a} \frac{|\langle \gamma_b \, , J_b, M_{Jb} \,|d_z|\, \gamma_a \, , J_a, M_{Ja} \rangle|^2}{E_a - E_b} \ . \tag{12.37}$$

Here, γ_a and γ_b refer not only to electronic configurations, but also the quantum numbers that make up the zero-order level (whether it is an LS-coupled fine-structure level, or something more appropriately described in another scheme).

We can exclude from the summation all states $|\,\gamma_b, J_b, M_{Jb}\rangle$ that have the same parity as $|\,\gamma_a, J_a, M_{Ja}\rangle$. In other words, the perturbing (mixing) state must belong to a different configuration than the perturbed one. This also means that in absence of configuration mixing, a non-degenerate perturbation analysis will suffice.

For the quantum number M_J, we can use the same arguments that we applied in order to derive (12.28). Regardless of how we couple all the individual electronic angular momenta, it will be generally true that:

$$\mathbf{J} = \sum_i (\mathbf{L}_i + \mathbf{S}_i) \ , \tag{12.38}$$

with the sum taken over *all* electrons. Thus, the quantum number M_J must be:

$$M_J = \sum_i (m_{li} + m_{si}) \ . \tag{12.39}$$

The Stark effect Hamiltonian does not act on the spins, and the extension of (12.28) becomes the constraint that the only non-zero matrix elements are those for which $M_{Jb} = M_{Ja}$. Analogously, for J we can follow a line of reasoning similar to that leading up to (12.29). A more common route is to make the analogy with light-atom interactions in the electric dipole limit, and to use the Wigner-Eckart theorem [3]. Either way, the resulting condition for finite matrix elements is that $|J_b - J_a| \leq 1$.

Taking the above rules into account, (12.37) can be stated as:

$$E^{(2)}_{\mathrm{St}}(\gamma J M_J) = \mathscr{E}^2 \sum_{\gamma_b \neq \gamma} \sum_{J_b = J-1}^{J_b = J+1} \frac{|\langle \gamma_b, J_b, M_J \,|d_z|\, \gamma, J, M_J \rangle|^2}{E_{\gamma J} - E_{\gamma_b J_b}} \ . \tag{12.40}$$

In this expression, we have assumed an energy degeneracy in M_J for the unperturbed states. The explicit dependence of the matrix elements on M_J can be shown to be (see, for example, [5]):

$$|\langle \gamma_b, J_b, M_J \,|d_z|\, \gamma, J, M_J \rangle|^2 \propto \begin{cases} J^2 - M_J^2 & \text{for} \quad J_b = J-1 \\ M_J^2 & \text{for} \quad J_b = J \\ (J+1)^2 - M_J^2 & \text{for} \quad J_b = J+1 \end{cases} \ . \tag{12.41}$$

From (12.41) and (12.40), we see that the Stark effect will break the degeneracy in M_J, but not entirely. States with opposite signs of the projection quantum number will have the same shift. We can also note that the Stark shift will be of the form:

$$E^{(2)}_{\mathrm{St}}(\gamma J M_J) = \mathscr{E}^2 (A_{\gamma J} + B_{\gamma J} M_J^2) \ , \tag{12.42}$$

where $A_{\gamma J}$ and $B_{\gamma J}$ are parameters that depend of the coupling scheme and the zero-order state.

Also for the multielectron atom, the static electric field will induce a dipole moment. Negative charges will on average be displaced towards the negative field direction, relative to the nucleus. This can be illustrated from the viewpoint of the mixing of states of opposite parity. As an example, consider the admixture of an s-orbital and a p-orbital, for a state $m_l = 0$. The moduli squared of the corresponding probability distributions are shown in the upper left of table 1.2.

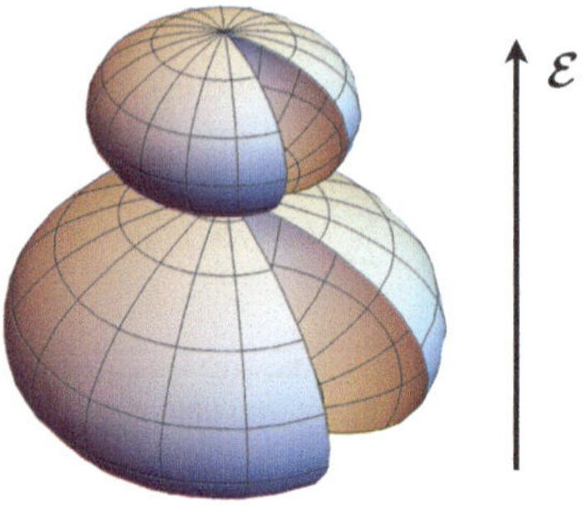

Fig. 12.3 Pictorial representation of the electron probability density resulting from a mixing of a hydrogenic s-state with a p-state, both with $m_l = 0$ (see also table 1.2). The outcome is an average displacement of probability towards the negative field direction. The direction of the static electric field responsible for the mixing is indicated by the arrow.

If we sum the spherical harmonics $Y_{1,0}$ and $Y_{0,0}$ with some proper weight, we obtain an electronic probability density as shown in figure 12.3. That is, a multi-electron atom in a static electric field will have its average charge density displaced from the position of the nucleus, and spherical symmetry will be broken. We can define the static polarisability, just as in (12.33), with d_z now used as a quantity, rather than as an operator. The polarisability is:

$$\alpha_{\mathrm{sp}}(\gamma J M_J) = 2 \sum_{\gamma_b \neq \gamma} \sum_{J_b = J-1}^{J_b = J+1} \frac{|\langle \gamma_b, J_b, M_J \,| d_z |\, \gamma, J, M_J \rangle|^2}{E_{\gamma J} - E_{\gamma_b J_b}} , \tag{12.43}$$

and in terms of α_{sp}, the Stark shift can be written as:

$$E_{\mathrm{St}}^{(2)}(\gamma J M_J) = -\frac{1}{2}\,\alpha_{\mathrm{sp}}(\gamma J M_J)\,\mathcal{E} . \tag{12.44}$$

12.3 Atoms in Static Electric Fields

Our survey of the Stark effect has been made at a low order of approximation and for reasonably weak fields. For very strong fields, the effect of a static electric field on atomic structure changes. For example, for a multielectron atom — for which the linear Stark effect is absent for weak fields — the energy shift may get so large that levels of different principal quantum numbers become degenerate. Thus, the first-order effect comes into play and energies as functions of field amplitude will be highly non-linear.

Also for very weak fields, our approximations are less than perfect. This is mentioned in the beginning of section 12.1. In that section, we ignore spin–orbit effects, or rather we assume that the Stark shift is large enough to render the fine-structure levels quasi-degenerate. Thereby, and at that limit, the first-order perturbation treatment can be applied.

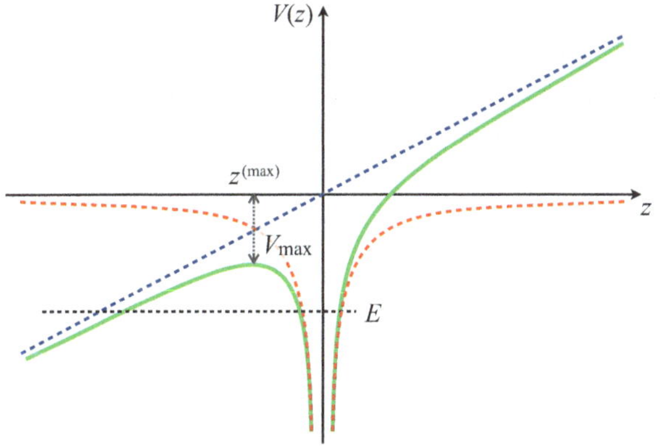

Fig. 12.4 Illustration of (12.45); the sum of a Coulomb attraction and a linear Stark potential, plotted along the field direction. The dotted, horizontal line represents a quasi-bound state, with a finite tunnelling probability towards the negative field direction. One consequence of this is that the effective ionisation potential of the atom is lowered, as in (12.47).

When the electric field strength is even lower, a refined approach must be applied (see, for example, [6]). Take some low lying excited fine-structure states in hydrogen as an example. The states $2s\,^2S_{1/2}$ and $2p\,^2P_{1/2}$ are degenerate, with relativistic effects taken into account but disregarding the small Lamb shift. This means that also for a very weak field, these states will experience a linear Stark effect, without $2p\,^2P_{3/2}$ being involved. With increasingly stronger field, when the Stark shift becomes comparable to the fine-structure splitting of $2p\,^2P$, the energies will asymptotically merge into ones commensurate with the analysis in section 12.1, illustrated in figure 12.1.

As was mentioned in the discussion following (12.4), a Stark shifted atom lacks truly bound states. For a one-electron atom with a static electric field parallel with $\hat{e}_z$, the electronic potential energy has the form of:

$$V(r) = -\frac{Z}{r} + \mathscr{E}z \ . \tag{12.45}$$

In figure 12.4, this is plotted as function of z. For negative z, there is a local maximum at:

$$z^{(\mathrm{max})} = -\sqrt{\frac{Z}{\mathscr{E}}} \ . \tag{12.46}$$

At that point, the potential energy is:

$$V_{\mathrm{max}} = -2\sqrt{Z\mathscr{E}} \ , \tag{12.47}$$

and consequently, the ionisation energy is lowered by that same amount. This means that while low energy states will see but a slight widening due to the finite probability for tunnelling, states close to the series limit will be strongly modified and highly susceptible to auto-ionisation.

A quantum mechanical calculation shows that a quasi-discrete energy spectrum extends a little further in energy than the potential maximum in (12.47) — see, for example, [6]. The existence of discrete energies above the ionisation limit can be explained by a classical analogue. The rim of the three-dimensional potential is lowered only along one direction. A classical harmonic potential with a low rim only around one point may well sustain oscillations with higher energies along some other direction.

Further Reading

Atomic spectra, by Kuhn [7]
The theory of atomic structure and spectra, by Cowan [8]
Quantum Mechanics (non-relativistic theory), by Landau & Lifshitz [2]
Atomic Spectra and Radiative Transitions, by Sobelman [5]
Physics of Atoms and Molecules, by Bransden & Joachain [3]

References

1. G.B. Arfken, H.J. Weber, F.E. Harris, *Mathematical Methods for Physicists: A Comprehensive Guide*, 7th edn. (Academic Press, Amsterdam, 2012)
2. L.D. Landau, E.M. Lifshitz, *Quantum Mechanics — Course of Theoretical Physics, volume 3*, 3rd edn. (Butterworth-Heinemann, Amsterdam, 1981)
3. B.H. Bransden, C.J. Joachain, *Physics of Atoms and Molecules*, 2nd edn. (Prentice Hall, Harlow, England, 2003)
4. H.A. Bethe, E.E. Salpeter, *Quantum Mechanics of One- and Two-Electron Atoms* (Springer-Verlag, Berlin, 1957)
5. I.I. Sobelman, *Atomic Spectra and Radiative Transitions*, 2nd edn. (Springer, Berlin, 1992)
6. E.U. Condon, G.H. Shortley, *The theory of atomic spectra* (Cambridge University Press, Cambridge, 1935)
7. H.G. Kuhn, *Atomic spectra* (Longmans, London, 1969)
8. R.D. Cowan, *The theory of atomic structure and spectra* (University of California press, Berkeley, 1981)

Chapter 13
Complex and Exotic Excitations

In the preceding twelve chapters, we have concentrated on the least complicated issues in atomic structures, which are arguably also those that are most readily observed in nature. Concretely, what this means is atomic ground configurations and ground states, as well as atomic states where just one electron is excited (from the viewpoint of the CFA). Moreover, this said excitation should not be above, or too close to the ionisation limit.

It is natural to begin studies of atomic structure with the above stated constraints. This provides a platform for further studies of more complex states, and also of larger systems such as molecules. For the most part, we will leave the detailed studies of more complicated atomic systems to other volumes, see for example [1]. In this chapter, we will just introduce a few extensions to what has hitherto been treated in this book, and we give them a cursory treatment.

One case that we have deferred treatment of is that when more than one electron is excited. Such atomic states tend to lie above the first ionisation limit in energy, and hence they are often *auto-ionising*. That makes it natural to discuss such atoms in concert with a general treatment of atomic states coupled to a continuum. We do this in section 13.2. However, there are some examples of truly bound doubly excited states, and we give one example towards the end of this introductory section of chapter 13.

Another important omission so far is atoms for which one electron is excited to a very high principal quantum number, while still being in a bound state. Such *Rydberg atoms* have some characteristics similar to alkalis (see section 5.4), and we will look at these large atoms in section 13.1.

In section 13.3 we will cover atoms for whom the excitation is not that of the valence electron, but rather comes from an inner shell. This is relevant for X-ray physics. However, this book is strictly about *atomic structure*, and not about radiative transitions or interactions. With that restriction, there is not so much to add here and now, but we will give a brief introduction. Similar things can be said for highly

A. Kastberg, *Structure of Multielectron Atoms*, Springer Series on Atomic, Optical, and Plasma Physics 112, https://doi.org/10.1007/978-3-030-36420-5_13

ionised atoms. The main difference from atomic states covered so far is that the excitation energies involved are far above the optical region of the electromagnetic radiation spectrum.

Doubly Excited States

If we crassly take the Schrödinger equation for a multielectron atom, in the CFA, it is clear that this has solutions for which more than one atom is excited. This does not necessarily have to be two valence electrons. There will also be solutions that involve electrons from the inner core of orbitals. Doubly excited states almost always have all eigenenergies higher than the binding energy of the most loosely bound electron. This makes it highly probable for them to couple to a continuum state of a singly ionised atom plus a free electron. That in turn is likely to cause an ejection of one electron (or indeed more than one), with the atom relaxing into a bound ionic state.

The most direct example for which this can be illustrated is neutral He. We ignore the electron–electron interaction and use the hydrogenic energy equation of (1.24). In that approximation, the atom has the energy:

$$E = E_1 + E_2 = -\frac{Z^2}{2}\left(\frac{1}{n_1^2} + \frac{1}{n_2^2}\right) . \tag{13.1}$$

We use this to calculate the energy of the doubly excited state $2s^2$, and find that it is $E = -1\,E_h$, which is much higher that the value obtained by taking $n_1 = 1$ and letting n_2 go to infinity ($E_{ion} = -2\,E_h$). The mutual Coulomb repulsion and other perturbations cannot compensate for such a difference in energy (see also table D.2), and indeed all doubly excited states of neutral He are auto-ionising.

For a heavy and highly ionised atom, it is more likely to find doubly excited states below the first ionisation limit, because of the strong nuclear biding potential. However, for He-like ions, there are no known states for which this is enough. To take an example, barium ($Z = 56$) 54 times ionised has a first ionisation limit of about 43 keV, whereas the energy of the state $2s^2$ is as high as 66 keV above the ground state [2].

In contrast, there are a few neutral atoms with more than two electrons that have doubly excited bound states. To give an example, see figure 13.1 for a partial Grotrian diagram for Be. The doubly excited configuration $2p^2$ has its highest energy term, ^1S above the ionisation limit, but ^3P and ^1D are well below.

For a bound doubly excited state, there is nothing fundamentally different concerning how angular momentum couplings can be constructed, and how energies can be estimated. The main difference lies in an increased complexity for all analyses.

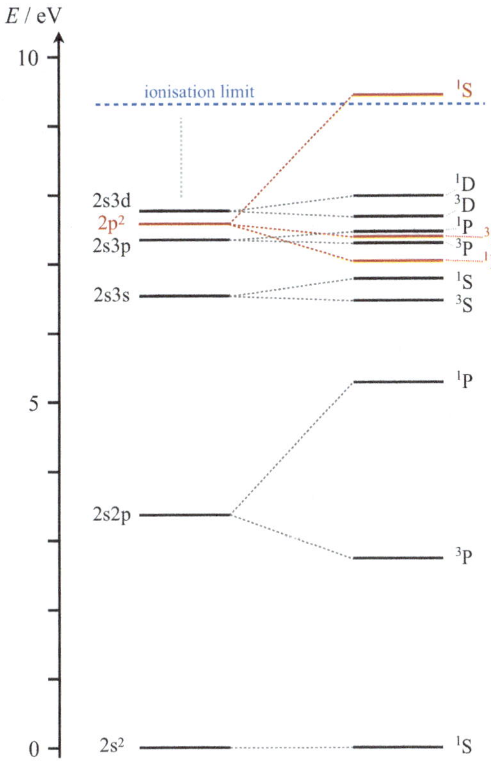

Fig. 13.1 Partial energy level diagram for Be, with empiric energies from [2]. The doubly excited configuration $2p^2$, and its *LS*-coupling terms, are in red. Zero energy has been taken as the energy of the $2s^2$ ground state. The energies of the configurations — in the left part of the figure — are averages over the number of states for the respective configurations. All energy levels between 2s3s ^1D and the ionisation limit have been omitted from the diagram.

13.1 Rydberg Atoms

The somewhat vague, but generally accepted, definition of a *Rydberg atom* — or in an analogue way a Rydberg molecule — is an atom where at least one electron is excited to a high principle quantum number, n. That electron is said to be in a *Rydberg state*. There is no rigid rule for exactly how high n has to be. The point is that a substantial part of the electron probability density should be located outside the electronic core so that the latter together with the nucleus can be treated as a single positive charge. That will leave the Rydberg electron in a semi-pure central potential, at some level of approximation.

The study of Rydberg states is highly relevant for a wide range of subjects in fundamental physics. However, most of these have to do with the way Rydberg atoms interact with other systems and with each other, and less with the direct specificities of how they are built up. For example, Rydberg atoms have a large polarisability, leading to large dipole moments. That, in turn, signals the onset of long range interactions and enables studies of fundamental scattering phenomena. The large size of Rydberg atoms — see (13.4) below — also makes them interesting platforms for studies of QED and of the interface between classical and quantum domains. For a deeper survey of Rydberg atoms, we refer to [3] or [4]. Here follows an overview of a few of the salient structural properties.

The central point of the structure of a Rydberg atom is that it will be akin to a one-electron system (as long as just one electron is highly excited). This means that energies can be reasonably well characterised by (1.24), and a lot of what has been detailed in chapter 1 and appendix B apply. In a more detailed analysis, quantum defect theory (see section 5.4) should be used. As is the case for alkali atoms — described by (5.11) and (5.13) and illustrated in table 5.3 — also Rydberg atoms become more hydrogen-like for a high orbital angular momentum number, l. The reason is that for a higher l, the centrifugal barrier becomes more important and there is less probability for penetration of the core (see also figure 1.3).

From (1.24), revealing a hydrogenic energy scaling with n^{-2}, it is clear that an atom with a large n will be close to its ionisation limit. So as to say, the excited electron is weakly bound. A scaling with n^2 will also emerge for the radial extension of the probability density. In terms of classical physics, this amount to saying that the atomic radius will have that dependence. If we apply the Bohr model, and take the charge of the nucleus (really the ionic core left behind by the Rydberg electron) as plus one, the radius of the classical orbit is, in SI-units:

$$r = \frac{n\hbar}{m_e v} \ , \tag{13.2}$$

with v as the classical orbital velocity. The latter we can take from classical centripetal motion of a particle with the mass and charge of an electron, given the Coulomb attraction from the nucleus and the Bohr model energy:

$$\frac{m_e v^2}{r} = \frac{e^2}{4\pi\varepsilon_0 r^2} \ . \tag{13.3}$$

The two relations above combine to:

$$r = \frac{4\pi\varepsilon_0 \hbar^2}{m_e e^2} n^2 \ , \tag{13.4}$$

which in atomic units is simply $r = n^2$. To take a numerical example, a principal quantum number of $n = 50$ leads to an atomic radius of about 130 nm — enormous on a typical atomic scale.

A somewhat counter-intuitive property of Rydberg atoms is that they have extraordinarily long radiative lifetimes, compared to most low energy excited states. The reason is that the wave function of a Rydberg state has a small overlap with the ground state, which from the point of view of the Fermi golden rule will lead to a small transition amplitude.

A consequence of the weak binding of the highly excited electron is a large polarisability, a corresponding big dipole moment, and an acute sensibility to external fields. Even a rather small external field will typically lead to a more important interaction than the hyperfine structure does. Thus, for a Rydberg atom, it often suffices to describe an atom merely by its fine-structure states and not involve the nuclear spin.

For Rydberg atoms with at least two valence electrons, whereof one is in a Rydberg state, the relative importance of *jj*-coupling over *LS*-coupling increases. This is another effect of the small overlap of the probability density of the Rydberg electron with its neighbours. The latter are distant enough to reduce the effect of the interelectronic repulsion, and accordingly, the intrinsic spin–orbit interaction for the highly excited electron may take precedence.

13.2 Continuum States

A system of an atomic nucleus and an ensemble of electrons typically has bound states. If we define zero energy as that for which the least bound electron of the atom has been removed to infinity — the first ionisation potential — it is clear that the atomic Schrödinger equation has solutions also for positive energies. These kinds of solutions can then include a continuous energy distribution for at least one electron. From one point of view, this is an extension of a Rydberg series (see section 13.1).

We can initiate an analysis of atomic continuum states by considering, within the CFA, an atom in a bound state which sees one of its electrons being excited to an energy above the ionisation limit, as in:

$$\gamma(nl)^N \to \gamma(nl)^{N-1}\left[E_\mathrm{e}l_\mathrm{e}\right] \ . \tag{13.5}$$

An atom with closed orbitals configuration γ has N valence electrons in the orbital (nl). One of these gets excited to an energy above its ionisation limit. The outcome is a system in a product state of a *parent ion* and an electron with the positive kinetic energy E_e and the orbital angular momentum l_e. Since the excited electron is no longer bound, it will eventually leave the system, and we will be left with a bound state of an ion. That brings us back to the focal point of the book — a bound atomic state. However, the transient state described by the right part of (13.5) is interesting in its own right.

To start with, an atomic state with a substantial amount of configuration interaction (see section 9.5) may have to be described by an expansion involving some configurations above the ionisation limit. This necessitates an understanding also of the latter. An atom with two excited electrons, or one with an excited core electron, may have a discrete solution with an energy above the first ionisation limit. Such a state may interact with a continuum state. That will lead to an ejection of an electron — the process of *auto-ionisation*, discussed in section 13.2.1 — and a broadening of the *quasi-discrete state*. These are a couple of reasons for which an understanding of continuum states may be necessary also for a deeper understanding of bound states.

The Schrödinger equation for a continuum state electron is exactly the same as for a bound state. The single free-electron eigenstate, however, will be of the form:

$$\Psi(\mathbf{q}) = R_{E_{el}l}(r)\, Y_{lm_l}(\theta,\varphi)\, \zeta(s,m_s) \quad , \tag{13.6}$$

with the radial function labelled by the kinetic energy, rather than with a discrete quantum number. This does not have to be seen as an excitation of bound state. It is just a solution to the Schrödinger equation and if it is at all necessary to invoke a preceding history, this could just as well have been a free electron coming into contact with an ion. In that case, one would be studying a scattering problem and/or collisional physics. These are subjects worthy of their own books, but they will not be further deliberated on here. Some more details about the mathematical form of scattering states, in the context of atomic structure, can be found, for example, in [5] or [6].

13.2.1 Auto-Ionising States

An auto-ionising state involves a discrete solution to the CFA Schrödinger equation wherein one or more electrons are excited so as to leave the atom in an energy state higher than that of the ionisation limit of the originally most loosely bound electron. As previously mentioned, this semi-discrete state may couple to a solution of the Schrödinger equation, for which one electron has a continuous positive energy, but with some conditions. The two states must have the same parity and the same J. If LS-coupling applies, then the quantum numbers L must also correspond, for angular momentum to be conserved.

The consequence is that the atom will decay to the ion parent state, while the superfluous electron parts with the excess energy. The occurrence of these kinds of radiationless transitions is known as the *Auger effect*. There exists an extensive literature on this, see for example [7] or [8] for an introduction.

The Auger effect is important in many fields of collisional physics. For example, the presence of these states can give rise to *radiative recombination* in atoms, which is the process where an electron colliding with an atom may be captured by the latter. The existence of auto-ionising channels also profoundly effects atomic spectra of doubly excited and core excited states. Emission lines from such quasi-discrete states become substantially broadened and weakened.

13.3 Inner Orbital Excitations and Highly Ionised Atoms

We have so far exclusively dealt with excited states where the valence electrons are responsible for the excitation. Atomic states for which one of the closed orbit electrons is excited are also important, especially in X-ray physics. However, since we adhere to the principle of treating structure rather than radiation and radiative transitions, this section will be brief. For more details, we refer to other works, for example [9]. In terms of structure, most of what has been described so far in the book still apply. The same is true for ions that have been stripped of many of its electrons.

13.3.1 Inner Excitations

If, in a multielectron atom, an electron in one of the innermost shells is excited, it will leave behind a vacancy in its initial orbital — a hole. When this hole is subsequently filled by the decay of an electron further out, this is accompanied by radiation in a typical wavelength range of 0.1–10 nm. Conversely, the binding energies of inner orbital electrons are in the corresponding range.

An inner electron will be exposed to more or less the full nuclear charge, with very little effective screening. This will increase the relative importance of the nuclear Coulomb Hamiltonian, and as a first approximation of the energies, the principal quantum number can be used alone. The convention for this, in X-ray physics, is to use the typical shell denominations from chemistry, that is principal quantum numbers $n = 1, 2, 3, 4, \ldots$ are labelled K, L, M, N … and so on.

As a first approximation of the binding energy for an inner shell electron, the following formula can be used:

$$E_n = -\frac{1}{2}\frac{(Z-\sigma_n)^2}{n^2} \quad . \tag{13.7}$$

n refers to the principal quantum number, Z to the nuclear charge and $\sigma_n \ll Z$ is a *screening term*. The numerator is thus the square of an effective charge. As far as structure is concerned, the equation is more pragmatic than informative, but for the purpose of calculating X-ray emission energies, it is a decent first approximation. Rough approximations, based on empirical data, for the K and L-shells are $\sigma_1 \approx 1$ and $\sigma_2 \approx 7.4$.

13.3.2 Highly Charged Ions

From a structural point of view, the main difference between an atom that has lost many of its electrons, and a neutral one, is that for highly charged ions, the relative importance of the nuclear Coulomb potential increases also for valence electrons. The charge screening will be less effective than what it is for a neutral. Another important difference is that relativistic effects become more pronounced.

The increased weight of the nuclear attraction term in the Hamiltonian renders a highly charged ion more hydrogenic in character. As a consequence, approximating the energy with a simple function of the principal quantum number becomes more pertinent. A version of (13.7), with an adjusted screening constant, can be used as a first approximation [9].

13.4 Further Complexity

There are many more examples of complex atoms than the ones given in this chapter, and many are a lot more complex. For example, one of the electrons could be replaced by some other negatively charged particle, such as a muon. A bound system of one positron and one electron — *positronium* — could be seen as an atomic system, and if it is an anti-proton and a positron, we have an anti-hydrogen atom. One could also easily envisage regular atoms in very complex excited states.

All of this, and the systems that are briefly treated in this chapter, deserves their own texts, and their extensive treatments fall beyond the scope of this book. The present descriptions should be seen as introductions to such subjects. One good source for further studies is [1] and references therein.

Further Reading

The Theory of Auger Transitions, by Chattarji [7]
The theory of atomic structure and spectra, by Cowan [5]
Rydberg Atoms, by Gallagher [3]
Atoms, by Kleinpoppen [9]
Physics of Atoms and Molecules, by Bransden & Joachain [10]
Springer Handbook of Atomic, Molecular, and Optical Physics, by Drake [1]

References

1. G.W.F. Drake (ed.), *Springer Handbook of Atomic, Molecular, and Optical Physics* (Springer-Verlag, New York, 2006)
2. A. Kramida, Y. Ralchenko, J. Reader, and NIST ASD Team. NIST Atomic Spectra Database (ver. 5.3). [Online]. Available: http://physics.nist.gov/asd (2018). Accessed: 2019-07-14
3. T.F. Gallagher, *Rydberg Atoms* (Cambridge University Press, Cambridge, 1994)
4. T.F. Gallagher, in *Springer Handbook of Atomic, Molecular, and Optical Physics*, ed. by G.W.F. Drake (Springer-Verlag, New York, 2006), p. 235
5. R.D. Cowan, *The theory of atomic structure and spectra* (University of California press, Berkeley, 1981)
6. P. Burke, *Electron-Atom, Electron-Ion, and Electron-Molecule Collisions* (Springer-Verlag, New York, 2006), p. 705
7. D. Chattarji, *The Theory of Auger Transitions* (Academic Press, London, 1976)
8. A. Temkin, A.K. Bhatia, *Autoionization* (Springer-Verlag, New York, 2006), p. 391
9. H. Kleinpoppen, in *Constituents of Matter*, ed. by W. Raith (Walter de Gruyter, Berlin, 1997), p. 1
10. B.H. Bransden, C.J. Joachain, *Physics of Atoms and Molecules*, 2nd edn. (Prentice Hall, Harlow, England, 2003)

Chapter 14
Numerical Solutions of the Atomic Schrödinger Equation

The methods for calculation of energies and wave functions presented in this book are mostly first approximations, aimed primarily at providing a physical understanding of atomic structure, and less at exact results. In order to achieve more precise theoretical values for atomic quantities, more sophisticated methods, involving ample computational resources, are needed. A detailed description of the associated methods is a subject that merits its own literature, and several such works exist (see suggested further reading). It should also be noted that at the moment of writing this book, numerical calculations of atomic wave functions is an active field of research, and every year brings progress and new developments.

In what follows, we will give short accounts of the underlying ideas for a few numerical methods used for multielectron atoms. This type of computational techniques is actually a part of a more general field, which could be called many-body quantum mechanics or quantum chemistry. Adaptations of the same methods that we will describe in sections 14.1 and 14.2 are used also for analyses of, for example, nuclear or molecular systems.

14.1 Self-Consistent Field Methods

The essence of self-consistent field (SCF) methods, when applied to atomic structure, is that for each electron in a multielectron atom the overall Coulomb potential it feels is taken as the time average of that produced by all other charges — that is, all other electrons and the nucleus. We will soon see what the meaning is of self-consistency of this average field in this context.

The great advantage of this is that the total wave function can be expressed as a product state. Thus, the SCF method is essentially the same as the central-field approximation, except that this time it is applied with the goal of obtaining more accurate quantitative results. This particular terminology is the choice used in this book, albeit it is not universally applied.

A. Kastberg, *Structure of Multielectron Atoms*, Springer Series on Atomic, Optical, and Plasma Physics 112, https://doi.org/10.1007/978-3-030-36420-5_14

If we would start with a guess, as well informed as possible, of the average potentials felt by each individual electron, the subsequent step would then be to apply all those potentials in a Schrödinger equation and to solve the latter. In most cases, this would be a numerical solution. Since we have made the crude approximation of a product function, we will, in fact, have N separate equations — each with an individual wave function and its corresponding probability density. This new charge distribution translates into N new average potentials, and then the Schrödinger equation can be solved anew for *that* potential.

If the initial average potential has been chosen wisely, this loop of generating charge distributions and using them to find improved expressions for stationary states should not be infinite. It should converge, and at the point where the solution of the Schrödinger equation leads to the same potential as the input one, one has reached *self-consistency*.

This described method is in fact an extension of the variational method in quantum mechanics (see chapter 3). The search for extrema in the variational method is mathematically equivalent to the iterative SCF method [1].

14.1.1 The Hartree Equation

We will outline one of the original forms of the self-consistent field method, originally introduced by Hartree. We start by writing the total non-relativistic wave function of an atom, with N electrons and nuclear charge Z, as the product function:

$$\psi(\mathbf{r}_1 \dots \mathbf{r}_N) = \prod_{i=1}^{N} \psi_{n_i l_i m_{l_i}}(\mathbf{r}_i) = \prod_{i=1}^{N} R_{n_i l_i}(r_i)\, Y_{l_i m_{l_i}}(\theta_i, \varphi_i)\,. \tag{14.1}$$

At this stage, we do not bother about making the wave function antisymmetric, but we do make sure that the Pauli principle is obeyed. That is, for each quantum number combination $n_i l_i m_{l_i}$ there can at most be two electrons. Other than allowing for this, spin does not play a role for the solution.

Since we will take spherical averages of the Coulomb interactions, the angular function in (14.1) will be exactly the spherical harmonics used for hydrogenic solutions. The radial function, however, will be quite different from the hydrogen one, albeit still obeying the quantisation rules of central potentials. The goal of the exercise is to derive as good an approximation to this radial function as possible.

The full Hamiltonian is the one in (2.8):

$$H = -\sum_{i=1}^{N} \frac{\nabla_i^2}{2} - \sum_{i=1}^{N} \frac{Z}{r_i} + \sum_{i=1}^{N} \sum_{j>i}^{N} \frac{1}{r_{ij}}\,. \tag{14.2}$$

However, we are going to make individual spherical averages of the last term for each of the N electrons. Thus, the full solution will be one corresponding to every electron moving in its own specific potential. Since the latter is strongly coupled to *all* electrons, this statement is in a literal sense inherently inconsistent, but as we shall see, the approach can still lead to self-consistent solutions.

As for many issues with multielectron atoms, it is the cross term that causes difficulties. It is responsible both for entanglement and for a breakdown of spherical symmetry. It will need special treatment, and we therefore divide the expectation value of the Hamiltonian in (14.2) into two parts. We also make the assumption inherent in (14.1) that we can represent the total atomic wave function with a product state. If we also take the individual electron wave functions as normalised, the expectation value of the Hamiltonian reduces to, first of all, a sum of single-electron integrals, plus one double sum of integrals taken over just two electron coordinates. That is, we have:

$$\begin{aligned}\langle H\rangle &= \left\langle \psi(\mathbf{r}_1\ldots\mathbf{r}_N)\left|-\sum_{i=1}^{N}\frac{\nabla_i^2}{2}-\sum_{i=1}^{N}\frac{Z}{r_i}+\sum_{i=1}^{N}\sum_{j>i}^{N}\frac{1}{r_{ij}}\right|\psi(\mathbf{r}_1\ldots\mathbf{r}_N)\right\rangle\\ &=\sum_{i=1}^{N}\left\langle \psi_{n_il_im_{li}}(\mathbf{r}_i)\left|-\frac{\nabla_i^2}{2}-\frac{Z}{r_i}\right|\psi_{n_il_im_{li}}(\mathbf{r}_i)\right\rangle\\ &\quad+\sum_{i=1}^{N}\sum_{j>i}^{N}\left\langle \psi_{n_il_im_{li}}(\mathbf{r}_i)\,\psi_{n_jl_jm_{lj}}(\mathbf{r}_j)\left|\frac{1}{r_{ij}}\right|\psi_{n_il_im_{li}}(\mathbf{r}_i)\,\psi_{n_jl_jm_{lj}}(\mathbf{r}_j)\right\rangle\\ &=\sum_{i=1}^{N}\mathscr{I}_{\text{single}}^{(i)}+\sum_{i=1}^{N}\sum_{j>i}^{N}\mathscr{I}_{\text{pair}}^{(i,j)}\,. \end{aligned}\tag{14.3}$$

The single-electron integral has a purely central potential term. Thus, the angular part of this can be factorised out, and with normalised spherical harmonics:

$$\begin{aligned}\mathscr{I}_{\text{single}}^{(i)} &= \int_0^\infty R_{n_il_i}^*(r_i)\left\{-\frac{1}{2r_i^2}\frac{\mathrm{d}}{\mathrm{d}r_i}\left[r_i^2\frac{\mathrm{d}}{\mathrm{d}r_i}\right]+\frac{l_i(l_i+1)}{2r_i^2}-\frac{Z}{r_i}\right\}\\ &\qquad\times R_{n_il_i}(r_i)\,r_i^2\,\mathrm{d}r_i = I(n_il_i)\,. \end{aligned}\tag{14.4}$$

This is formally identical with the one-electron integral defined in (7.16).

Next step is to find the spherical average of the pair integral in (14.3). The physical meaning of the term is the potential energy emerging from the Coulomb repulsion between the two charge distributions $|\psi_{n_il_im_{li}}(\mathbf{r}_i)|^2$ and $|\psi_{n_jl_jm_{lj}}(\mathbf{r}_j)|^2$. However, in order to be able to handle it mathematically, we will approach it less directly.

From (1.17), we know that the charge density (in atomic units) of a spherical shell between r_j and $r_j+\mathrm{d}r_j$ of a hydrogenic wave function is:

$$|R_{n_jl_j}(r_j)|^2\,r_j^2\,\mathrm{d}r_j\,. \tag{14.5}$$

To get the spherically averaged potential energy resulting from the interaction between this charge and the electron i, we think of the total charge distribution of electron j as a uniformly charged sphere, centred at the origin and whose radius is one of the integration variables. This means that for all $r_i<r_j$ in the double integral

in $\mathscr{I}_{\text{pair}}^{(i,j)}$, the Coulomb potential from electron j is constant, and equal to the value at the surface of the sphere. For $r_i > r_j$, the interaction will be the same as for an electron j located at the origin. This gives the potential energy:

$$V_{\mathscr{E}} = \frac{1}{r_i}\int_0^{r_i} |R_{n_jl_j}(r_j)|^2\, r_j^2\,\mathrm{d}r_j + \int_{r_i}^{\infty} |R_{n_jl_j}(r_j)|^2\, r_j\,\mathrm{d}r_j \equiv \frac{1}{r_i}\,Y_0(n_il_i:n_jl_j/r_i)\ , \qquad (14.6)$$

where with the last step we have defined the *Hartree potential*, $Y_0(n_il_i:n_jl_j/r_i)$.

From the Hartree potential in (14.6), the pair integral in (14.3) is now found by integrating over r_i:

$$\begin{aligned}\mathscr{I}_{\text{pair}}^{(i,j)} &= \int_0^{\infty} |R_{n_il_i}(r_i)|^2\, Y_0(n_il_i:n_jl_j/r_i)\, r_i\,\mathrm{d}r_i \\ &= \int_0^{\infty}\int_0^{\infty} |R_{n_il_i}(r_i)|^2\, |R_{n_jl_j}(r_j)|^2\, \frac{1}{r_>}\, r_i^2 r_j^2\,\mathrm{d}r_i\,\mathrm{d}r_j = F^{(0)}(n_il_i:n_jl_j)\ . \qquad (14.7)\end{aligned}$$

The denominator $r_>$ refers to the largest of the two radial coordinates in each integration. In the last line of (14.7), we note that the pair integral equals a zero-order Slater F-integral (see section 7.2.3 and appendix D). Together with (14.4), this means that we can summarise the expectation value in (14.3) as:

$$\langle H\rangle = \sum_{i=1}^{N} I(n_il_i) + \sum_{i=1}^{N}\sum_{j>i}^{N} F^{(0)}(n_il_i:n_jl_j)\ . \qquad (14.8)$$

We want to have a set of N differential equations — one for the radial wave function of each electron. In order to get there, we can vary the radial function (all the time maintaining normalisation), and search for a minimum of the expectation value in (14.8). For an explicit demonstration of this exercise, we refer to the suggested further reading (for example, [2]). The resulting *Hartree equation* is:

$$\begin{aligned}&\left[-\frac{1}{2}\frac{\mathrm{d}^2}{\mathrm{d}r_i^2} + \frac{l_i(l_i+1)}{2r_i^2} - \frac{Z}{r_i} + \sum_{j=0}^{N}\delta(i,j)\,\frac{1}{r_i}\,Y_0(n_il_i:n_jl_j/r_i)\right] U_{n_il_i}(r_i) \\ &\qquad = \varepsilon_{n_il_i}\, U_{n_il_i}(r_i)\ . \qquad (14.9)\end{aligned}$$

In (14.9), $U_{n_il_i}(r_i) = r_i R_{n_il_i}(r_i)$, and $\varepsilon_{n_il_i}$ is the eigenvalue corresponding to that function. Since the potential $Y_0(n_il_i{:}n_jl_j/r_i)$ itself contains integrals of $R_{n_il_i}$ (and thus of $U_{n_il_i}$), this is an *integrodifferential equation*, and analytical solutions will be difficult to find. Typically, the only way to reach a self-consistent solution is to begin with as good a guess as possible of all the potentials Y_0, and to solve the N equations (14.9) numerically. The solution can then be used to calculate a new Y_0, and so on. In a successful attempt, this iterative process will converge towards a decent approximation of the total wave function.

It is important to emphasise that a slightly different Schrödinger equation will be solved for each electron. A consequence of this is that the N different wave functions will not necessarily come out mutually orthogonal. The angular parts of the wave functions have been factored out, and the spherical harmonics are in themselves orthogonal. However, two different radial functions with the same value of l_i may violate the fundamental requirement of orthogonality. This caveat led to a further development of the Hartree method, which will be treated in the following section.

14.1.2 The Hartree-Fock Method

The inconvenience of the Hartree method, of not giving wave functions that are strictly exchange antisymmetric, has led to it being rarely used in its original form. Another reason for this is that not long after its conception, a patch of the Hartree method was introduced. In this, the *Hartree-Fock method*, the resulting wave functions fulfil the symmetry requirements, and as a consequence calculated energies — as well as other quantities — better reproduce empiric observations.

In a nutshell, the difference between the Hartree and the Hartree-Fock methods is that the latter uses Slater determinant wave functions. The single-electron wave functions will still be derived using single-electron equations, but when the expectation value of the energy is optimised, or the expectation value of the charge distributions, this has to be done with Slater determinants. That means that care has to be taken when calculating the $1/r_{12}$ part of the matrix element integrals. The rules outlined in section 2.3 have to be adhered to (see also [3]).

The procedure needed to derive the *Hartree-Fock equations* — the analogy of (14.9) — follows essentially the same lines as those demonstrated in the preceding section. For the details, we refer to the suggested further reading, for example, [2, 4] or [5]. It amounts to varying all the individual electronic wave functions and thereby finding the Slater determinant (alternatively a superposition of Slater determinants) that best represents the atomic state. All the while, normalisation and mutual orthogonality must be maintained.

14.2 Correlation Methods

The Hartree-Fock method described in section 14.1 has been of great importance in the development of atomic structure theory. For the most precise computations, it is however necessary to develop the techniques further.

The Hartree-Fock wave function is a Slater determinant. That means that electron–electron correlation has been taken into account as so far as that the wave function is antisymmetric. It is however still a sum of product functions, and the interelectronic Coulomb repulsion is not explicitly taken into account. In order to remedy this, correlation method extensions of the SCF and Hartree-Fock methods have been developed. The classical Hartree-Fock method is also non-relativistic, and for accurate results for multielectron atoms, this will not be enough.

One successful technique that takes into account correlation effects is the *coupled cluster* method. Detailed descriptions of this can be found in, for example, [3] and [6]. The coupled cluster method is based on constructing a multielectron wave function in which a basis function is multiplied by an exponential operator:

$$\Psi(\mathbf{q}) = e^S \Psi_0(\mathbf{q}) . \tag{14.10}$$

$\mathbf{q}$ stands for the complete set of spatial and spin coordinates for all electrons. The function $\Psi_0(\mathbf{q})$ can be a Slater determinant obtained from a preceding analysis, and the electron correlation will be contained in the cluster operator S. The latter is formulated as an infinite series:

$$S \equiv S_1 + S_2 + S_3 + S_4 + \ldots \quad , \tag{14.11}$$

and the component operators S_n describe successive multiparticle effects. S_1 correspond to the creation of *one* excitation, according to:

$$S_1 \equiv \sum_{i=1}^{N_{\text{occ}}} \sum_{p=1}^{N_{\text{emp}}} t_i^{(p)} \hat{a}_p^\dagger \hat{a}_i \quad . \tag{14.12}$$

N_{occ} is the number of electrons and the corresponding sum is over all states occupied by an electron. The second sum goes over all empty states, to which an electron may transition. N_{emp} is thus the number of empty states included in the calculation. A larger set of states gives better precision.

In (14.12) the annihilation and creation operators promote one electron from an occupied orbital to one that has a vacancy. The coefficients $t_i^{(p)}$ are to be determined in the computation. In an analogue fashion, S_2 is an operator which excites *two* electrons from occupied orbitals to excited ones:

$$S_2 \equiv \sum_{i=1}^{N_{\text{occ}}} \sum_{j>i}^{N_{\text{occ}}} \sum_{p=1}^{N_{\text{emp}}} \sum_{q>p}^{N_{\text{emp}}} t_{i,j}^{(p,q)} \hat{a}_p^\dagger \hat{a}_q^\dagger \hat{a}_j \hat{a}_i \quad , \tag{14.13}$$

and so on. The number of terms in (14.11) is thus limited to the overall number of electrons, N_{occ}.

14.3 The Starting Point for Further Studies of Atomic Structure

The title of this chapter is 'Numerical solutions of the atomic Schrödinger equation'. While numerical methods have indeed been introduced, the methods that we have outlined, such as the self-consistent field, Hartree-Fock, correlation methods and coupled clusters, are much more than just 'numerical methods'. They are all theoretical developments in their own rights. They do provide results in terms of expressions for wave functions, and quantitative values for observables such as energies, wavelengths, multipole coupling and radiative lifetimes, but besides that, they give important insights into the fundamentals of the structure of atoms, beyond the introductions given in this book, and frequently beyond empirical data.

The methods that we have discussed in the chapter are more general than being applicable only to atomic systems. They could also be described as components of general quantum many-body physics, and this includes many-body perturbation theory and the incorporation of relativistic effects. At the forefront of the field, it combines QED and many-body physics. While still being pursued and further developed at leading-edge research institutions, the field has also led to a plethora of software resources for making advanced atomic structure computation, both commercial and open-source ones.

This last chapter has been kept intentionally brief. The present section marks the end of the book, but the intention of this book is that rather than an end, this should for the reader mark a starting point for more profound studies of multielectron atoms. There are many sources for further reading on the subject — good examples are [3, 7, 8] and [9] — and others may be expected to emerge, along with further developments of methodologies and of the understanding of the structure of atoms.

Further Reading

Quantum theory of atomic structure, by Slater [2]
Physics of Atoms and Molecules, by Bransden & Joachain [4]
The theory of atomic structure and spectra, by Cowan [7]
Atomic Many-Body Theory, by Lindgren & Morrison [3]
Springer Handbook of Atomic, Molecular, and Optical Physics, by Drake [8]
Relativistic many-body theory, a new field-theoretical approach, by Lindgren [9]

References

1. G.W.F. Drake, in *Springer Handbook of Atomic, Molecular, and Optical Physics*, ed. by G.W.F. Drake (Springer-Verlag, New York, 2006), p. 199
2. J.C. Slater, *Quantum theory of atomic structure* (McGraw-Hill, New York, 1960)
3. I. Lindgren, J. Morrison, *Atomic Many-Body Theory*, 2nd edn. (Springer Verlag, Berlin, 1986)
4. B.H. Bransden, C.J. Joachain, *Physics of Atoms and Molecules*, 2nd edn. (Prentice Hall, Harlow, England, 2003)
5. C.F. Fischer, in *Springer Handbook of Atomic, Molecular, and Optical Physics*, ed. by G.W.F. Drake (Springer-Verlag, New York, 2006), p. 307
6. J. Paldus, in *Springer Handbook of Atomic, Molecular, and Optical Physics*, ed. by G.W.F. Drake (Springer-Verlag, New York, 2006), p. 101

7. R.D. Cowan, *The theory of atomic structure and spectra* (University of California press, Berkeley, 1981)
8. G.W.F. Drake (ed.), *Springer Handbook of Atomic, Molecular, and Optical Physics* (Springer-Verlag, New York, 2006)
9. I. Lindgren, *Relativistic Many-Body Theory, A New Field-Theoretical Approach*, 2nd edn. (Springer-Verlag, Switzerland, 2016)

Appendix A
Atomic Units

Atomic units (a.u.) are used extensively in atomic physics, see, for example, [1]. The benefit this brings is that mathematical expressions can be written in a way that makes many natural constants inherently included. Thus, it becomes technically easier to deal with formulae and ideally the mathematics becomes more transparent and easier to interpret.

The starting point for atomic units is to set the following natural constants equal to unity:

$$e = m_{\mathrm{e}} = \hbar = \frac{1}{4\pi\varepsilon_0} = 1 \ . \qquad \text{(A.1)}$$

From there, other constants can be derived, and units for physical quantities can be expressed accordingly. The atomic units for a selection of quantities, relevant for atomic physics, are given in table A.1. Among these are the a.u. for energy, E_{h} (one *hartree*), and that for length, a_0 (the *Bohr radius*). The values are commensurate with the table of constants in the preamble of the book, and as sources we have used [2] and [3].

A.1 Constants in Atomic Units

The conversion of constants to atomic units is typically straightforward from their definitions (see the table in the preamble of the book), and their relations with the quantities included in (A.1). In table A.2, we present the a.u. value of some of the more important constants for atomic physics.

A. Kastberg, *Structure of Multielectron Atoms*, Springer Series on Atomic, Optical, and Plasma Physics 112, https://doi.org/10.1007/978-3-030-36420-5

Table A.1 Some quantities important for the study of atomic structure, expressed in atomic units, together with their corresponding values in SI-units (see also the table of constants).

Quantity	atomic unit	value in SI-units
Charge	e	$1.602\,176\,634 \times 10^{-19}$ C
Mass	m_e	$9.109\,383\,7015 \times 10^{-31}$ kg
Angular momentum	$\hbar$	$1.054\,571\,817 \times 10^{-34}$ J s
Length	a_0	$0.529\,177\,210\,903 \times 10^{-10}$ m
Energy	E_h	$4.359\,744\,722\,2071 \times 10^{-18}$ J
Time	$\hbar/E_h$	$2.418\,884\,325\,1036 \times 10^{-17}$ s
Velocity	$c\,\alpha$	$2.187\,691\,263\,64 \times 10^{6}$ m s^{-1}
Momentum	$\hbar/a_0$	$1.992\,851\,912\,88 \times 10^{-24}$ kg m s^{-1}
Force	E_h/a_0	$8.238\,723\,4983 \times 10^{-8}$ N
Charge density	e/a_0^3	$1.081\,202\,384\,57 \times 10^{12}$ C m^{-3}
Electric potential	E_h/e	$27.211\,386\,245\,988$ V
Electric field	$E_h/(e\,a_0)$	$5.142\,206\,747\,63 \times 10^{11}$ V m^{-1}
Electric dipole moment	$e\,a_0$	$8.478\,353\,6255 \times 10^{-30}$ C m
Polarisability	$e^2\,a_0^2/E_h$	$1.648\,777\,274\,36 \times 10^{-41}$ C^2 m^2 J^{-1}
Magnetic flux density	$\hbar/(e\,a_0^2)$	$2.350\,517\,567 \times 10^{5}$ T
Electromagnetic vector potential	$\hbar/(e\,a_0)$	$1.243\,840\,3298 \times 10^{-5}$ T m
Temperature	E_h/k_B	$3.157\,750\,2480 \times 10^{5}$ K

Table A.2 A selection of constants expressed in atomic units.

c	$=$	$1/\alpha$
μ_0	$=$	$4\pi\,\alpha^2$
ε_0	$=$	$1/(4\pi)$
h	$=$	2π
m_p	$=$	$1\,836.152\,673\,43$
m_n	$=$	$1\,838.683\,661\,73$
m_u	$=$	$1\,822.888\,486\,21$
μ_B	$=$	$1/2$
μ_N	$=$	$e\hbar/(2m_p)$
R_∞	$=$	$\alpha/4\pi$
a_0	$=$	1
α	$=$	$0.007\,297\,352\,5693$
α^{-1}	$=$	$137.035\,999\,084$

A.1.1 The Fine-Structure Constant

The two last entries in table A.2 are the *fine-structure constant* and its inverse. Its definition in SI-units is:

$$\alpha \equiv \frac{e^2}{4\pi\varepsilon_0 \hbar c}\,. \tag{A.2}$$

This renders it dimensionless, and thus it has the same value in all units.

The explanation for the name 'fine-structure constant' is historic. Its first specified value was the ratio of the velocity of an electron in the first Bohr orbit to the speed of light. This turned out to be a measure of the relative magnitude of the fine-structure splitting in hydrogen. Albeit the fine-structure now has a solid relativistic quantum mechanical explanation (see appendix E), the latter fact still essentially holds.

The modern interpretation is that α is the coupling constant for the electromagnetic force, and as such it can be compared to the other fundamental forces in the standard model of particle physics [4]. The fine-structure constant occurs abundantly in atomic physics, and its invariability with the choice of units makes it most useful.

References

1. W.E. Baylis, G.W.F. Drake, in *Springer Handbook of Atomic, Molecular, and Optical Physics*, ed. by G.W.F. Drake (Springer-Verlag, New York, 2006), p. 1
2. P.J. Mohr, D.B. Newell, B.N. Taylor, Journal of Physical and Chemical Reference Data **45**, 043102 (2016)
3. Physical Measurement Laboratory of NIST. The NIST reference on constants, units and uncertainties. [Online]. Available: https://physics.nist.gov/cuu/Constants/index.html (2019). Accessed: 2019-05-22
4. E.D. Commins, *Quantum mechanics: an experimentalist's approach* (Cambridge University Press, New York, 2014)

Appendix B
Radial Hydrogenic Wave Functions

In this appendix, we will begin by deriving the solution to the radial part of the non-relativistic Schrödinger equation for hydrogen, $R(r)$, in spherical coordinates. More detailed and rigorous versions of this can be found in many texts on fundamental quantum mechanics and atomic physics, for example, [1, 2] or [3]. After that, we tackle the solution to the hydrogenic Schrödinger equation in parabolic coordinates. It is included in this appendix, albeit it does not involve a symmetric radial parameter and in spite of involving some angular parameters (see appendix C). It is nevertheless relevant to compare this with the solution in polar spherical coordinates, and the formalism is useful when an external field breaks the spherical symmetry of the free atom (see chapter 12). In the last section of the appendix, we table a series of integrals over hydrogenic radial wave functions and positive and negative powers of the radial parameter r.

B.1 The Radial Solutions to the Hydrogenic Schrödinger Equation

When the Schrödinger equation for a hydrogenic system has been factorised into radial and angular parts, the radial equation, in atomic units, is the one in (1.9) in chapter 1:

$$\frac{1}{r^2}\frac{\mathrm{d}}{\mathrm{d}r}\left[r^2\frac{\mathrm{d}}{\mathrm{d}r}R(r)\right]+\left[\frac{2Z}{r}+2E-\frac{l(l+1)}{r^2}\right]R(r)=0\,. \tag{B.1}$$

We reformulate this in terms of a function defined as $U(r)\equiv rR(r)$. This gives us:

$$\frac{\mathrm{d}^2}{\mathrm{d}r^2}U(r)+\left[\frac{2Z}{r}+2E-\frac{l(l+1)}{r^2}\right]U(r)=0\,, \tag{B.2}$$

which is equivalent to (1.11).

A. Kastberg, *Structure of Multielectron Atoms*, Springer Series on Atomic, Optical, and Plasma Physics 112, https://doi.org/10.1007/978-3-030-36420-5

To identify solutions to (B.2), we will begin by studying the asymptotic regions, where $r \to 0$ and $r \to \infty$. The respective special cases, $U^{(0)}(r)$ and $U^{(\infty)}(r)$, will then be used in order to derive general solutions.

For $r \to 0$, the terms $2Z/r$ and $2E$ in (B.2) can be neglected and we have:

$$\frac{d^2}{dr^2} U^{(0)}(r) - \frac{l(l+1)}{r^2} U^{(0)}(r) = 0 . \tag{B.3}$$

This equation has the following two solutions:

$$\begin{aligned} U^{(0)}(r) &= r^{l+1} \\ U^{(0)}(r) &= r^{-l} \quad . \end{aligned} \tag{B.4}$$

From the definition of $U(r)$ we can see that this function must be finite also as $r \to 0$. Therefore, the latter of the two solutions above can be discarded.

We then consider the other limit, where $r \to \infty$. This time it is the two terms proportional to $1/r$ and $1/r^2$ that are negligible, and the differential equation becomes:

$$\frac{d^2}{dr^2} U^{(\infty)}(r) + 2E\, U^{(\infty)}(r) = 0 . \tag{B.5}$$

Also here we get two solutions:

$$U^{(\infty)}(r) = e^{\pm r\sqrt{-2E}} \quad . \tag{B.6}$$

For bound states, $E < 0$, this gives real solutions. We cannot allow $U(r)$ to diverge, and thus we retain only the negative exponential.

With the form of $U(r)$ determined for $r \to 0$ and $r \to \infty$, we introduce a trial solution which is a product of the limiting cases above and a polynomial function:

$$U(r) = r^{l+1} e^{-r\sqrt{-2E}} \left(A_0 + A_1 r + A_2 r^2 + A_3 r^3 + \ldots\right) \; . \tag{B.7}$$

With properly chosen polynomial coefficients, this will have the correct behaviour for very small and very large r. The trial solution is then substituted into (B.2), which will give us a recursion formula for the A's in (B.7):

$$A_k = -2A_{k-1} \frac{Z - (l+k)\sqrt{-2E}}{(l+k)(l+k+1) - l(l+1)} \, . \tag{B.8}$$

The problem with this is that when r goes to infinity, the infinite series in (B.7) increases exponentially, and thus $U(r)$ will no longer be finite. The remedy to this is to force the series in (B.7) to break off at some point, and form a limited polynomial. This can be achieved if the numerator in (B.8) becomes zero for some k. Thus, we get a limiting condition for k:

$$(l+k)\sqrt{-2E} \leq Z \; . \tag{B.9}$$

Since k is a series of integers, a particular value of k_{max} must correspond to a particular energy eigenvalue. This means that we can directly find the hydrogenic bound state energies (in a.u.) from (B.9):

$$E = -\frac{Z^2}{2\,(l+k_{\mathrm{max}})^2} \equiv -\frac{Z^2}{2\,n^2} \; . \tag{B.10}$$

In the second step, we have defined the *principal quantum number*, n.

B.1.1 Wave Functions

To find the mathematical expressions for the stationary states, we use the definition of n above and we rescale the radial parameter as:

$$\rho = \frac{2Zr}{n} \; . \tag{B.11}$$

Using the recursion formula in (B.8), this allows us to express (B.7) in terms of an associated Laguerre polynomial (see, for example, [4] or [5]), with ρ as parameter:

$$U_{nl}(\rho) = -\sqrt{\frac{Z\,(n-l-1)!}{n^2\,[(n+l)!]^3}}\;\rho^{l+1}\,\mathrm{e}^{-\rho/2}\,\mathrm{L}_{n+l}^{(2l+1)}(\rho) \; , \tag{B.12}$$

or, as expressed in $R(r)$:

$$R_{nl}(r) = -\frac{2Z^{3/2}}{n^2}\sqrt{\frac{(n-l-1)!}{[(n+l)!]^3}}\;\rho^{l}\,\mathrm{e}^{-\rho/2}\,\mathrm{L}_{n+l}^{(2l+1)}(\rho) \; . \tag{B.13}$$

The associated Laguerre polynomial in (B.12) and (B.13) can be written as:

$$\mathrm{L}_{n+l}^{(2l+1)}(\rho) = B_0 + B_1\,\rho + B_2\,\rho^2 + B_3\,\rho^3 + \cdots + B_{n-l-1}\,\rho^{n-l-1} \; , \tag{B.14}$$

with the coefficients

$$B_k = -B_{k-1}\,\frac{n-l-k}{(l+k)\,(l+k+1) - l(l+1)} \; . \tag{B.15}$$

The series must be finite, and the last coefficient is:

$$B_{k_{\mathrm{max}}} = B_{n-l-1} = (-1)^{n+l}\,\frac{(n+l)!}{(n-l-1)!} \; . \tag{B.16}$$

The denominator factorial $(n-l-1)$ must be zero or positive, and thus we get the condition that l must be a positive integer or zero, and also that l has a maximal allowed value:

$$l = 0, 1, 2, \ldots, n-1 \, . \tag{B.17}$$

Equation B.13 is the radial solution to the $1/r$ potential of hydrogenic systems. It is periodic and discretised in the quantum numbers n and l. A list of the lowest order functions $R_{nl}(r)$ is provided in (1.16). Charge densities resulting from the lowest order solutions are shown in figure 1.3.

B.2 The One-Electron Schrödinger equation in Parabolic Coordinates

The hydrogenic Schrödinger equation in parabolic coordinates is used in section 12.1.3 to formulate a Stark shift Hamiltonian. Here, we will derive the field free wave function in this format, show that it is separable, and that the solutions are identical with the ones obtained with spherical coordinates.

We define the parabolic coordinates as follows [4]:

$$\begin{aligned} \xi &\equiv r+z \\ \eta &\equiv r-z \\ \varphi &\equiv \arctan\left(\frac{y}{x}\right) \, . \end{aligned} \tag{B.18}$$

ξ and η range from 0 to ∞, and φ from 0 to 2π. Inverse transformations are:

$$\begin{aligned} x &= \sqrt{\xi\eta}\cos\theta \\ y &= \sqrt{\xi\eta}\sin\theta \\ z &= \frac{\xi-\eta}{2} \\ r &= \frac{\xi+\eta}{2} \, . \end{aligned} \tag{B.19}$$

The volume element is:

$$\mathbf{dr} = \frac{\xi+\eta}{4}\, \mathrm{d}\xi \mathrm{d}\eta \mathrm{d}\varphi \, , \tag{B.20}$$

and the Laplacian:

$$\nabla^2 = \frac{4}{\xi+\eta}\left[\frac{\partial}{\partial\xi}\left(\xi\frac{\partial}{\partial\xi}\right) + \frac{\partial}{\partial\eta}\left(\eta\frac{\partial}{\partial\eta}\right)\right] + \frac{1}{\xi\eta}\frac{\partial^2}{\partial\varphi^2} \, . \tag{B.21}$$

This taken into account, the Schrödinger equation in parabolic coordinates for a one-electron atom becomes (in a.u.):

$$\frac{2}{\xi+\eta}\left[\frac{\partial}{\partial\xi}\left(\xi\frac{\partial}{\partial\xi}\right)+\frac{\partial}{\partial\eta}\left(\eta\frac{\partial}{\partial\eta}\right)\right]\psi(\xi,\eta,\varphi)+\frac{1}{2\xi\eta}\frac{\partial^2}{\partial\varphi^2}\psi(\xi,\eta,\varphi)$$
$$+\left(\frac{2Z}{\xi+\eta}+E\right)\psi(\xi,\eta,\varphi)=0\,, \quad \text{(B.22)}$$

with Z being the charge state of the nucleus.

To proceed, we need to factorise (B.22). The azimuthal angle coordinate is the same as that used for polar spherical coordinates, and the corresponding function should be the same as the one for the standard solution to the hydrogen angular Schrödinger equation (C.30), discretised in the projection quantum number m_l. The solutions sought after are therefore of the form:

$$\psi(\xi,\eta,\varphi)=\Xi(\xi)\,H(\eta)\,\mathrm{e}^{\mathrm{i}m_l\varphi}\,. \quad \text{(B.23)}$$

We introduce two separation parameters, defined by their sum:

$$\beta_1+\beta_2\equiv Z\ . \quad \text{(B.24)}$$

With this, we can write the separate equations:

$$\frac{\mathrm{d}}{\mathrm{d}\xi}\left[\xi\frac{\mathrm{d}\Xi(\xi)}{\mathrm{d}\xi}\right]+\left(\frac{E\,\xi}{2}-\frac{m_l^2}{4\xi}+\beta_1\right)\Xi(\xi)=0$$
$$\frac{\mathrm{d}}{\mathrm{d}\eta}\left[\eta\frac{\mathrm{d}H(\eta)}{\mathrm{d}\eta}\right]+\left(\frac{E\,\eta}{2}-\frac{m_l^2}{4\eta}+\beta_2\right)H(\eta)=0\ . \quad \text{(B.25)}$$

We limit the study to bound states ($E<0$). The procedure that follows then takes a form similar to that deployed in section B.1 (see, for example, [3, 6] or [7] for more details). We use the result in (B.10) for the energy expressed in terms of the spherical coordinates principal quantum number:

$$n=\frac{Z}{\sqrt{-2E}}\,, \quad \text{(B.26)}$$

and adopt the rescaled spatial parameters:

$$\rho_1\equiv\frac{Z\xi}{n}\quad,\quad\rho_2\equiv\frac{Z\eta}{n}\ . \quad \text{(B.27)}$$

The two differential equations are now:

$$\frac{\mathrm{d}^2\Xi(\rho_1)}{\mathrm{d}\rho_1^2}+\frac{1}{\rho_1}\frac{\mathrm{d}\Xi(\rho_1)}{\mathrm{d}\rho_1}+\left[-\frac{1}{4}+\frac{\beta_1 n}{Z\rho_1}-\frac{m_l^2}{4\rho_1^2}\right]\Xi(\rho_1)=0$$
$$\frac{\mathrm{d}^2H(\rho_2)}{\mathrm{d}\rho_2^2}+\frac{1}{\rho_2}\frac{\mathrm{d}H(\rho_2)}{\mathrm{d}\rho_2}+\left[-\frac{1}{4}+\frac{\beta_2 n}{Z\rho_2}-\frac{m_l^2}{4\rho_2^2}\right]H(\rho_2)=0\,. \quad \text{(B.28)}$$

We look for solutions in the limits $\rho_1, \rho_2 \to 0$ and $\rho_1, \rho_2 \to \infty$. For the first of the equations in (B.28) these are:

$$\begin{aligned} \Xi^{(0)}(\rho_1) &= \rho_1^{|m_l|/2} \\ \Xi^{(\infty)}(\rho_1) &= \mathrm{e}^{-\rho_1/2} . \end{aligned} \tag{B.29}$$

The respective general solution should then be of the form:

$$\Xi(\rho_1) = \rho_1^{|m_l|/2} \, \mathrm{e}^{-\rho_1/2} \, w_1(\rho_1) , \tag{B.30}$$

where we have also introduced the function $w_1(\rho_1)$. Substitution of (B.30) in (B.28) results in a differential equation for $w_1(\rho_1)$, and the same procedure may be followed for $H(\eta)$ — introducing $w_2(\rho_2)$. The two emerging equations are:

$$\begin{aligned} \left[\rho_1 \frac{\mathrm{d}^2}{\mathrm{d}\rho_1^2} + (|m_l| + 1 - \rho_1) \frac{\mathrm{d}}{\mathrm{d}\rho_1} + \left(\frac{n\beta_1}{Z} - \frac{|m_l|}{2} - \frac{1}{2}\right)\right] w_1(\rho_1) = 0 \\ \left[\rho_2 \frac{\mathrm{d}^2}{\mathrm{d}\rho_2^2} + (|m_l| + 1 - \rho_2) \frac{\mathrm{d}}{\mathrm{d}\rho_2} + \left(\frac{n\beta_2}{Z} - \frac{|m_l|}{2} - \frac{1}{2}\right)\right] w_2(\rho_2) = 0 . \end{aligned} \tag{B.31}$$

We define the *parabolic quantum numbers* as:

$$\begin{aligned} n_1 &\equiv \frac{n\beta_1}{Z} - \frac{|m_l|}{2} - \frac{1}{2} \\ n_2 &\equiv \frac{n\beta_2}{Z} - \frac{|m_l|}{2} - \frac{1}{2} , \end{aligned} \tag{B.32}$$

and in terms of these, a solution to (B.31), substituted in (B.30) and (B.23) and properly normalised, results in the wave functions:

$$\begin{aligned} \psi_{n_1 n_2 m_l}(\xi, \eta, \varphi) = \frac{Z^{3/2}}{\sqrt{\pi}\, n^2} \sqrt{\frac{n_1!\, n_2!}{\left[(n_1 + |m_l|)!\, (n_2 + |m_l|)!\right]^3}} \, (\rho_1 \rho_2)^{|m_l|/2} \\ \times \mathrm{e}^{-(\rho_1 + \rho_2)/2} \, \mathrm{L}_{n_1 + |m_l|}^{(|m_l|)}(\rho_1) \, \mathrm{L}_{n_2 + |m_l|}^{(|m_l|)}(\rho_2) \, \mathrm{e}^{\mathrm{i} m_l \varphi} . \end{aligned} \tag{B.33}$$

This includes the spherical principal quantum number, which is related to n_1 and n_2 through the relation:

$$n = n_1 + n_2 + |m_l| + 1 . \tag{B.34}$$

The associated Laguerre polynomials are those defined in (B.14), and (B.33) and (B.13) are functionally equivalent.

B.3 Radial Integrals and Radial Functions

In this section, we give the hydrogenic expectation values for a number of functions r^k, with k being a positive or negative integer. These integrals are taken over the radial one-electron atom wave functions, such as those presented in (1.14), (1.16) and (B.13). Knowledge of expectation values for powers of r is essential for the determination of a great number of quantities and perturbation energies. For example, $k=2$ is used in analyses of diamagnetism, $k=-1$ for Coulomb potential energy, and $k=-3$ for the spin–orbit interaction.

We give the value of integrals as functions of hydrogenic radial function quantum numbers n and l, and the nuclear charge Z. The formalism is the standard one for expectation values:

$$\left\langle \psi_{nl} \left| r^k \right| \psi_{nl} \right\rangle \equiv \iiint \psi^*_{nl}(\mathbf{r})\, r^k\, \psi_{nl}(\mathbf{r})\, \mathrm{d}V = \int_0^\infty |R_{nl}(r)|^2\, r^{k+2}\, \mathrm{d}r\,. \tag{B.35}$$

The wave function $\psi_{nl}(\mathbf{r})$ has been separated in radial and angular coordinates, and the spherical harmonics (see appendix C.3) are chosen as normalised. For a review of the detailed steps needed for the derivations, and a few further orders, we refer to [1, 8, 9] or [3].

The solutions to the expectation values below are given in atomic units. Note the constraints: $l>0$ for (B.38) and (B.39), and $l>1$ for (B.37) and (B.36). A couple of remarks are that the positive powers depend heavily on n, but are less sensitive to an increasing l. In contrast, the negative powers have a strong dependence on angular momentum. The reasons for this can be traced to (B.4) and (B.6). Positive powers are dominated by the behaviour of $R(r)$ for large r, where it is essentially a pure exponential in r. For r close to the origin, $R(r)$ will scale approximately as r^l, which explains the pronounced l dependence for the negative powers in (B.35).

$$\begin{aligned}\left\langle \psi_{nl} \left| r^{-6} \right| \psi_{nl} \right\rangle &= Z^6 \{35n^4 - n^2[30l(l+1)-25] + 3(l-1)l(l+1)(l+2)\} \\ &\times \left[8n^7(l-3/2)(l-1)(l-1/2)l(l+1/2)(l+1)(l+3/2)(l+2)(l+5/2)\right]^{-1}\end{aligned} \tag{B.36}$$

$$\left\langle \psi_{nl} \left| r^{-5} \right| \psi_{nl} \right\rangle = \frac{Z^5\left[5n^2 - 3l(l+1) + 1\right]}{2n^5(l-1)(l-1/2)l(l+1/2)(l+1)(l+3/2)(l+2)} \tag{B.37}$$

$$\langle \psi_{nl} | r^{-4} | \psi_{nl} \rangle = \frac{Z^4\left[3n^2 - l(l+1)\right]}{2n^5(l-1/2)l(l+1/2)(l+1)(l+3/2)} \tag{B.38}$$

$$\langle \psi_{nl} | r^{-3} | \psi_{nl} \rangle = \frac{Z^3}{n^3 l(l+1/2)(l+1)} \tag{B.39}$$

$$\langle \psi_{nl} | r^{-2} | \psi_{nl} \rangle = \frac{Z^2}{n^3(l+1/2)} \tag{B.40}$$

$$\langle \psi_{nl} | r^{-1} | \psi_{nl} \rangle = \frac{Z}{n^2} \tag{B.41}$$

$$\langle \psi_{nl} | r | \psi_{nl} \rangle = \frac{1}{2Z} \left[3n^2 - l(l+1)\right] \tag{B.42}$$

$$\langle \psi_{nl} | r^2 | \psi_{nl} \rangle = \frac{n^2}{2Z^2} \left[5n^2 - 3l(l+1) + 1\right] \tag{B.43}$$

$$\begin{aligned} \langle \psi_{nl} | r^3 | \psi_{nl} \rangle &= \frac{n^2}{8Z^3} \{35n^4 - n^2 [30l(l+1) - 25] \\ &\quad + 3(l-1)l(l+1)(l+2)\} \end{aligned} \tag{B.44}$$

$$\begin{aligned} \langle \psi_{nl} | r^4 | \psi_{nl} \rangle &= \frac{n^4}{8Z^4} \{63n^4 - n^2 [70l(l+1) - 105] \\ &\quad + 15(l-1)l(l+1)(l+2) - 20l(l+1) + 12\} \end{aligned} \tag{B.45}$$

$$\begin{aligned} \left\langle \psi_{nl} \middle| r^5 \middle| \psi_{nl} \right\rangle &= \frac{n^4}{16Z^5} \Big\{231n^6 - n^4 [315l(l+1) - 735] \\ &\quad + n^2 [105(l-1)l(l+1)(l+2) - 315l(l+1) + 294] \\ &\quad - 5(l-2)(l-1)l(l+1)(l+2)(l+3)\Big\} \end{aligned} \tag{B.46}$$

Further Reading

The theory of atomic spectra, by Condon & Shortley [2]
Quantum Mechanics of One- and Two-Electron Atoms, by Bethe & Salpeter [1]
Quantum theory of atomic structure, by Slater [5]
Quantum mechanics of atomic spectra and atomic structure, by Mizushima [9]
Physics of Atoms and Molecules, by Bransden & Joachain [3]
Hydrogenic Wave Functions, by Hill [7]

References

1. H.A. Bethe, E.E. Salpeter, *Quantum Mechanics of One- and Two-Electron Atoms* (Springer-Verlag, Berlin, 1957)
2. E.U. Condon, G.H. Shortley, *The theory of atomic spectra* (Cambridge University Press, Cambridge, 1935)
3. B.H. Bransden, C.J. Joachain, *Physics of Atoms and Molecules*, 2nd edn. (Prentice Hall, Harlow, England, 2003)
4. G.B. Arfken, H.J. Weber, F.E. Harris, *Mathematical Methods for Physicists: A Comprehensive Guide*, 7th edn. (Academic Press, Amsterdam, 2012)
5. J.C. Slater, *Quantum theory of atomic structure* (McGraw-Hill, New York, 1960)

6. L.D. Landau, E.M. Lifshitz, *Quantum Mechanics — Course of Theoretical Physics, volume 3*, 3rd edn. (Butterworth-Heinemann, Amsterdam, 1981)
7. R.N. Hill, in *Springer Handbook of Atomic, Molecular, and Optical Physics*, ed. by G.W.F. Drake (Springer-Verlag, New York, 2006), p. 153
8. K. Bockasten, Phys. Rev. A **9**, 1087 (1974)
9. M. Mizushima, *Quantum mechanics of atomic spectra and atomic structure* (W. A. Benjamin, New York, 1970)

Appendix C
Angular Momentum

Angular momentum plays a central role in the theory of atomic structure. The first three sections of this appendix contain a general introduction to quantum mechanical angular momentum. This includes the spherical harmonics, which are identical to the solution for the angular non-relativistic hydrogenic Schrödinger equation. The latter part of the appendix deals with more specific applications of angular momentum in the formalism of multielectron atoms. The electron spin will be treated separately in appendix E.

The treatment of angular momentum in this appendix is somewhat brief, and is limited to what is needed for the principal chapters of the book. For more complete coverage of the subject of angular momentum, we refer to other more detailed sources, and there are many. Some notable examples are [1–4] and [5].

C.1 General Angular Momentum

In classical mechanics, the orbital angular momentum is defined as $\mathbf{L} = \mathbf{r} \times \mathbf{p}$. In Cartesian coordinates, the components of this vector are:

$$
\begin{aligned}
L_x &= y p_z - z p_y \\
L_y &= z p_x - x p_z \\
L_z &= x p_y - y p_x \,.
\end{aligned}
\tag{C.1}
$$

Using the quantum mechanical operator forms for the linear momenta, we have:

A. Kastberg, *Structure of Multielectron Atoms*, Springer Series on Atomic, Optical, and Plasma Physics 112, https://doi.org/10.1007/978-3-030-36420-5

$$L_x = -\mathrm{i}\hbar \left(y\frac{\partial}{\partial z} - z\frac{\partial}{\partial y} \right)$$
$$L_y = -\mathrm{i}\hbar \left(z\frac{\partial}{\partial x} - x\frac{\partial}{\partial z} \right)$$
$$L_z = -\mathrm{i}\hbar \left(x\frac{\partial}{\partial y} - y\frac{\partial}{\partial x} \right) \quad . \tag{C.2}$$

This is in SI-units, and in this section we will stick to this. Changing to atomic units would here simply mean discarding the factors of $\hbar$.

From (C.2), we can calculate the commutators of the components of $\mathbf{L}$. The resulting rules are general; they will be the same for any type of angular momentum (see, for example, [6]), and we can write them in terms of a generalised one, $\mathbf{J}$:

$$[J_x, J_y] = \mathrm{i}\hbar J_z$$
$$[J_y, J_z] = \mathrm{i}\hbar J_x$$
$$[J_z, J_x] = \mathrm{i}\hbar J_y \quad . \tag{C.3}$$

Next, we define the ladder operators: the operators that increment (or decrement) the projection of the angular momentum $\mathbf{J}$ along the $\hat{e}_z$-axis by one unit of $\hbar$:

$$J_+ \equiv J_x + \mathrm{i}J_y$$
$$J_- \equiv J_x - \mathrm{i}J_y \quad . \tag{C.4}$$

By taking the products of these raising and lowering operators, we find that they do not commute:

$$J_+J_- = J_x^2 + J_y^2 + \hbar J_z$$
$$J_-J_+ = J_x^2 + J_y^2 - \hbar J_z$$
$$[J_+, J_-] = 2\hbar J_z \quad . \tag{C.5}$$

C.1.1 Eigenvalues

Since orthogonal Cartesian components of an angular momentum never commute, the least ambiguous way in which one such vector can be specified is by the combination of the square of its absolute value $\mathbf{J}^2$, and its projection along the $\hat{e}_z$-axis, J_z. The square of the vector is:

$$\mathbf{J}^2 = \mathbf{J} \cdot \mathbf{J} = J_x^2 + J_y^2 + J_z^2 \quad . \tag{C.6}$$

$\mathbf{J}^2$ commutes with J_z, and thus these operators have common eigenfunctions ψ_{ab}:

$$\mathbf{J}^2 \psi_{ab} = a\,\psi_{ab}$$
$$J_z \psi_{ab} = b\,\psi_{ab} \quad . \tag{C.7}$$

Here a and b are the eigenvalues of the respective operators, and together a and b provide a unique label for the state. From (C.6) and (C.7):

$$(J_x^2 + J_y^2)\,\psi_{ab} = (\mathbf{J}^2 - J_z^2)\,\psi_{ab} = (a - b^2)\,\psi_{ab} \quad , \tag{C.8}$$

and since this sum of two squares necessarily has to be positive or zero:

$$a \geq b^2 \ . \tag{C.9}$$

Next, we apply the ladder operators (C.4) on ψ_{ab}. From the fact that $\mathbf{J}^2$ commutes with all the components of $\mathbf{J}$ follows that the functions $J_\pm\, \psi_{ab}$ are also eigenfunctions of $\mathbf{J}^2$, with the same eigenvalue a. Then we let J_z operate on $J_\pm\, \psi_{ab}$ and, using the commutation relations in (C.3), we find:

$$\begin{aligned} J_z J_\pm\, \psi_{ab} &= (J_z J_x \pm \mathrm{i} J_z J_y)\psi_{ab} = [(J_x J_z + \mathrm{i}\hbar J_y) \pm \mathrm{i}(J_y J_z - \mathrm{i}\hbar J_x)]\, \psi_{ab} \\ &= [(J_x \pm \mathrm{i} J_y)(J_z \pm \hbar)]\, \psi_{ab} = (b \pm \hbar)\, J_\pm \psi_{ab} \ . \end{aligned} \tag{C.10}$$

Thus, unless $J_\pm\, \psi_{ab}$ is zero, it must be an eigenfunction of J_z, with eigenvalue $b \pm \hbar$. If we now apply $J_\pm$ repeatedly to ψ_{ab}, we find that the eigenvalues of $\mathbf{J}^2$ and J_z to the resulting functions are:

$$\begin{aligned} \mathbf{J}^2\, (J_\pm)^n\, \psi_{ab} &= a\, (J_\pm)^n\, \psi_{ab} \\ J_z\, (J_\pm)^n\, \psi_{ab} &= (b \pm n\hbar)\, (J_\pm)^n\, \psi_{ab}\ , \end{aligned} \tag{C.11}$$

except for the cases where $(J_\pm)^n \psi_{ab}$ is zero.

The equations in (C.11) show that the ladder operators do indeed either increase or decrease the projection of the angular momentum $\mathbf{J}$ along the z-axis with units of $\hbar$. They also show that for a given eigenvalue of $\mathbf{J}^2$, a, there is a discrete spectrum of eigenvalues for J_z:

$$b = \ldots, b' - 2\hbar\,, b' - \hbar\,, b'\,, b' + \hbar\,, b' + 2\hbar\,, \ldots\ . \tag{C.12}$$

Because of the restriction in (C.9), this spectrum cannot be infinite, but must have lower and upper bounds, set by $\pm\sqrt{a}$.

This means that the eigenfunctions corresponding to these limits in the spectrum of J_z must return zero if they are acted on by appropriate ladder operators. For example, we have:

$$\begin{aligned} J_- \left[J_+ \psi_{ab}^{(\max)} \right] &= 0 \\ J_+ \left[J_- \psi_{ab}^{(\min)} \right] &= 0\,. \end{aligned} \tag{C.13}$$

Using (C.5), this can be written as:

$$\begin{aligned} (J_x^2 + J_y^2 - \hbar J_z)\, \psi_{ab}^{(\max)} &= (\mathbf{J}^2 - J_z^2 - \hbar J_z)\, \psi_{ab}^{(\max)} \\ &= (a - b_{\max}^2 - \hbar\, b_{\max})\, \psi_{ab}^{(\max)} = 0 \\ (J_x^2 + J_y^2 + \hbar J_z)\, \psi_{ab}^{(\min)} &= (\mathbf{J}^2 - J_z^2 + \hbar J_z)\, \psi_{ab}^{(\min)} \\ &= (a - b_{\min}^2 + \hbar\, b_{\min})\, \psi_{ab}^{(\min)} = 0\ . \end{aligned} \tag{C.14}$$

Since the eigenfunctions $\psi_{ab}^{(\max)}$ and $\psi_{ab}^{(\min)}$ are non-zero, albeit limiting cases, we can deduce:

$$(a - b_{\min}^2 + \hbar b_{\min}) = (a - b_{\max}^2 - \hbar b_{\max}) = 0 \ . \tag{C.15}$$

From that we can derive the relation:

$$(b_{\max} + b_{\min})(\hbar + b_{\max} - b_{\min}) = 0 \ . \tag{C.16}$$

The second parenthesis in (C.16) is necessarily non-zero. Therefore $b_{\min} = -b_{\max}$, and thus the limits to the spectrum of J_z must be symmetrically placed around zero. This together with (C.12) means that all values of b are either integers or half integers of $\hbar$. This characteristic holds for all quantum mechanical angular momenta.

From (C.15), we also get a condition for the eigenvalues of $\mathbf{J}^2$:

$$a = b_{\max}(b_{\max} + \hbar) \ . \tag{C.17}$$

Introducing the quantum numbers $j \equiv b_{\max}/\hbar$, and $m_j \equiv b/\hbar$, we have now shown that:

$$\begin{aligned} \mathbf{J}^2 \, \psi_{jm_j} &= j(j+1)\hbar^2 \, \psi_{jm_j} \\ J_z \, \psi_{jm_j} &= m_j \hbar \, \psi_{jm_j} \ . \end{aligned} \tag{C.18}$$

We end this section by computing a normalisation constant, $c_\pm$, for the ladder operators. We take:

$$J_\pm \, \psi_{jm_j} = c_\pm \, \psi_{j,m_j \pm 1} \ . \tag{C.19}$$

It is convenient to express this in Dirac notation, using state vectors rather than wave functions:

$$J_\pm \, | \, j, m_j \rangle = c_\pm \, | \, j, m_j \pm 1 \rangle \ . \tag{C.20}$$

We consistently assume normalised wave functions and thus, by using (C.5), we get:

$$\begin{aligned} |c_\pm|^2 &= \langle \, j, m_j \pm 1 \, | \, (c_\pm)^* c_\pm \, | \, j, m_j \pm 1 \, \rangle = \langle \, j, m_j \, | \, J_\mp J_\pm \, | \, j, m_j \, \rangle \\ &= \langle \, j, m_j \, | \, (J_x^2 + J_y^2 \mp \hbar J_z) \, | \, j, m_j \, \rangle = \langle \, j, m_j \, | \, (\mathbf{J}^2 - J_z^2 \mp \hbar J_z) \, | \, j, m_j \, \rangle \\ &= \hbar^2 \, [\, j(j+1) - m_j(m_j \pm 1) \,] \ . \end{aligned} \tag{C.21}$$

The phase is irrelevant so, without loss of generality, we can choose:

$$c_\pm = \hbar \sqrt{j(j+1) - m_j(m_j \pm 1)} = \hbar \sqrt{(j \pm m_j + 1)(j \mp m_j)} \ . \tag{C.22}$$

C.2 Orbital Angular Momentum

To get explicit expressions for the orbital angular momentum, we first choose spherical coordinates, with θ and φ respectively as the zenith and azimuthal angles:

$$\begin{cases} x = r\sin\theta\cos\varphi \\ y = r\sin\theta\sin\varphi \\ z = r\cos\theta \end{cases} , \qquad \begin{cases} r = (x^2+y^2+z^2)^{1/2} \\ \cos\theta = z\,(x^2+y^2+z^2)^{-1/2} \\ \tan\varphi = y/x \end{cases} . \tag{C.23}$$

From this, we can find spherical coordinate forms of the partial derivatives in (C.2):

$$\begin{aligned}
\frac{\partial r}{\partial x} &= x\,(x^2+y^2+z^2)^{-1/2} = \frac{x}{r} = \sin\theta\cos\varphi \\
\frac{\partial r}{\partial y} &= \sin\theta\sin\varphi \\
\frac{\partial r}{\partial z} &= \cos\theta \\
\frac{\partial}{\partial x}(\cos\theta) = -\sin\theta\,\frac{\partial\theta}{\partial x} &= -zx\,(x^2+y^2+z^2)^{-3/2} = -\frac{z}{r^2}\sin\theta\cos\varphi \\
&= -\frac{\sin\theta\cos\theta\cos\varphi}{r} \\
-\sin\theta\,\frac{\partial\theta}{\partial y} &= -\frac{z}{r^2}\sin\theta\sin\varphi = -\frac{\sin\theta\cos\theta\sin\varphi}{r} \\
-\sin\theta\,\frac{\partial\theta}{\partial z} &= -\frac{z}{r^2}\cos\theta + \frac{1}{r} = \frac{\sin^2\theta}{r} \\
\frac{\partial}{\partial x}(\tan\varphi) = \frac{1}{\cos^2\varphi}\,\frac{\partial\varphi}{\partial x} &= -\frac{y}{x^2} = -\frac{\tan\varphi}{x} = -\frac{\sin\varphi}{r\sin\theta\cos^2\varphi} \\
\frac{1}{\cos^2\varphi}\,\frac{\partial\varphi}{\partial y} &= \frac{1}{x} = \frac{1}{r\sin\theta\cos\varphi} \\
\frac{\partial\varphi}{\partial z} &= 0 \quad ,
\end{aligned} \tag{C.24}$$

which leads to:

$$\begin{aligned}
\frac{\partial}{\partial x} &= \sin\theta\cos\varphi\,\frac{\partial}{\partial r} + \frac{\cos\theta\cos\varphi}{r}\,\frac{\partial}{\partial\theta} - \frac{\sin\varphi}{r\sin\theta}\,\frac{\partial}{\partial\varphi} \\
\frac{\partial}{\partial y} &= \sin\theta\sin\varphi\,\frac{\partial}{\partial r} + \frac{\cos\theta\sin\varphi}{r}\,\frac{\partial}{\partial\theta} + \frac{\cos\varphi}{r\sin\theta}\,\frac{\partial}{\partial\varphi} \\
\frac{\partial}{\partial z} &= \cos\theta\,\frac{\partial}{\partial r} - \frac{\sin\theta}{r}\,\frac{\partial}{\partial\theta} \quad .
\end{aligned} \tag{C.25}$$

Combining (C.25) with (C.2), we get the operator expressions for the orbital angular momentum components:

$$
\begin{aligned}
L_x &= \mathrm{i}\hbar\left(\sin\varphi\,\frac{\partial}{\partial\theta} + \frac{\cos\varphi}{\tan\theta}\,\frac{\partial}{\partial\varphi}\right)\\
L_y &= \mathrm{i}\hbar\left(-\cos\varphi\,\frac{\partial}{\partial\theta} + \frac{\sin\varphi}{\tan\theta}\,\frac{\partial}{\partial\varphi}\right)\\
L_z &= -\mathrm{i}\hbar\,\frac{\partial}{\partial\varphi} \quad .
\end{aligned} \tag{C.26}
$$

With these explicit expressions for all components of $\mathbf{L}$, we can derive differential forms of its ladder operators:

$$
L_\pm = \hbar\,\mathrm{e}^{\pm\mathrm{i}\varphi}\left(\pm\frac{\partial}{\partial\theta} + \mathrm{i}\,\frac{1}{\tan\theta}\,\frac{\partial}{\partial\varphi}\right) \quad . \tag{C.27}
$$

Using (C.5), the operator for $\mathbf{L}^2$ becomes:

$$
\begin{aligned}
\mathbf{L}^2 &= L_x^2 + L_y^2 + L_z^2 = L_z^2 + L_-L_+ + \hbar L_z\\
&= L_z^2 + \hbar L_z - \hbar^2\left(\frac{\partial^2}{\partial\theta^2}\,\frac{1}{\tan\theta}\,\frac{\partial}{\partial\theta} - \mathrm{i}\,\frac{\partial}{\partial\varphi}\,\frac{1}{\tan^2\theta}\,\frac{\partial^2}{\partial\varphi^2}\right)\\
&= -\hbar^2\left[\frac{1}{\sin\theta}\,\frac{\partial}{\partial\theta}\left(\sin\theta\,\frac{\partial}{\partial\theta}\right) + \frac{1}{\sin^2\theta}\,\frac{\partial^2}{\partial\varphi^2}\right] \quad .
\end{aligned} \tag{C.28}
$$

The general properties of angular momenta — described in section C.1 — are valid. With l and m_l as the respective quantum numbers for the square and the projection along $\hat{\mathrm{e}}_z$ of $\mathbf{L}$, we have:

$$
\begin{aligned}
\mathbf{L}^2\,\psi_{lm_l} &= l(l+1)\,\hbar^2\,\psi_{lm_l}\\
L_z\,\psi_{lm_l} &= m_l\,\hbar\,\psi_{lm_l} \quad .
\end{aligned} \tag{C.29}
$$

In the case of $\mathbf{L}$, however, we have an extra constraint that was not present for the general angular momentum $\mathbf{J}$. The equation for L_z in (C.26) shows that the solution to the eigenvalue equation for L_z must be of the form:

$$
\psi_{lm_l}(r,\theta,\varphi) = f(r,\theta)\,\mathrm{e}^{\mathrm{i}\varphi m_l} \quad . \tag{C.30}
$$

Since this function has to be periodic, with the periodicity 2π, we get the condition:

$$
\mathrm{e}^{2\pi\mathrm{i}m_l} = 1 \quad , \tag{C.31}
$$

and thus the projection quantum number m_l for orbital angular momentum must be a whole integer (positive or negative). As a consequence, the quantum number l has to be a positive integer.

C.2.1 Further Commutation Relations

We here list a few more commutation relations between angular momentum operators and linear position operators, which will prove useful. These can all be derived from general commutator algebra, the canonical commutation relations, and (C.1) and (C.3).

First, we have:

$$\begin{aligned} [L_x, y] &= 2\mathrm{i}\hbar z \\ [L_y, z] &= 2\mathrm{i}\hbar x \\ [L_z, x] &= 2\mathrm{i}\hbar y \end{aligned} \tag{C.32}$$

and

$$\begin{aligned} [L_y, x] &= -2\mathrm{i}\hbar z \\ [L_z, y] &= -2\mathrm{i}\hbar x \\ [L_x, z] &= -2\mathrm{i}\hbar y \ . \end{aligned} \tag{C.33}$$

For the square operator, commutations with any of the Cartesian coordinates are as:

$$[\mathbf{L}^2, z] = 2\mathrm{i}\hbar\,(xL_y - L_x y)\,, \tag{C.34}$$

and

$$\left[\mathbf{L}^2, [\mathbf{L}^2, z]\right] = 2\hbar^2\,(\mathbf{L}^2 z + z\mathbf{L}^2) \ . \tag{C.35}$$

C.3 Spherical Harmonics

For an atom with a single electron, the spherical harmonics are exact solutions to the angular part of its Schrödinger equation. This solution will be the same for any quantum mechanical system having a potential $V(r)$ that only depends on the radial parameter (see, for example, [7]). Stated more generally, the spherical harmonics are the angular parts of a solution to the Laplace equation:

$$\nabla^2 \psi = 0\,. \tag{C.36}$$

In spherical coordinates, with standard definitions of the zenith and azimuthal angles, and of the Laplacian, and calling the angular part of the wave function $Y(\theta,\varphi) = Y_{lm_l}$, the equation to solve is — see (1.5) and (1.19):

$$\begin{aligned} -\frac{1}{Y(\theta,\varphi)}\left[\frac{1}{\sin\theta}\frac{\partial}{\partial\theta}\left(\sin\theta\,\frac{\partial}{\partial\theta}\right) + \frac{1}{\sin^2\theta}\frac{\partial^2}{\partial\varphi^2}\right] Y(\theta,\varphi) \\ = \frac{1}{Y(\theta,\varphi)}\mathbf{L}^2\,Y(\theta,\varphi) = \text{const.}\ . \end{aligned} \tag{C.37}$$

We have used the definition of the square of the orbital angular momentum $\mathbf{L}^2$ from (C.28) and eliminated factors of $\hbar$ by using atomic units.

The differential equation (C.37) can be integrated directly, but we instead take an algebraic route, based on the eigenvalue relation (1.7):

$$\mathbf{L}^2\, Y(\theta,\varphi) = l(l+1)\, Y(\theta,\varphi) \ . \tag{C.38}$$

The quantum numbers l and m_l are those used in (C.29).

The initial step is to separate the total angular function Y_{lm_l} into two components:

$$Y_{lm_l}(\theta,\varphi) = \Theta_{lm_l}(\theta)\, \Phi_{m_l}(\varphi) \ . \tag{C.39}$$

We start with the solution that has the minimum projection of L_z, which is $Y_{l,-l} = \Theta_{l,-l}\Phi_{-l}$. We then let the lowering ladder operator L_- — see (C.4) — act on $Y_{l,-l}$. This should yield zero, which gives us a solvable differential equation. From (C.27) we have:

$$L_-\, Y_{l,-l}(\theta,\varphi) = \mathrm{e}^{-\mathrm{i}\varphi}\left(-\frac{\partial}{\partial\theta} + \mathrm{i}\cot\theta\,\frac{\partial}{\partial\varphi}\right)\Theta_{l,-l}(\theta)\,\Phi_{-l}(\varphi) = 0 \ . \tag{C.40}$$

In the second term within the parenthesis, we can identify the expression for L_z from (C.26). Substituting this, we can eliminate $\Phi_{-l}(\varphi)$, as well as the initial exponential:

$$-\frac{\partial\Theta_{l,-l}}{\partial\theta} + l\cot\theta\;\Theta_{l,-l}(\theta) = 0 \ . \tag{C.41}$$

The solution to this is a sine function to the power of l. We chose an integration constant such that the zenith function becomes normalised:

$$\langle\Theta_{l,-l}|\Theta_{l,-l}\rangle = \int_0^\pi \Theta_{l,-l}(\theta)^*\,\Theta_{l,-l}(\theta)\,\sin\theta\,\mathrm{d}\theta = 1 \ , \tag{C.42}$$

with the result:

$$\Theta_{l,-l}(\theta) = \sqrt{\frac{(2l+1)!}{2}}\,\frac{\sin^l\theta}{2^l\, l!} \ . \tag{C.43}$$

For all other functions $\Theta_{lm_l}(\theta)$, with $m_l > -l$, we can repeatedly use the raising ladder operator. This will yield a recursion equation:

$$\Theta_{l,m_l+1}(\theta)\,\Phi_{m_l+1}(\varphi) = \frac{L_+\,\Theta_{l,m_l}(\theta)\,\Phi_{m_l}(\varphi)}{\sqrt{l(l+1)-m_l(m_l+1)}} \ , \tag{C.44}$$

where we have taken the prefactor from (C.22). The azimuthal function can be had from (C.30), and we take the rising operator from (C.27):

$$\Theta_{l,m_l+1}(\theta)\,\mathrm{e}^{\mathrm{i}(m_l+1)\varphi} = \frac{\mathrm{e}^{\mathrm{i}\varphi}\left(\frac{\partial}{\partial\theta} + \mathrm{i}\cot\theta\,\frac{\partial}{\partial\varphi}\right)}{\sqrt{l(l+1)-m_l(m_l+1)}}\;\Theta_{l,m_l}(\theta)\,\mathrm{e}^{\mathrm{i}m_l\varphi} \ . \tag{C.45}$$

We identify the expression (C.26) for L_z, and we get:

$$\Theta_{l,m_l+1}(\theta) = \frac{\left(\frac{\partial}{\partial\theta} - m_l \cot\theta\right)}{\sqrt{l(l+1) - m_l(m_l+1)}} \; \Theta_{l,m_l}(\theta) \; . \tag{C.46}$$

With (C.46) and (C.43), we can find normalised zenith wave functions for any allowed combination of l and m_l (with the constraint $|m_l| \leq l$):

$$\begin{aligned}
\Theta_{l,m_l}(\theta) &= \frac{(-1)^{l+m_l}}{2^l\, l!} \sqrt{\frac{(2l+1)\,(l-m_l)!}{2\,(l+m_l)!}} \; \sin^{m_l}\theta \; \frac{\mathrm{d}^{l+m_l}}{\mathrm{d}(\cos\theta)^{l+m_l}} \; \sin^{2l}\theta \\
&= (-1)^{(m_l+|m_l|)/2} \sqrt{\frac{(2l+1)\,(l-|m_l|)!}{2\,(l+|m_l|)!}} \; \mathrm{P}_l^{(|m_l|)}(\cos\theta) \; .
\end{aligned} \tag{C.47}$$

In the second line of (C.47), $\mathrm{P}_l^{(|m_l|)}(\cos\theta)$ is an associated Legendre polynomial of degree l and order m_l, with argument $\cos\theta$. By including the azimuthal function from (C.30), we get the complete expression for the spherical harmonics:

$$Y_{l,m_l}(\theta,\varphi) = (-1)^{(m_l+|m_l|)/2} \sqrt{\frac{(2l+1)\,(l-|m_l|)!}{4\pi\,(l+|m_l|)!}} \; \mathrm{P}_l^{(|m_l|)}(\cos\theta)\; \mathrm{e}^{\mathrm{i}\varphi m_l} \; . \tag{C.48}$$

The lowest orders of this are tabulated in (1.22) in chapter 1.

Integration of products of associated Legendre polynomials yields the following relation [8, 9]:

$$\begin{aligned}
\int_{-1}^{1} \mathrm{P}_{l_a}^{(m)}(x)\; \mathrm{P}_{l_b}^{(m)}(x)\, \mathrm{d}x &= \int_{0}^{\pi} \mathrm{P}_{l_a}^{(m)}(\cos\theta)\; \mathrm{P}_{l_b}^{(m)}(\cos\theta)\, \sin\theta\, \mathrm{d}\theta \\
&= \begin{cases} 0 & \text{if} \quad l_a \neq l_b \\ \dfrac{2}{2l_a+1}\, \dfrac{(l_a+m)!}{(l_a-m)!} & \text{if} \quad l_a = l_b \end{cases} \; .
\end{aligned} \tag{C.49}$$

This means that the functions $Y_{l,m_l}(\theta,\varphi)$ are all normalised, and mutually orthogonal. Thus, the spherical harmonics in (C.48) provide a set of orthonormal functions:

$$\left\langle Y_{lm_l} \,|\, Y_{l',m_l'} \right\rangle = \int_0^{2\pi}\!\! \int_0^{\pi} Y_{lm_l}^{*}\, Y_{l',m_l'} \, \sin\theta \; \mathrm{d}\theta \mathrm{d}\varphi = \delta_{ll'}\, \delta_{m_l m_l'} \; . \tag{C.50}$$

C.3.1 The Addition Theorem for Spherical Harmonics

In problems involving two electrons, at coordinates $\mathbf{r}_1$ and $\mathbf{r}_2$, it is useful to be able to express a function of the angle between these two vectors in terms of spherical harmonics. If the unit vectors along $\mathbf{r}_1$ and $\mathbf{r}_2$ are defined by the zenith and azimuthal angles θ_1, φ_1, θ_2 and φ_2, an angle ω defined as that *between* the two vectors can be written as:

$$\cos\omega = \cos\theta_1\,\cos\theta_2 + \sin\theta_1\,\sin\theta_2\,\cos(\varphi_1-\varphi_2)\,. \tag{C.51}$$

The addition theorem for spherical harmonics can be defined in terms of angles interrelated as in (C.51). The statement of the theorem is (see, for example, [7] or [10]):

$$\mathrm{P}_l(\cos\omega) = \frac{4\pi}{2l+1}\sum_{m_l=-l}^{l} Y^*_{lm_l}(\theta_1,\varphi_1)\,Y_{lm_l}(\theta_2,\varphi_2)\ , \tag{C.52}$$

where $\mathrm{P}_l(x)$ is a *Legendre polynomial* of order l.

From the theorem (C.52) we can deduce that for the special case with $\theta_1=\theta_2$ and $\varphi_1=\varphi_2$ we have:

$$\sum_{m_l=-l}^{l}\left|Y_{lm_l}(\theta,\varphi)\right|^2 = \frac{2l+1}{4\pi}\,, \tag{C.53}$$

which is known as *Unsölds theorem*. This is independent of the angle ω, which confirms that the sum over all m_l of any spherical harmonic is spherically symmetric, regardless of l.

The addition theorem in (C.52) provides a means for expressing the inverse distance between two electrons in terms of spherical harmonics [3]. This is important when we need to solve integrals involving the Hamiltonian of the Coulomb repulsion between two electrons. Expressed in terms of r_1, r_2 and the angle ω — defined in (C.51) — we can write:

$$\begin{aligned}\frac{1}{r_{12}} \equiv \frac{1}{|\mathbf{r}_1-\mathbf{r}_2|} &= (r_1^2+r_2^2-2r_1r_2\cos\omega)^{-1/2}\\ &= r_>^{-1}\left[1+(r_</r_>)^2-2(r_</r_>)\cos\omega\right]^{-1/2}.\end{aligned} \tag{C.54}$$

In this equation, $r_<$ represents the smaller radial parameter of r_1 and r_2, and $r_>$ the greater one.

The expression (C.54) can be developed in Legendre polynomials, or associated Legendre polynomials, as:

$$\begin{aligned}\frac{1}{r_{12}} &= \sum_{l=0}^{\infty}\frac{(r_<)^l}{(r_>)^{l+1}}\,\mathrm{P}_l(\cos\omega)\\ &= \sum_{l=0}^{\infty}\sum_{m_l=-l}^{l}\frac{(l-|m_l|)!}{(l+|m_l|)!}\,\frac{(r_<)^l}{(r_>)^{l+1}}\,\mathrm{P}_l^{(|m_l|)}(\cos\omega_1)\,\mathrm{P}_l^{(|m_l|)}(\cos\omega_2)\,\mathrm{e}^{im_l(\varphi_1-\varphi_2)}\,.\end{aligned} \tag{C.55}$$

Using (C.52), we can also write (C.55) in terms of the spherical harmonics:

$$\frac{1}{r_{12}} = \sum_{l=0}^{\infty} \sum_{m_l=-l}^{l} \frac{4\pi}{2l+1} \frac{(r_<)^l}{(r_>)^{l+1}} Y^*_{lm_l}(\theta_1, \varphi_1)\, Y_{lm_l}(\theta_2, \varphi_2) \ , \qquad \text{(C.56)}$$

which is the expression used in, for example, chapter 7 in order to calculate energies of *LS*-coupling terms.

C.4 Wigner *nj*-Symbols

The Wigner *nj*-symbols are a set of standardised mathematical tools, which are convenient for the treatment of a great variety of angular functions. Typical examples in atomic structure are the coupling of angular momentum vectors and the conversion between different representations of states. Other applications include products of spherical harmonics.

The *n* in *nj* should be a multiple of three. The most common symbol is the 3*j*, which plays essentially the same role as the Clebsch-Gordan coefficients (to be properly defined in section C.5). However, the 3*j*-symbol is more practical to use than the Clebsch–Gordan coefficient because of a higher degree of symmetry. A similar comparison can be made between 6*j*-symbols and so-called Racah coefficients [10]. Higher values of *n*, such as in 9*j* and 12*j*-symbols, are less common, but also find their use. The symmetry properties of the *nj*-symbols greatly facilitate tabulation and implementation in algorithms. Before the advent of desktop computers, the values of different 3*j*, 6*j* and 9*j*-symbols were typically sought in printed tables in books, whereas today they are frequently predefined as functions in mathematical software and in mathematical programming libraries.

This section will be limited to the definition of the symbols, and to a few of the most fundamental properties. For more details of the mathematics of *nj*-symbols, and of how they may be applied in atomic physics, examples of suitable sources are [1, 2, 11] and [10].

C.4.1 3j-Symbols

The definition of the 3*j*-symbol is:

$$\begin{pmatrix} j_1 & j_2 & j_3 \\ m_1 & m_2 & m_3 \end{pmatrix} \equiv \delta(m_1+m_2+m_3\,, 0)\, \Delta(j_1, j_2, j_3)$$
$$\times\, (-1)^{j_1-j_2-m_3} \sqrt{\Lambda(j_1, j_2, j_3)}\, \sqrt{\mathscr{T}_1} \sum_{k=k_{\min}}^{k_{\max}} \frac{(-1)^k}{k!\,(j_1+j_2-j_3-k)!\,\mathscr{T}_2} \ . \qquad \text{(C.57)}$$

The three arguments in the top row are angular momentum quantum numbers, and the three in the bottom row are corresponding projection quantum numbers. The summation over k has the limits:

$$\begin{aligned} k_{\min} &\equiv \max(0\,,\, j_2-j_3-m_1\,,\, j_1-j_3+m_2) \\ k_{\max} &\equiv \min(j_1+j_2-j_3\,,\, j_1-m_1\,,\, j_2+m_2) \ . \end{aligned} \tag{C.58}$$

The function $\Lambda(x,y,z)$ is a *triangular coefficient*, defined as:

$$\Lambda(x,y,z) \equiv \frac{(x+y-z)!\,(y+z-x)!\,(z+x-y)!}{(x+y+z+1)!} \ , \tag{C.59}$$

while the meanings of $\mathscr{T}_1$ and $\mathscr{T}_2$ are:

$$\begin{aligned} \mathscr{T}_1 &\equiv (j_1+m_1)!\,(j_1-m_1)!\,(j_2+m_2)!\,(j_2-m_2)!\,(j_3+m_3)!\,(j_3-m_3)! \\ \mathscr{T}_2 &\equiv (j_1-m_1-k)!\,(j_2+m_2-k)!\,(j_3-j_2+m_1+k)!\,(j_3-j_1-m_2+k)! \ . \end{aligned} \tag{C.60}$$

The function $\Delta(j_1,j_2,j_3)$ is a *triangular condition*, which is zero unless the three arguments fulfil the inequalities:

$$\begin{aligned} j_1+j_2 &\geq j_3 \\ j_2+j_3 &\geq j_1 \\ j_3+j_1 &\geq j_2 \ . \end{aligned} \tag{C.61}$$

This is equivalent to a requirement that the numbers j_1, j_2 and j_3 conform to the generic summation rule for angular momentum quantum numbers. That is, for any permutation of them, the ranges of their values are restricted as:

$$|j_1-j_2| \leq j_3 \leq j_1+j_2 \ . \tag{C.62}$$

If the condition (C.61) is adhered to, $\Delta(j_1,j_2,j_3)$ takes the value of unity.

The values of the numbers j_i and m_i should also follow other standard constraints for angular momenta. That is, they must be integers or half-integers, and:

$$|m_i| \leq j_i \quad , \tag{C.63}$$

for all three i. In addition, the two sums $(j_1+j_2+j_3)$ and $(m_1+m_2+m_3)$ must have integer values. These requirements assure that all factorial arguments in (C.57) are positive integers, and thus that the value of the $3j$-symbol is real.

A permutation of two columns in a $3j$-symbol can only change its sign. Mathematically, the relation is:

$$\begin{aligned} \begin{pmatrix} j_1 & j_2 & j_3 \\ m_1 & m_2 & m_3 \end{pmatrix} &= (-1)^{j_1+j_2+j_3} \begin{pmatrix} j_2 & j_1 & j_3 \\ m_2 & m_1 & m_3 \end{pmatrix} \\ &= (-1)^{j_1+j_2+j_3} \begin{pmatrix} j_1 & j_3 & j_2 \\ m_1 & m_3 & m_2 \end{pmatrix} \ . \end{aligned} \tag{C.64}$$

A parity inversion can also only affect the sign, according to:

$$\begin{pmatrix} j_1 & j_2 & j_3 \\ m_1 & m_2 & m_3 \end{pmatrix} = (-1)^{j_1+j_2+j_3} \begin{pmatrix} j_1 & j_2 & j_3 \\ -m_1 & -m_2 & -m_3 \end{pmatrix} . \tag{C.65}$$

C.4.2 6j-Symbols

A $6j$-symbol has the following definition:

$$\begin{aligned} \begin{Bmatrix} j_1 & j_2 & j_3 \\ j_4 & j_5 & j_6 \end{Bmatrix} &\equiv \Delta(j_1, j_2, j_3)\, \Delta(j_1, j_5, j_6)\, \Delta(j_2, j_4, j_6)\, \Delta(j_3, j_4, j_5) \\ &\times \sqrt{\Lambda(j_1, j_2, j_3)\, \Lambda(j_1, j_5, j_6)\, \Lambda(j_2, j_4, j_6)\, \Lambda(j_3, j_4, j_5)} \\ &\times \sum_{q=q_{\min}}^{q_{\max}} \frac{(-1)^q\,(q+1)!}{\mathscr{U}_1\,\mathscr{U}_2} . \end{aligned} \tag{C.66}$$

The summation limits are here:

$$\begin{aligned} q_{\min} &\equiv \max(j_1+j_2+j_3\,,\, j_1+j_5+j_6\,,\, j_2+j_4+j_6\,,\, j_3+j_4+j_5) \\ q_{\max} &\equiv \min(j_1+j_2+j_4+j_5\,,\, j_2+j_3+j_5+j_6\,,\, j_1+j_3+j_4+j_6) \ , \end{aligned} \tag{C.67}$$

and all six arguments must be angular momentum quantum numbers (not projections). The expressions $\mathscr{U}_1$ and $\mathscr{U}_2$ in the denominator in the sum are:

$$\begin{aligned} \mathscr{U}_1 &\equiv (q-j_1-j_2-j_3)!\,(q-j_1-j_5-j_6)!\,(q-j_2-j_4-j_6)!\,(q-j_3-j_4-j_5)! \\ \mathscr{U}_2 &\equiv (j_1+j_2+j_4+j_5-k)!\,(j_2+j_3+j_5+j_6-k)!\,(j_1+j_3+j_4+j_6-k)! \ . \end{aligned} \tag{C.68}$$

The symmetries of $6j$-symbols are such that any two columns may interchange, without the value being altered. An exchange of any two arguments in the top row of the symbol, with the corresponding numbers in the bottom row, also leaves the symbol unchanged.

C.4.3 Higher Order nj-Symbols

$3j$- and $6j$-symbols are extensively used in essentially all fields of quantum physics, while higher order Wigner nj-symbols are less common. This is because they can always be defined as a comparatively compact sum of lower order symbols, and also because they are only needed for relatively complex vector coupling considerations. Their main utility is to facilitate a more compact notation.

The definition of the $9j$-symbol is the following sum of $6j$-symbols:

$$\begin{Bmatrix} j_{11} & j_{12} & j_{13} \\ j_{21} & j_{22} & j_{23} \\ j_{31} & j_{32} & j_{33} \end{Bmatrix} \equiv \sum_{j=j_{\min}}^{j_{\max}} (-1)^{2j}\,(2j+1)$$
$$\times \begin{Bmatrix} j_{11} & j_{21} & j_{31} \\ j_{32} & j_{33} & j \end{Bmatrix} \begin{Bmatrix} j_{12} & j_{22} & j_{32} \\ j_{21} & j & j_{23} \end{Bmatrix} \begin{Bmatrix} j_{13} & j_{23} & j_{33} \\ j & j_{11} & j_{12} \end{Bmatrix} . \quad \text{(C.69)}$$

There are a number of triangular conditions, which all follow from those included in the definition of the $6j$-symbol in (C.66). This also imposes a finiteness of the sum over j. The limits of the latter can be derived as:

$$\begin{aligned} j_{\min} &\equiv \max(|j_{11}-j_{33}|\,,\,|j_{12}-j_{23}|\,,\,|j_{21}-j_{32}|) \\ j_{\max} &\equiv \min(j_{11}+j_{33}\,,\,j_{12}+j_{23}\,,\,j_{21}+j_{32}) \ . \end{aligned} \quad \text{(C.70)}$$

In a similar fashion, a $12j$-symbol can be written as an expansion of $9j$-symbols, a $15j$-symbol as one of $12j$-symbols, and so on [10].

C.5 Addition of Angular Momenta and Vector Coupling Coefficients

Whenever two or more angular momenta interact, neither one nor the other will remain a constant of the motion. However, in the absence of other external interactions, their sum will be conserved (it will commute with the interaction Hamiltonian). Having defined a nomenclature and the basic formalism for angular momenta in earlier sections, we will now consider how these rules apply to vector sums such as the one described above.

To give examples of applicability: to describe the effect on the atomic wave function of the interaction between angular momenta, we need to add the vectors describing the latter and deal with their sum as another quantum mechanical angular momentum. It could be a spin, $\mathbf{S}_i$, interacting with an orbital angular momentum, $\mathbf{L}_i$ via the spin–orbit interaction, forming the sum $\mathbf{J}_i = \mathbf{S}_i + \mathbf{L}_i$. It could also be the orbital angular momenta of two different electrons interacting through the Coulomb repulsion: $\mathbf{L} = \mathbf{L}_1 + \mathbf{L}_2$. It could be any interaction between angular momenta. As described in chapter 6, empirical rules for this were developed before quantum mechanics was invented, based on observations of atomic spectra (the *vector model*). Here, we will give these rules a better justification.

When we add two angular momenta, $\mathbf{J}_1$ and $\mathbf{J}_2$, to form the sum $\mathbf{J}$, we get two natural choices for sets of quantum numbers, or *representations*:

$$\begin{cases} |\,j_1, j_2, m_1, m_2\,\rangle_{m_1 m_2} \\ |\,j_1, j_2, J, M\,\rangle_{JM} \end{cases} . \quad \text{(C.71)}$$

Here, and for the rest of this section, we use the notations m_1, m_2 and M, rather than m_{j1}, m_{j2} and M_J, in order to clean up the notation, and for clarity we add subscripts to the state vectors. The conversion between the two representations above is mediated by *Clebsch–Gordan coefficients*. The theory behind this and equations for the determination of such coefficients are presented in the final part of the section.

C.5.1 Addition of Quantum Mechanical Angular Momenta

Consider an atomic state, having among its degrees of freedom two angular momenta, $\mathbf{J}_1$ and $\mathbf{J}_2$. The state can be described by eigenvectors to the operators $\mathbf{J}_1^2$, J_{1z}, $\mathbf{J}_2^2$ and J_{1z}, with associated quantum numbers j_1, m_1, j_2 and m_2. We assume that the two angular momenta couple, but that nothing else is affected by this interaction between $\mathbf{J}_1$ and $\mathbf{J}_2$. The wave function considered is:

$$\psi_{j_1, j_2, m_1, m_2} \quad , \tag{C.72}$$

corresponding to the state vector:

$$|\, j_1, j_2, m_1, m_2 \,\rangle_{m_1 m_2} \ . \tag{C.73}$$

Since the projection quantum numbers m_i can take any values between $-j_i$ and j_i, the function (C.72) spans a state space with $(2j_1+1)(2j_2+1)$ levels (degenerate or not).

We define the sum of the two vectors as:

$$\mathbf{J} \equiv \mathbf{J}_1 + \mathbf{J}_2 \ . \tag{C.74}$$

The here introduced vector $\mathbf{J}$ follows the same commutation rules as all quantum mechanical angular momenta, and it will best be represented by the operators $\mathbf{J}^2$ and J_z, and their associated quantum numbers J and M.

The introduction of the sum (C.74) comes with the possibility of an alternative representation:

$$|\, j_1, j_2, J, M \,\rangle_{JM} \ . \tag{C.75}$$

The vectors (C.73) and (C.75) span the same Hilbert space and they are both complete sets — described by four quantum numbers. We now need to determine which values of J and M that are possible for a given pair of j_1 and j_2, and we also want to unravel a procedure for the conversion between the representations.

From the definition in (C.74) follows that that each Cartesian component of $\mathbf{J}$ is the sum of the corresponding components of $\mathbf{J}_1$ and $\mathbf{J}_2$. From $J_z = J_{1z} + J_{2z}$ we have:

$$M = m_1 + m_2 \ . \tag{C.76}$$

We know that the maximum value of M is $J^{(\max)}$ and from (C.76) we can then deduce that this maximum must equal j_1+j_2, for any given combination of values for j_1 and j_2:

$$J^{(\max)} = j_1 + j_2 \; . \tag{C.77}$$

With this, we have identified one possible state in the basis $\{j_1, j_2, J, M\}$, namely:

$$|\, j_1, j_2, j_1+j_2, j_1+j_2 \,\rangle_{JM} \; . \tag{C.78}$$

States with $M = j_1+j_2-1$ can be formed in two different ways: the combinations $\{m_1 = j_1-1\, , m_2 = j_2\}$ or $\{m_1 = j_1\, , m_2 = j_2-1\}$. This must correspond to the two values of J for which $M = j_1+j_2-1$ are possible, $J = j_1+j_2$ and $J = j_1+j_2-1$. We have thus identified two more states:

$$|\, j_1, j_2, j_1+j_2-1, j_1+j_2-1 \,\rangle_{JM} \quad , \quad |\, j_1, j_2, j_1+j_2, j_1+j_2-1 \,\rangle_{JM} \, . \tag{C.79}$$

If we go on, we find that the projection quantum number $M = j_1+j_2-2$ is possible for three levels and so on. Eventually, the number of possible states corresponding to incrementally smaller values of J will cease to grow. When we get to either of the values $m_1 = -j_1$ or $m_2 = -j_2$, whichever comes first, we will reach a minimum value for J:

$$J^{(\min)} = |j_1 - j_2| \, . \tag{C.80}$$

This proves what was postulated in chapter 6, namely that:

$$\begin{aligned} J &= j_1+j_2\, , \, j_1+j_2-1\, , \, \dots\, , \, |j_1-j_2| \\ M &= J\, , J-1\, , \, \dots\, , \, -J \; . \end{aligned} \tag{C.81}$$

These rules are valid for any sum of two angular momenta. If three angular momenta need to be summed, we first add two of them, and then we add the third to the intermediate sum. We will always get vectors following the same rules of discretisation.

C.5.1.1 Matrix Elements of $\mathbf{J}^2$

Before we look into the coupling coefficients between the two representations (C.73) and (C.75), we shall express the operator $\mathbf{J}^2$ in the basis of (C.73). Using (C.4), we can write:

$$\begin{aligned} \mathbf{J}^2 &= \mathbf{J}_1^2 + \mathbf{J}_2^2 + 2\,(J_{1x}J_{2x} + J_{1y}J_{2y} + J_{1z}J_{2z}) \\ &= \mathbf{J}_1^2 + \mathbf{J}_2^2 + 2J_{1z}J_{2z} + J_{1+}J_{2-} + J_{1-}J_{2+} \; . \end{aligned} \tag{C.82}$$

From (C.20) and (C.22):

$$\begin{aligned}\mathbf{J}^2 \,|\, j_1, j_2, m_1, m_2 \rangle_{m_1 m_2} &= [\, j_1(j_1+1) + j_2(j_1+2) + 2m_1 m_2] \,|\, j_1, j_2, m_1, m_2 \rangle_{m_1 m_2} \\ &+ \sqrt{j_1(j_1+1) - m_1(m_1+1)} \\ &\quad \times \sqrt{j_2(j_2+1) - m_2(m_2-1)} \,|\, j_1, j_2, m_1+1, m_2-1 \rangle_{m_1 m_2} \\ &+ \sqrt{j_1(j_1+1) - m_1(m_1-1)} \\ &\quad \times \sqrt{j_2(j_2+1) - m_2(m_2+1)} \,|\, j_1, j_2, m_1-1, m_2+1 \rangle_{m_1 m_2} \ , \end{aligned} \tag{C.83}$$

and from this, matrix elements can be found.

C.5.2 Transformation Between Representations

To find out how to express a state vector $|\, j_1, j_2, J, M \rangle_{JM}$ as superpositions of the vectors $|\, j_1, j_2, m_1, m_2 \rangle_{m_1 m_2}$, we begin by applying the closure theorem:

$$\begin{aligned} &|\, j_1, j_2, J, M \rangle_{JM} \\ &= \sum_{m_1=-j_1}^{j_1} \sum_{m_2=-j_2}^{j_2} \left[|\, j_1, j_2, m_1, m_2 \rangle_{m_1 m_2} \langle\, j_1, j_2, m_1, m_2 \,|_{m_1 m_2} \right] |\, j_1, j_2, J, M \rangle_{JM} \\ &= \sum_{m_1=-j_1}^{j_1} \sum_{m_2=-j_2}^{j_2} \langle\, j_1, j_2, m_1, m_2 \,|\, j_1, j_2, J, M \rangle \,|\, j_1, j_2, m_1, m_2 \rangle_{m_1 m_2} \ , \end{aligned} \tag{C.84}$$

where $\langle\, j_1, j_2, m_1, m_2 \,|\, j_1, j_2, J, M \rangle$ are the *Clebsch–Gordan coefficients* (CGC) (we have dropped the indices from the CGC). For (C.84) to be an eigenstate of J_z, we have to have the constraint $M = m_1 + m_2$, which means that we can simplify the expression by eliminating one of the summations:

$$\begin{aligned} &|\, j_1, j_2, J, M \rangle_{JM} \\ &= \sum_{m_1=-j_1}^{j_1} \langle\, j_1, j_2, m_1, M - m_1 \,|\, j_1, j_2, J, M \rangle \,|\, j_1, j_2, m_1, M - m_1 \rangle_{m_1 m_2} \ . \end{aligned} \tag{C.85}$$

The Clebsch-Gordan coefficients can be calculated analytically, but as we will see in the following, the generating formula is daunting. Therefore, the coefficients have been listed in long tables — in the form of books (see, for example, [12]). Using instead Wigner nj-formalism (see section C.4) calculation and tabulation are facilitated. Today, access to vector coupling coefficients is available in mathematical software, and the need for long tables have diminished. However, in the upcoming section, we will demonstrate how to derive the CGCs.

C.5.2.1 Derivation of Clebsch-Gordan Coefficients

To find a formula for the CGCs, we begin by a tentative choice for the coefficient $\langle j_1, j_2, m_1, m_2 \,|\, j_1, j_2, J, M \rangle$ for the case with the extreme value of the projection quantum number, $M = J$. We then justify the choice by operating on the state vector with the raising ladder operator $J_+ = J_{1+} + J_{2+}$ and showing that it yields zero.

Our trial expression for this CGC is included in:

$$A \,|\, j_1, j_2, J, J \rangle_{JM} = \sum_{\mu_1 = J - j_2}^{j_1} \delta(J, \mu_1 + \mu_2)\, (-1)^{j_1 - \mu_1} \\ \times \left[\frac{(j_1 + \mu_1)!\, (j_2 + \mu_2)!}{(j_1 - \mu_1)!\, (j_2 - \mu_2)!} \right]^{1/2} |\, j_1, \mu_1, j_2, \mu_2 \rangle_{m_1 m_2} \,. \tag{C.86}$$

The lower bound on the summation variable μ_1 is $-j_1$ or $J - j_2$, whichever is the smallest, and in accordance with (C.85) we have $\mu_2 = J - \mu_1$. The normalising constant A is yet to be determined.

We now use (C.20) and (C.22) and apply J_+ to (C.86):

$$\begin{aligned} A J_+ \,|\, j_1, j_2, J, J \rangle_{JM} = & \sum_{\mu_1 = J - j_2}^{j_1} \delta(J, \mu_1 + \mu_2)\, (-1)^{j_1 - \mu_1} \left[\frac{(j_1 + \mu_1)!\, (j_2 + \mu_2)!}{(j_1 - \mu_1)!\, (j_2 - \mu_2)!} \right]^{1/2} \\ & \times \Big[\sqrt{(j_1 + \mu_1 + 1)(j_1 - \mu_1)}\; |\, j_1, j_2, \mu_1 + 1, \mu_2 \rangle_{m_1 m_2} \\ & + \sqrt{(j_2 + \mu_2 + 1)(j_2 - \mu_2)}\; |\, j_1, j_2, \mu_1, \mu_2 + 1 \rangle_{m_1 m_2} \Big] \\ = & \sum_{\mu_1 = J - j_2}^{j_1} \delta(J, \mu_1 + \mu_2)\, (-1)^{j_1 - \mu_1} \\ & \times \left[\frac{(j_1 + \mu_1 + 1)!\, (j_2 + \mu_2)!}{(j_1 - \mu_1 - 1)!\, (j_2 - \mu_2)!} \right]^{1/2} |\, j_1, j_2, \mu_1 + 1, \mu_2 \rangle_{m_1 m_2} \\ & - \sum_{\nu_1 = J - j_2 - 1}^{j_1 - 1} \delta(J, \nu_1 + \nu_2)\, (-1)^{j_1 - \nu_1} \\ & \times \left[\frac{(j_1 + \nu_1 + 1)!\, (j_2 + \nu_2)!}{(j_1 - \nu_1 - 1)!\, (j_2 - \nu_2)!} \right]^{1/2} |\, j_1, j_2, \nu_1 + 1, \nu_2 \rangle_{m_1 m_2} \,. \end{aligned} \tag{C.87}$$

In the last sum we have made the substitutions $\nu_1 = \mu_1 - 1$ and $\nu_2 = \mu_2 + 1$. The two sums after the last equality in (C.87) are identical, except that each of them has a term that the other one has not. In the first of the two, the extra term is the one for $\mu_1 = j_1$. However, this term includes the component $|\, j_1, j_1 + 1 \rangle_{m_1}$, which is zero. In the second sum, the additional term is that for $\nu_1 = J - j_2 - 1$. Since we have chosen the state with the maximum value of M, we have $J = M = \nu_1 + \nu_2$, and thus $\nu_2 = j_2 + 1$. The component $|\, j_2, j_2 + 1 \rangle_{m_2}$ is also zero and thus the two sums in the final stage of (C.87) are identical, and the expression as a whole cancels. Thereby, we have indeed showed that the expression in (C.86) can be used as the CGC for $M = J$.

The constant A can be determined using the binomial theorem. The derivation is quite involved, and for the details we refer to [10].

$$\begin{aligned}|A|^2 &= \sum_{\mu_1=J-j_2}^{j_1} \frac{(j_1+\mu_1)!\,(j_2+J-\mu_1)!}{(j_1-\mu_1)!\,(j_2-J+\mu_1)!} \\ &= \sum_{n=0}^{j_1+j_2-J} \frac{(j_1+J-j_2+n)!\,(2j_2-n)!}{n!\,(j_1+j_2-J-n)!} \\ &= \frac{(j_1-j_2+J)!\,(J-j_1+j_2)!\,(j_1+j_2+J+1)!}{(j_1+j_2-J)!\,(2J+1)!} \quad . \end{aligned} \tag{C.88}$$

A must be real, and it is practical to choose it to be positive in order to end up with a consistent phase factor.

With (C.86) and (C.88), we have a complete expression for the CGC for $M=J$. To find the coefficients for all other allowed values of M, we make use of the lowering ladder operator $J_-=J_{1-}+J_{2-}$. From (C.20) and (C.22), we can deduce:

$$(J_-)^n\,|\,j_1,j_2,J,M\,\rangle_{JM} = \left[\frac{(J+M)!\,(J-M+n)!}{(J-M)!\,(J+M-n)!}\right]^{1/2} |\,j_1,j_2,J,M\,\rangle_{JM} \quad . \tag{C.89}$$

With the binomial theorem, we can also express a particular power of J_- as:

$$(J-)^{J-M} = (J_{1-}+J_{2-})^{J-M} = \sum_n \binom{J-M}{n} (J_{1-})^n\,(J_{2-})^{J-M-n} , \tag{C.90}$$

and if we let (C.90) operate on both sides of (C.86) (the transformation relation for $M=J$) we get:

$$\begin{aligned} & A\left[\frac{2J!\,(J-m)!}{(J+m)!}\right]^{1/2} |\,j_1.j_2,J,M\,\rangle_{JM} \\ &= \sum_{\mu_1}\delta(J,\mu_1+\mu_2)\,(-1)^{j_1-\mu_1}\left[\frac{(j_1+\mu_1)!\,(j_2+\mu_2)!}{(j_1-\mu_1)!\,(j_2-\mu_2)!}\right]^{1/2} \\ &\times\sum_n\binom{J-M}{n}\left[\frac{(j_1+\mu_1)!\,(j_1-\mu_1+n)!\,(j_2+\mu_2)!\,(j_2-\mu_2+J-M-n)!}{(j_1-\mu_1)!\,(j_1+\mu_1-n)!\,(j_2-\mu_2)!\,(j_2+\mu_2-J+M+n)!}\right]^{1/2} \\ &\times |\,j_1,j_2,\mu_1-n,\mu_2-J+M+n\,\rangle_{m_1m_2} \\ &= \sum_{m_1}\delta(M,m_1+m_2)\,(-1)^{j_1-m_1}\left[\frac{(j_1-m_1)!\,(j_2-m_2)!}{(j_1+m_1)!\,(j_2+m_2)!}\right]^{1/2} \\ &\times B\,|\,j_1,j_2,m_1,m_2\,\rangle_{m_1m_2} \ . \end{aligned} \tag{C.91}$$

In the last line, we have introduced a new constant, B, and we have made the substitutions $\mu_1=m_1+n$ and $\mu_2=J-M-n+m_2$. B is defined by (C.91) as:

$$B \equiv \sum_n (-1)^n \frac{(J-M)!\,(j_i+m_1+n)!\,(j_2+J-m_1-n)!}{n!\,(j_2-J+m_1+n)!\,(J-M-n)!\,(j_1-m_1-n)!} \ . \tag{C.92}$$

Taking again advantage of the binomial theorem [10], we can find that:

$$B = \sum_k (-1)^{j_1-m_1+k} \frac{(J-M)!\,(j_1+m_1)!\,(j_2+m_2)!}{k!\,(j_2+m_2-k)!\,(j_1-m_1-k)!} \times \frac{(J-j_1+j_2)!\,(J+j_1-j_2)!}{(j_1+j_2-J-k)!\,(J-j_1-m_2+k)!\,(J-j_2+m_1+k)!} . \tag{C.93}$$

With all this, it is now possible to write down the general formula for the CGCs by substituting (C.93) and (C.88) into (C.91). A comparison with (C.84) then yields:

$$\begin{aligned} \langle j_1, j_2, m_1, m_2 \,|\, j_1, j_2, J, M \rangle &= \delta(M, m_1+m_2) \\ &\times \Bigg[(2J+1)\,(j_1-m_1)!\,(j_1+m_1)!\,(j_2-m_2)!\,(j_2+m_2)!\,(J-M)!\,(J+M)! \\ &\quad \times \frac{(j_1+j_2-J)!\,(j_1-j_2+J)!\,(J-j_1+j_2)!}{j_1+j_2+J+1} \Bigg]^{1/2} \\ &\times \sum_k (-1)^k\, [k!\,(j_2+m_2-k)!\,(j-j_1-m_2+k)! \\ &\times (j_1-m_1-k)!\,(j_1+j_2-J-k)!\,(J-j_2+m_1+k)!]^{-1} . \end{aligned} \tag{C.94}$$

Equation (C.94) is generally valid, but not attractive. However, it can be greatly shortened with Wigner $3j$-symbols, as defined in (C.57). Using this notation, we can write the CGC as:

$$\langle j_1, j_2, m_1, m_2 \,|\, j_1, j_2, J, M \rangle = (-1)^{j_1-j_2+M} \sqrt{2J+1} \begin{pmatrix} j_1 & j_2 & J \\ m_1 & m_2 & -M \end{pmatrix} . \tag{C.95}$$

When we derived (C.95), we chose a phase convention. However, if we had chosen $\mathbf{J} \equiv \mathbf{J}_2 + \mathbf{J}_1$ instead of $\mathbf{J} \equiv \mathbf{J}_1 + \mathbf{J}_2$, this would have amounted to interchanging the first two columns in the $3j$-symbol. Using the symmetry relation of $3j$-symbols (C.64), and using the closure theorem, this leads to:

$$\begin{aligned} |\, j_2, j_1, J, M \rangle_{JM} &= \sum_{m_1, m_2} (-1)^{j_2-j_1+M} \sqrt{2J+1} \begin{pmatrix} j_2 & j_1 & J \\ m_2 & m_1 & -M \end{pmatrix} |\, j_1, j_2, m_1, m_2 \rangle_{m_1 m_2} \\ &= (-1)^{-j_1+3j_2+J} \sum_{m_1, m_2} (-1)^{j_1-j_2+M} \sqrt{2J+1} \begin{pmatrix} j_1 & j_2 & J \\ m_1 & m_2 & -M \end{pmatrix} |\, j_1, j_2, m_1, m_2 \rangle_{m_1 m_2} \\ &= (-1)^{-j_1-j_2+J}\, |\, j_1, j_2, J, M \rangle_{JM} . \end{aligned} \tag{C.96}$$

In the last line, we have used the fact that $4j_2$ is necessarily an even integer. The phase can equally well be written $(-1)^{j_1+j_2-J}$, since the sum in the exponent must be an integer.

One conclusion from this is that the definition of the total angular momentum $\mathbf{J}$ made in chapters 4 and 6 is done in a cavalier way. When we casually take $\mathbf{J} = \mathbf{L} + \mathbf{S}$

to mean the same thing as $\mathbf{J}=\mathbf{S}+\mathbf{L}$, we introduce a phase error of $(-1)^{L+S-J}$. However, we are saved from this inconvenience by the fact that when we calculate matrix elements of the electron–electron interaction Hamiltonian H_{ee} in chapter 7, H_{ee} is diagonal in the quantum numbers L, S and J. This will cancel any possible phase errors, and thus we can follow the common standard in the literature and call the coupling scheme in chapter 7 *LS*-coupling, even if a more appropriate name would be *SL*-coupling.

C.5.2.2 Recursion Relation

Step operators can also be used to derive a recursion relation. That is, an expression relating CGCs differing by one unit of the projection quantum number m. Suppose that the values of j_1, j_2 and J are fixed. We take the double sum formula in (C.84) and apply the ladder operators $J_\pm$ on both sides. Using (C.22), we may write:

$$\sqrt{(J\pm M+1)(J\mp M)}\,|\,j_1,j_2,J,M\pm 1\,\rangle = \sum_{\mu_1=-j_1}^{j_1}\sum_{\mu_2=-j_2}^{j_2}\langle\, j_1,j_2,\mu_1,\mu_2\,|\,j_1,j_2,J,M\,\rangle$$
$$\times\left(\sqrt{(j_1\pm\mu_1+1)(j_1\mp\mu_1)}\,|\,j_1,j_2,\mu_1+1,\mu_2\,\rangle_{\mu_1\mu_2}\right.$$
$$\left.+\sqrt{(j_2\pm\mu_2+1)(j_2\mp\mu_2)}\,|\,j_1,j_2,\mu_1,\mu_2+1\,\rangle_{\mu_1\mu_2}\right)\,. \quad \text{(C.97)}$$

For the summation variables and the corresponding individual electronic projection quantum numbers, we have used μ_1 and μ_2 in order to be able to multiply both sides from the left with $\langle\, j_1,j_2,m_1,m_2\,|$. The outcome of the latter is that most terms in the sum may be eliminated, due to the orthogonality of the state vectors. We are left with the recursion relation:

$$\sqrt{(J\pm M+1)(J\mp M)}\;\langle\, j_1,j_2,m_1,m_2\,|\,j_1,j_2,J,M\pm 1\,\rangle$$
$$=\sqrt{(j_1\mp m_1+1)(j_1\pm m_1)}\;\langle\, j_1,j_2,m_1\mp 1,m_2\,|\,j_1,j_2,J,M\,\rangle$$
$$+\sqrt{(j_2\mp m_2+1)(j_2\pm m_2)}\;\langle\, j_1,j_2,m_1,m_2\mp 1\,|\,j_1,j_2,J,M\,\rangle\,. \quad \text{(C.98)}$$

With the exception for an arbitrary sign convention, (C.98) can be used together with the normalisation condition as a generator for the CGC coefficients.

C.5.3 Coupling of Three Angular Momenta

When there are three or more angular momentum vectors that need to be added — through various interactions — the difference between different summations orders will no longer be restricted to just a phase factor. When we are transforming an interaction Hamiltonian matrix between two different vector coupling representations, for example, from *LS*-coupling to *jj*-coupling, it will not suffice to use only CGCs, or 3*j*-symbols. It will, however, be enough to formulate a general formula for the coupling of three angular momenta. If we have four or more, a three-vector equation can be applied several times in succession.

Suppose we have three mutually interacting angular momenta, $\mathbf{J}_1$, $\mathbf{J}_2$ and $\mathbf{J}_3$, and we want to compare different ways to add them to a resultant, $\mathbf{J}$. The summation order may correspond to successive layers of perturbation theory, and a corresponding set of basis vectors. To be more concrete, we will compare the two schemes:

$$\begin{aligned} \mathbf{J}_1+\mathbf{J}_2=\mathbf{K}_\alpha \;\rightarrow\; \mathbf{K}_\alpha+\mathbf{J}_3=\mathbf{J} \quad &; \quad \big|\,[(j_1,j_2)K_\alpha,j_3]JM_J\,\big\rangle \\ \mathbf{J}_2+\mathbf{J}_3=\mathbf{K}_\beta \;\rightarrow\; \mathbf{K}_\beta+\mathbf{J}_1=\mathbf{J} \quad &; \quad \big|\,[(j_2,j_3)K_\beta,j_1]JM_J\,\big\rangle \ . \end{aligned} \tag{C.99}$$

To transform to these from $|\,j_1,j_2,j_3,m_1,m_2,m_3\,\rangle$, we need to uncouple the vectors by applying the transformations in (C.84) and (C.95), twice for each case. This yields the following superpositions:

$$\begin{aligned} \big|\,[(j_1,j_2)K_\alpha,j_3]JM_J\,\big\rangle = &\sum_{M_{K_\alpha}=-K_\alpha}^{K_\alpha}\ \sum_{m_3=-j_3}^{j_3} (-1)^{K_\alpha-j_3+M_J}\sqrt{2J+1}\begin{pmatrix} K_\alpha & j_3 & J \\ M_{K_\alpha} & m_3 & -M_J \end{pmatrix} \\ &\times \sum_{m_1=-j_1}^{j_1}\ \sum_{m_2=-j_2}^{j_2} (-1)^{j_1-j_2+M_{K_\alpha}}\sqrt{2K_\alpha+1}\begin{pmatrix} j_1 & j_2 & K_\alpha \\ m_1 & m_2 & -M_{K_\alpha} \end{pmatrix} \\ &\times |\,j_1,j_2,j_3,m_1,m_2,m_3\,\rangle \quad , \end{aligned} \tag{C.100}$$

and

$$\begin{aligned} \big|\,[(j_2,j_3)K_\beta,j_1]JM_J\,\big\rangle = &\sum_{M_{K_\beta}=-K_\beta}^{K_\beta}\ \sum_{m_1=-j_1}^{j_1} (-1)^{K_\beta-j_1+M_J}\sqrt{2J+1}\begin{pmatrix} K_\beta & j_1 & J \\ M_{K_\beta} & m_1 & -M_J \end{pmatrix} \\ &\times \sum_{m_2=-j_2}^{j_2}\ \sum_{m_3=-j_3}^{j_3} (-1)^{j_2-j_3+M_{K_\beta}}\sqrt{2K_\beta+1}\begin{pmatrix} j_2 & j_3 & K_\beta \\ m_2 & m_3 & -M_{K_\beta} \end{pmatrix} \\ &\times |\,j_1,j_2,j_3,m_1,m_2,m_3\,\rangle \quad . \end{aligned} \tag{C.101}$$

These two transformations are different. This can be seen by taking the scalar product between (C.100) and (C.101). The same will also provide the transformation matrix elements for a conversion between these two different ways to couple three angular momenta.

The scalar product is an eight-fold sum. However, since the uncoupled wave functions are orthogonal in m_1, m_2 and m_3, this reduces to five sums. Using the symmetry constraints and definitions for 3*j*-symbols and 6*j*-symbols, the transformation can be written (see also [10]):

$$\left\langle [(j_1, j_2)K_\alpha, j_3]JM_J \,\middle|\, [(j_2, j_3)K_\beta, j_1]JM_J \right\rangle$$
$$= (-1)^{j_2+j_3-K_\beta+2J} \sqrt{(2K_\alpha+1)(2K_\beta+1)} \begin{Bmatrix} j_1 & j_2 & K_\alpha \\ j_3 & J & K_\beta \end{Bmatrix} . \quad \text{(C.102)}$$

By using (C.102) in combination with (C.96), a corresponding transformation matrix can be found for any permutation of the sums of the vectors $\mathbf{J}_1$, $\mathbf{J}_2$ and $\mathbf{J}_3$.

C.6 The Wigner-Eckart Theorem

A crude definition of a tensor is an object which is geometrically transformed following a certain set of rules, for example under rotation. Tensors come in different ranks or orders. A zero-order tensor is a scalar. A tensor of rank one can be said to be equivalent to a vector. It will have a number of components, specified by one index. A rank two tensor is akin to a matrix, and has two indices, and so on. Any tensor can be broken up into irreducible parts, which for our purpose will mean that under rotation, these parts will behave like components of the spherical harmonics (see section C.3).

The above definition is general, albeit lacking in stringency, and tensors are employed as mathematical tools in many fields of science. Tensor operators will here be taken to mean the quantum mechanical analogues, and for a profound study of angular momentum and atomic structure — beyond the coverage in this book — they are indispensable instruments. A seminal work on this topic is [2]. In the present volume, we employ tensors rather sparingly, and the main reason for introducing them here is to demonstrate the Wigner–Eckart theorem in the context of reduced matrix elements.

C.6.1 Scalar, Vector, and Tensor Operators

A scalar operator is one which is entirely unchanged under rotation. Mathematically, this can be stated as (see, for example, [2] or [13]):

$$[T, \mathbf{J}] = 0 \ , \quad \text{(C.103)}$$

where T is the operator and $\mathbf{J}$ the angular momentum vector.

A vector operator does change when it is rotated, but in a way that is equivalent to a classical vector in a three-dimensional space. Formulated in terms of commutation relations, this means that:

$$\begin{aligned} [T_i, J_i] &= 0 \\ [T_i, J_j] &= \mathrm{i}\hbar T_k \\ [T_i, J_k] &= -\mathrm{i}\hbar T_j \quad . \end{aligned} \quad \text{(C.104)}$$

The indices *ijk* are here the Cartesian coordinates *xyz* in cyclic order of the three-dimensional vector operator $\mathbf{T}$ and of $\mathbf{J}$. Rotations around the $\hat{e}_z$-axis can be represented by:

$$-\frac{1}{\sqrt{2}}(T_x+\mathrm{i}\,T_y) = -\frac{|T|}{\sqrt{2}}\sin\theta\,\mathrm{e}^{\mathrm{i}\varphi} \equiv \mathbf{T}_1^{(1)}$$
$$\frac{1}{\sqrt{2}}(T_x-\mathrm{i}\,T_y) = \frac{|T|}{\sqrt{2}}\sin\theta\,\mathrm{e}^{-\mathrm{i}\varphi} \equiv \mathbf{T}_{-1}^{(1)} \;, \tag{C.105}$$

where in the last step we have defined this as two components of an *irreducible spherical tensor operator of rank one*.

The expressions in (C.105) have the same functional form as the corresponding ones for the spherical harmonics $Y_{1,\pm1}$ — see (1.22) and section C.3. This logically leads to the definition of a third component of the same first order tensor operator as:

$$T_z = |T|\cos\theta \equiv \mathbf{T}_0^{(1)} \;. \tag{C.106}$$

The relationship with the spherical harmonics — equivalent transformation under rotation — holds also for higher order tensors. One example of a tensor operator of rank two is the nuclear electric quadrupole moment, employed in the study of hyperfine structure in section 10.2.5. When rotated, it will transform as the spherical harmonics $Y_{2,0}$, $Y_{2,\pm1}$ and $Y_{2,\pm2}$.

The commutation relations of an angular momentum and a general spherical tensor operator can be shown to be (see, for example, [14] or [15]):

$$\left[J_z,\mathbf{T}_q^{(k)}\right] = \hbar\, q\,\mathbf{T}_q^{(k)}$$
$$\left[J_\pm,\mathbf{T}_q^{(k)}\right] = \hbar\sqrt{(k\mp q)(k\pm q+1)}\;\mathbf{T}_{q\pm1}^{(k)} \;. \tag{C.107}$$

C.6.2 Statement of the Theorem

The Wigner-Eckart theorem is a mathematical tool with which an angular momentum matrix element may be reduced into one factor to describe the dynamics of the system, and another one containing the geometry. The former is called a *reduced matrix element* and is written as a usual matrix element in Dirac notation, albeit with double bars in the middle. It is void of the geometric quantum numbers *m*. These appear instead as components inside a Clebsch–Gordan coefficient or, with an alternative formalism, a 3*j*-symbol (see section C.4).

The statement of the theorem is:

$$\left\langle \alpha', j', m' \left| \mathbf{T}_q^{(k)} \right| \alpha, j, m \right\rangle = \frac{\langle j, k, m, q \,|\, j, k, j', m' \rangle}{\sqrt{2j+1}} \left\langle \alpha', j' \left\| \mathbf{T}^{(k)} \right\| \alpha, j \right\rangle$$
$$= (-1)^{j-k+m'} \begin{pmatrix} j & k & j' \\ m & q & -m' \end{pmatrix} \left\langle \alpha', j' \left\| \mathbf{T}^{(k)} \right\| \alpha, j \right\rangle . \quad \text{(C.108)}$$

α is here shorthand for all degrees of freedom other than the angular momentum $\mathbf{J}$. In order to prove the theorem, we begin with the commutation relations in (C.107). From there, we can use (C.20) and (C.22), which leads to:

$$\sqrt{(j' \mp m' + 1)(j' \pm m')} \left\langle \alpha', j', m' \mp 1 \left| \mathbf{T}_q^{(k)} \right| \alpha, j, m \right\rangle$$
$$= \sqrt{(j \pm m + 1)(j \mp m)} \left\langle \alpha', j', m' \left| \mathbf{T}_q^{(k)} \right| \alpha, j, m \pm 1 \right\rangle$$
$$+ \sqrt{(k \pm q + 1)(k \mp q)} \left\langle \alpha', j', m' \left| \mathbf{T}_{q \pm 1}^{(k)} \right| \alpha, j, m \right\rangle . \quad \text{(C.109)}$$

A comparison with (C.98), and identification of the CGC-coefficients therein with the tensor operator matrix elements in (C.109), leads directly to (C.108).

A special case of the Wigner-Eckart theorem is when it is applied to a vector operator $\mathbf{V}$, and the matrix element desired is diagonal, $j' = j$. This is known as the *projection theorem*. The statement of the projection theorem is:

$$\left\langle \alpha', j, m' \left| V_q \right| \alpha, j, m \right\rangle = \frac{\langle \alpha', j, m \,|\, \mathbf{J} \cdot \mathbf{V} \,|\, \alpha, j, m \rangle}{\hbar^2 \, j(j+1)} \left\langle j, m' \left| J_q \right| j, m \right\rangle . \quad \text{(C.110)}$$

For a proof of the projection theorem, we refer to the general literature, for example, [14] or [15].

Further Reading

The theory of atomic spectra, by Condon & Shortley [16]
Quantum theory of atomic structure, by Slater [9]
Quantum Mechanics (non-relativistic theory), by Landau & Lifshitz [4]
Atomic Many-Body Theory, by Lindgren & Morrison [2]
Angular Momentum in Quantum Mechanics, by Edmonds [6]
Quantum Mechanics, by Merzbacher [13]
Physics of Atoms and Molecules, by Bransden & Joachain [3]
Springer Handbook of Atomic, Molecular, and Optical Physics, by Drake [17]
Modern Quantum Mechanics, by Sakurai [14]
Quantum Mechanics: An Experimentalists Approach, by Commins [15]

References

1. J.D. Louck, *Angular Momentum Theory* (Springer-Verlag, New York, 2006)
2. I. Lindgren, J. Morrison, *Atomic Many-Body Theory*, 2nd edn. (Springer Verlag, Berlin, 1986)
3. B.H. Bransden, C.J. Joachain, *Physics of Atoms and Molecules*, 2nd edn. (Prentice Hall, Harlow, England, 2003)
4. L.D. Landau, E.M. Lifshitz, *Quantum Mechanics — Course of Theoretical Physics, volume 3*, 3rd edn. (Butterworth-Heinemann, Amsterdam, 1981)
5. U. Fanu, G. Racah, *Irreducible Tensorial Sets* (Academic Press, Cambridge, 1959)
6. A.R. Edmonds, *Angular Momentum in Quantum Mechanics* (Princeton University Press, Princeton, 1996)
7. E.W. Hobson, *The Theory of Spherical and Ellipsoidal Harmonics* (Cambridge University Press, Cambridge, 2012)
8. E.C. Kemble, *The Fundamental Principles of Quantum Mechanics: With Elementary Applications* (Dover Publications, Dover, 2012)
9. J.C. Slater, *Quantum theory of atomic structure* (McGraw-Hill, New York, 1960)
10. R.D. Cowan, *The theory of atomic structure and spectra* (University of California press, Berkeley, 1981)
11. I.I. Sobelman, *Atomic Spectra and Radiative Transitions*, 2nd edn. (Springer, Berlin, 1992)
12. M. Rotenberg, N. Metroplis, R. Bivens, J.K. Wooten, *The 3-j and 6-j Symbols* (Technology-Press/Massachusetts Institute Technology, Cambridge, 1959)
13. E. Merzbacher, *Quantum Mechanics*, 3rd edn. (Wiley, New York, 1997)
14. J.J. Sakurai, J.J. Napolitano, *Modern Quantum Mechanics*, 2nd edn. (Pearson, Harlow, 2010)
15. E.D. Commins, *Quantum mechanics: an experimentalist's approach* (Cambridge University Press, New York, 2014)
16. E.U. Condon, G.H. Shortley, *The theory of atomic spectra* (Cambridge University Press, Cambridge, 1935)
17. G.W.F. Drake (ed.), *Springer Handbook of Atomic, Molecular, and Optical Physics* (Springer-Verlag, New York, 2006)

Appendix D
Coulomb and Exchange Integrals

This appendix aims to demonstrate how to solve the Coulomb and exchange integrals involved in the electron–electron interaction Hamiltonian (see chapters 3, 6, 7 and 8). The wave functions we here consider are ones that are solutions to a Schrödinger equation with a purely central potential. This means that radial and angular parts can be factorised, and the angular functions will be the spherical harmonics (see appendix C.3). The spin parts of the wave functions are not taken into account in this appendix since the relevant Hamiltonian does not act on the spin degree of freedom.

We will provide specific numerical coefficients and solutions to certain limiting cases. This includes a table of Gaunt coefficients — used together with Slater integrals, solutions to the radial equation for pairs of hydrogenic wave functions, and the T-polynomials used for the treatment of jj-coupling. The terminology above will be explained in what follows and is also outlined in the chapters referenced to in the preceding paragraph.

D.1 Analytical Integration of the Coulomb Repulsion Potential

The two-electron Coulomb and exchange integrals are defined in (3.10):

$$\begin{aligned} J_{n_il_i:n_jl_j} &= \left\langle n_il_im_{li}(1)\,,\, n_jl_jm_{lj}(2) \,\middle|\, \frac{1}{r_{12}} \,\middle|\, n_il_im_{li}(1)\,,\, n_jl_jm_{lj}(2) \right\rangle \\ &= \iint |\psi_{n_il_im_{li}}(\mathbf{r}_1)|^2\, |\psi_{n_jl_jm_{lj}}(\mathbf{r}_2)|^2\, \frac{1}{r_{12}}\, \mathrm{d}\mathbf{r}_1 \mathrm{d}\mathbf{r}_2 \\ K_{n_il_i:n_jl_j} &= \left\langle n_il_im_{li}(1)\,,\, n_jl_jm_{lj}(2) \,\middle|\, \frac{1}{r_{12}} \,\middle|\, n_il_im_{li}(2)\,,\, n_jl_jm_{lj}(1) \right\rangle \\ &= \iint \psi^*_{n_il_im_{li}}(\mathbf{r}_1)\, \psi^*_{n_jl_jm_{lj}}(\mathbf{r}_2)\, \frac{1}{r_{12}}\, \psi_{n_il_im_{li}}(\mathbf{r}_2)\, \psi_{n_jl_jm_{lj}}(\mathbf{r}_1)\, \mathrm{d}\mathbf{r}_1 \mathrm{d}\mathbf{r}_2 \;. \qquad \text{(D.1)} \end{aligned}$$

A. Kastberg, *Structure of Multielectron Atoms*, Springer Series on Atomic, Optical, and Plasma Physics 112, https://doi.org/10.1007/978-3-030-36420-5

The labels (1) and (2) in the matrix elements refer to the respective electron coordinates, and all integrals are taken over the entire two-electron six-dimensional space.

The hydrogenic wave functions are the ones in (1.23), with the lower order radial, $R(r)$, and angular, $Y(\theta,\varphi)$, functions tabulated in (1.16) and (1.22). The interaction operator, $1/r_{12}$, is most conveniently expressed as an expansion of spherical harmonics, see (C.56). This substitution gives the following development for the integrals:

$$\begin{aligned} J_{n_il_i:n_jl_j} = &\sum_{k=0}^{\infty} \sum_{q=-k}^{k} \frac{4\pi}{2k+1} \int_0^{\infty}\!\!\int_0^{\infty} |R_{n_il_i}(r_1)|^2\, |R_{n_jl_j}(r_2)|^2\, \frac{(r_<)^k}{(r_>)^{k+1}}\, r_1^2 r_2^2\, \mathrm{d}r_1 \mathrm{d}r_2 \\ &\times \int_0^{2\pi}\!\!\int_0^{\pi} \left|Y_{l_im_{li}}(\theta_1,\varphi_1)\right|^2 Y_{kq}^*(\theta_1,\varphi_1)\, \sin\theta_1\, \mathrm{d}\theta_1 \mathrm{d}\varphi_1 \\ &\times \int_0^{2\pi}\!\!\int_0^{\pi} \left|Y_{l_jm_{lj}}(\theta_2,\varphi_2)\right|^2 Y_{kq}(\theta_2,\varphi_2)\, \sin\theta_2\, \mathrm{d}\theta_2 \mathrm{d}\varphi_2 \end{aligned} \tag{D.2}$$

$$\begin{aligned} K_{n_il_i:n_jl_j} = &\sum_{k=0}^{\infty} \sum_{q=-k}^{k} \frac{4\pi}{2k+1} \\ &\times \int_0^{\infty}\!\!\int_0^{\infty} R_{n_il_i}^*(r_1)\, R_{n_jl_j}^*(r_2)\, R_{n_il_i}(r_2)\, R_{n_jl_j}(r_1)\, \frac{(r_<)^k}{(r_>)^{k+1}}\, r_1^2 r_2^2\, \mathrm{d}r_1 \mathrm{d}r_2 \\ &\times \int_0^{2\pi}\!\!\int_0^{\pi} Y_{l_im_{li}}^*(\theta_1,\varphi_1)\, Y_{l_jm_{lj}}(\theta_1,\varphi_1)\, Y_{kq}^*(\theta_1,\varphi_1)\, \sin\theta_1\, \mathrm{d}\theta_1 \mathrm{d}\varphi_1 \\ &\times \int_0^{2\pi}\!\!\int_0^{\pi} Y_{l_jm_{lj}}^*(\theta_2,\varphi_2)\, Y_{l_im_{li}}(\theta_2,\varphi_2)\, Y_{kq}(\theta_2,\varphi_2)\, \sin\theta_2\, \mathrm{d}\theta_2 \mathrm{d}\varphi_2 \ . \end{aligned} \tag{D.3}$$

Here, k and q are summation variables. The radial parameters $r_<$ and $r_>$ refer, respectively, to the smaller and the larger of the two radial coordinates during integrations. The solutions to the angular equations will hold generally for central potentials. They are valid for purely hydrogenic wave functions as well as for any other central potential.

D.1.1 The Angular Integrals

Spherical harmonics, see (C.48), are products of associated Legendre polynomials of $\cos\theta$, functions of φ, and normalisation factors (see appendix C.3). This means that the angular integrals in (D.2) and (D.3) can be further factorised into ones over the zenith and azimuthal angles.

The integrands over φ_1 and φ_2 are exponential functions, and for $J_{n_il_i:n_jl_j}$ the two exponents are:

$$\begin{cases} -\mathrm{i}q\varphi_1 \\ \mathrm{i}q\varphi_2 \end{cases} . \tag{D.4}$$

For $K_{n_il_i:n_jl_j}$, the exponents are:

$$\begin{cases} \mathrm{i}(-m_{li}+m_{lj}-q)\,\varphi_1 \\ \mathrm{i}(-m_{lj}+m_{li}+q)\,\varphi_2 \end{cases} . \tag{D.5}$$

The integrals are taken from 0 to 2π, and the projection quantum numbers are necessarily positive or negative integers or zero. This means that in the sum over q, all terms will vanish except for:

$$\begin{cases} q=0 & \text{for} \quad J_{n_il_i:n_jl_j} \\ q=m_{lj}-m_{li} & \text{for} \quad K_{n_il_i:n_jl_j} \end{cases} . \tag{D.6}$$

For these remaining terms, the integrals over the exponentials return 2π, and this factor is included in the normalisation constants of the spherical harmonics.

The zenith integrals will be over products of three associated Legendre functions. This is a general mathematical problem, and it has a solution in the form of a serial expansion known as *Gaunt's formula* (see [1, 2] or [3]). The general form is:

$$\begin{aligned} &\frac{1}{2}\int_{-1}^{1} P_l^{(u)}(\mu)\,P_m^{(v)}(\mu)\,P_n^{(w)}(\mu)\,\mathrm{d}\mu \\ &\qquad = (-1)^{s-m-w}\,\frac{(m+v)!\,(n+w)!\,(2s-2n)!\,s!}{(m-v)!\,(s-l)!\,(s-m)!\,(s-n)!\,(2s+1)!} \\ &\qquad\quad \times \sum_{t=a}^{b}(-1)^t\,\frac{(l+u+t)!\,(m+n-u-t)!}{t!\,(l-u-t)!\,(m-n+u+t)!\,(n-w-t)!} . \end{aligned} \tag{D.7}$$

This equation comes with a list of definitions and conditions. To start with:

$$\begin{aligned} 2s &\equiv l+m+n \\ a &\equiv \max(0,\,-m+n-u) \\ b &\equiv \min(m+n-u\,,\,l-u\,,\,n-w) . \end{aligned} \tag{D.8}$$

The last two of these are necessary for the factorials under the sum over t to be defined. The conditions for the validity of (D.7) are:

$$\begin{aligned} &u\,,\,v\,,\,w \geq 0 \\ &u = v+w \\ &m \geq n \\ &s \ \text{ is a positive integer} . \end{aligned} \tag{D.9}$$

Moreover, in order for the integral to be non-zero, the following triangular condition has to be met:

$$m-n \leq l \leq m+n \ . \tag{D.10}$$

This formulation follows the original one by Gaunt. The equation can be made more accessible by expressing (D.7) with the sum and the fractions of factorials in terms of Wigner 3j-symbols (see appendix C.4). With this modification, use of the same notation for the coefficients as in (D.2) and (D.3), the constraints in (D.9) and those for the azimuthal integrals in (D.6), we can write an integral over a product of three spherical harmonics as [4]:

$$\begin{aligned}
&\sqrt{\frac{4\pi}{2k+1}} \int_0^{2\pi}\!\!\int_0^{\pi} Y_{l_i,m_{l_i}}(\theta,\varphi)\, Y_{l_j,m_{l_j}}(\theta,\varphi)\, Y_{k,m_{l_i}-m_{l_j}}(\theta,\varphi)\, \sin\theta\, \mathrm{d}\theta \mathrm{d}\varphi \\
&\quad = (-1)^{m_{li}} \sqrt{(2l_i+1)\,(2l_j+1)} \begin{pmatrix} l_i & k & l_j \\ 0 & 0 & 0 \end{pmatrix} \begin{pmatrix} l_i & k & l_j \\ -m_{li} & m_{li}-m_{lj} & m_{lj} \end{pmatrix} \\
&\quad \equiv c^{(k)}[l_i m_{li} : l_j m_{lj}] \quad .
\end{aligned} \tag{D.11}$$

In the last line of (D.11) we have defined the *Gaunt coefficient*. This is a function of the orbital angular momentum quantum numbers, and their corresponding projection quantum numbers, associated with two electron wave functions. The utility of the Gaunt coefficients is that they contain the entire angular parts of the Coulomb and exchange integrals. Using (D.11), (D.2) and (D.3), we can now write the angular parts of the integrals as:

$$\text{for} \quad J_{n_i l_i n_j l_j} \ : \ c^{(k)}[l_i m_{li} : l_i m_{li}]\, c^{(k)}[l_j m_{lj} : l_j m_{lj}] \tag{D.12}$$

$$\text{for} \quad K_{n_i l_i : n_j l_j} \ : \ \left\{ c^{(k)}[l_i m_{li} : l_j m_{lj}] \right\}^2 \ . \tag{D.13}$$

With modern personal computers, it is relatively straightforward to compute Gaunt coefficients. In some mathematical software and program languages, they are included as predefined functions. In older literature, Gaunt coefficients are presented in long tables, and in table D.1 we give a shortened example of this. We list values of the coefficients for combinations of angular momenta up to $l_i=2$ and $l_j=2$. We have used the following symmetry property, which facilitates a shorter table:

$$c^{(k)}[l_j m_{lj} : l_i m_{li}] = (-1)^{m_{li}-m_{lj}}\, c^{(k)}[l_i m_{li} : l_j m_{lj}] \ . \tag{D.14}$$

Table D.1 Gaunt coefficients $c^{(k)}[l_i m_{li} : l_j m_{lj}]$, see (D.11), tabulated for orbital angular momentum numbers up to $l_i = 2$ and $l_j = 2$, and the corresponding ranges for m_{li} and m_{lj}. l_i and l_j are given in terms of their spectroscopic symbols.

l_i	m_{li}	l_j	m_{lj}	k				
				0	1	2	3	4
s	0	s	0	1	0	0	0	0
s	0	p	0	0	$1/\sqrt{3}$	0	0	0
s	0	p	±1	0	$-1/\sqrt{3}$	0	0	0
s	0	d	0	0	0	$1/\sqrt{5}$	0	0
s	0	d	±1	0	0	$-1/\sqrt{5}$	0	0
s	0	d	±2	0	0	$1/\sqrt{5}$	0	0
p	0	p	0	1	0	$2/5$	0	0
p	0	p	±1	0	0	$\sqrt{3}/5$	0	0
p	±1	p	±1	1	0	$-1/5$	0	0
p	±1	p	∓1	0	0	$-\sqrt{6}/5$	0	0
p	0	d	0	0	$2/\sqrt{15}$	0	$\sqrt{27/245}$	0
p	0	d	±1	0	$-1/\sqrt{5}$	0	$-\sqrt{24/245}$	0
p	0	d	±2	0	0	0	$\sqrt{3}/7$	0
p	±1	d	0	0	$-1/\sqrt{15}$	0	$\sqrt{18/245}$	0
p	±1	d	±1	0	$1/\sqrt{5}$	0	$-3/\sqrt{245}$	0
p	±1	d	∓1	0	0	0	$-\sqrt{6}/7$	0
p	±1	d	±2	0	0	0	$3/7$	0
p	±1	d	∓2	0	$-\sqrt{2/5}$	0	$\sqrt{3/245}$	0
d	0	d	0	1	0	$2/7$	0	$2/7$
d	0	d	±1	0	0	$1/7$	0	$-\sqrt{10/147}$
d	0	d	±2	0	0	$-2/7$	0	$\sqrt{5/147}$
d	±1	d	±1	1	0	$1/7$	0	$-4/21$
d	±1	d	∓1	0	0	$-\sqrt{6}/7$	0	$-\sqrt{40}/21$
d	±1	d	±2	0	0	$-\sqrt{6}/7$	0	$\sqrt{5}/15$
d	±1	d	∓2	0	0	0	0	$\sqrt{5/63}$
d	±2	d	±2	1	0	$-2/7$	0	$1/21$
d	±2	d	∓2	0	0	0	0	$\sqrt{10/63}$

D.1.2 The Radial Integrals

The radial integrals in (D.2) and (D.3) are of the form:

$$R^{(k)}_{ab:cd}(r_1,r_2) \equiv \int_0^\infty\int_0^\infty R^*_{n_al_a}(r_1)\,R^*_{n_bl_b}(r_2)\,R_{n_cl_c}(r_1)\,R_{n_dl_d}(r_2)\,\frac{(r_<)^k}{(r_>)^{k+1}}\,r_1^2\,r_2^2\,\mathrm{d}r_1\mathrm{d}r_2\,. \quad \text{(D.15)}$$

This is the most general form of a so-called *Slater integral* (see, for example, [2] or [5]). In this integral, a, b, c and d are indices for four (possibly) different radial wave functions. Slater integrals will be further specified in section D.3, where we will see that we will typically use generalised forms for just two electrons.

For a solution of (D.15), we first integrate over one of the radial coordinates, say r_1, in two steps. We commence with the interval $0 \le r_1 \le r_2$, where $r_< = r_1$ and $r_> = r_2$. Then we proceed with $r_2 \le r_1 \le \infty$, where the roles of $r_<$ and $r_>$ are reversed. When that is done, the integration over r_2 can be performed. Stating the same thing mathematically:

$$R^{(k)}_{ab:cd}(r_1,r_2) = \int_0^\infty R^*_{n_bl_b}(r_2)\,R_{n_dl_d}(r_2)\left[r_2^{-k-1}\int_0^{r_2} R^*_{n_al_a}(r_1)\,R_{n_cl_c}(r_1)\,r_1^{k+2}\,\mathrm{d}r_1 + r_2^k\int_{r_2}^\infty R^*_{n_al_a}(r_1)\,R_{n_cl_c}(r_1)\,r_1^{1-k}\,\mathrm{d}r_1\right] r_2^2\,\mathrm{d}r_2\,. \quad \text{(D.16)}$$

We can combine this with the angular solutions from (D.12) and (D.13). The complete Coulomb and exchange integrals are then:

$$J_{n_il_i:n_jl_j} = \sum_{k=0}^{\infty} c^{(k)}[l_im_{li}:l_im_{li}]\;c^{(k)}[l_jm_{lj}:l_jm_{lj}]\;R^{(k)}_{ij:ij} \quad \text{(D.17)}$$

$$K_{n_il_i:n_jl_j} = \sum_{k=0}^{\infty}\left\{c^{(k)}[l_im_{li}:l_jm_{lj}]\right\}^2 R^{(k)}_{ij:ji}\,. \quad \text{(D.18)}$$

From this we can conclude that the general formulation of the Slater integral in (D.15), which includes four different electron radial functions, is simplified to two-electron integrals, when they are applied to the Coulomb and exchange integrals.

The sums in (D.17) and (D.18) are infinite only in principle. As can be seen in table D.1, only very few of the Gaunt coefficients are non-zero, which means that less than a handful of terms will have to be retained (a consequence of the conditions associated to Gaunt's formula), except for states with very large angular momenta. Note that great attention must be given to the order of the indices in the functions $R_{ij:ij}$ and $R_{ij:ji}$ in (D.17) and (D.18).

D.2 Solutions for Hydrogenic Wave Functions

For a perturbative calculation of energies for true two-electron atoms, one needs solutions to (D.17) and (D.18) for hydrogenic one-electron functions (see sections 3.1.1 and 3.1.2). In the absence of an additional core of closed-orbital electrons one can approach the integrals analytically, in the non-relativistic limit. This works since we have access to analytical expressions for both radial and angular parts of the wave functions — (1.16) and (1.22) — for the product of zero-order hydrogenic wave functions.

In the equations below, we give examples of Coulomb and exchange integrals for quantum numbers up to $l_i = 2$ for the first electron, and with $l_j \leq l_i$. Generalisation to higher angular momenta is straightforward. These integrals are independent of m_{li} and m_{lj}, which means that $m_{li} = m_{lj} = 0$ can be used for simplicity.

For $l_i = 0$ and $l_j = 0$:

$$\begin{aligned} J_{n_i\mathrm{s}:n_j\mathrm{s}} &= \left\{c^{(0)}[00:00]\right\}^2 R^{(0)}_{n_i\mathrm{s}n_j\mathrm{s}:n_i\mathrm{s}n_j\mathrm{s}} \\ K_{n_i\mathrm{s}:n_j\mathrm{s}} &= \left\{c^{(0)}[00:00]\right\}^2 R^{(0)}_{n_i\mathrm{s}n_j\mathrm{s}:n_j\mathrm{s}n_i\mathrm{s}} \quad . \end{aligned} \tag{D.19}$$

For $l_i = 1$ and $l_j \leq 1$:

$$\begin{aligned} J_{n_i\mathrm{p}:n_jl_j} &= c^{(0)}[10:10]\, c^{(0)}[l_j0:l_j0]\, R^{(0)}_{n_i\mathrm{p}n_jl_j:n_i\mathrm{p}n_jl_j} \\ &\quad + c^{(2)}[10:10]\, c^{(2)}[l_j0:l_j0]\, R^{(2)}_{n_i\mathrm{p}n_jl_j:n_i\mathrm{p}n_jl_j} \\ K_{n_i\mathrm{p}:n_jl_j} &= \left\{c^{(l_j-1)}[10:l_j0]\right\}^2 R^{(l_j-1)}_{n_i\mathrm{p}n_jl_j:n_jl_jn_i\mathrm{p}} \\ &\quad + \left\{c^{(l_j+1)}[10:l_j0]\right\}^2 R^{(l_j+1)}_{n_i\mathrm{p}n_jl_j:n_jl_jn_i\mathrm{p}} \quad . \end{aligned} \tag{D.20}$$

For $l_i = 2$ and $l_j \leq 2$:

$$\begin{aligned} J_{n_i\mathrm{d}:n_jl_j} &= c^{(0)}[20:20]\, c^{(0)}[l_j0:l_j0]\, R^{(0)}_{n_i\mathrm{d}n_jl_j:n_i\mathrm{d}n_jl_j} \\ &\quad + c^{(2)}[20:20]\, c^{(2)}[l_j0:l_j0]\, R^{(2)}_{n_i\mathrm{d}n_jl_j:n_i\mathrm{d}n_jl_j} \\ &\quad + c^{(4)}[20:20]\, c^{(4)}[l_j0:l_j0]\, R^{(4)}_{n_i\mathrm{d}n_jl_j:n_i\mathrm{d}n_jl_j} \\ K_{n_i\mathrm{d}:n_jl_j} &= \left\{c^{(l_j-2)}[20:l_j0]\right\}^2 R^{(l_j-2)}_{n_i\mathrm{d}n_jl_j:n_jl_jn_i\mathrm{d}} \\ &\quad + \left\{c^{(l_j)}[20:l_j0]\right\}^2 R^{(l_j)}_{n_i\mathrm{d}n_jl_j:n_jl_jn_i\mathrm{d}} \\ &\quad + \left\{c^{(l_j+2)}[20:l_j0]\right\}^2 R^{(l_j+2)}_{n_i\mathrm{d}n_jl_j:n_jl_jn_i\mathrm{d}} \quad . \end{aligned} \tag{D.21}$$

In table D.2 we provide quantitative values (linear functions of Z) for $J_{n_il_i:n_jl_j}$ and $K_{n_il_i:n_jl_j}$ for the ground state and the lowest excited states of a two-electron atom, as well as for a few doubly excited states. Note that the latter will be autoionising, since they have energies in the continuum, above the first ionisation limit. Computing integrals for higher angular momentum numbers is straightforward, but the resulting fractions become long and unsightly.

Table D.2 Numerical values, in a.u., of the electron pair overlap integrals J and K, computed for pairs of hydrogenic radial functions. This is applicable for perturbative calculations of energies for true two-electron systems. Note the linear scaling with Z.

$n_il_i:n_jl_j$	$J_{n_il_i:n_jl_j}$	$K_{n_il_i:n_jl_j}$
1s:1s	$\frac{5\,Z}{8}$	$\frac{5\,Z}{8}$
1s:2s	$\frac{17\,Z}{81}$	$\frac{16\,Z}{729}$
1s:2p	$\frac{59\,Z}{243}$	$\frac{112\,Z}{6651}$
1s:3s	$\frac{815\,Z}{8192}$	$\frac{189\,Z}{32768}$
1s:3p	$\frac{1783\,Z}{16384}$	$\frac{297\,Z}{65536}$
1s:3d	$\frac{1819\,Z}{16384}$	$\frac{81\,Z}{327680}$
2s:2s	$\frac{77\,Z}{512}$	$\frac{77\,Z}{512}$
2s:2p	$\frac{82\,Z}{512}$	$\frac{15\,Z}{512}$
2p:2p	$\frac{501\,Z}{2560}$	$\frac{501\,Z}{2560}$

D.3 Slater Integrals

In (D.15), we introduced the general form of a Slater integral, $R^{(k)}_{ab:cd}(r_1,r_2)$. This is a double integral over two electronic radial coordinates and each of the four indices a, b, c and d corresponds to a combination of one principal quantum number and one orbital angular momentum quantum number: n_al_a, n_bl_b and so on.

For the Coulomb and exchange integrals, $J_{n_il_i:n_jl_j}$ and $K_{n_il_i:n_jl_j}$, it is superfluous with four different orbitals (different combinations of nl), since they always involve one specific *electron pair*. This means that we can simplify the notation by introducing the *Slater F- and G-integrals*, which are generalisations of $R^{(k)}_{ab:cd}(r_1,r_2)$. The definitions of the F- and G-integrals are:

$$
\begin{aligned}
F^{(k)}_{n_il_i:n_jl_j} &\equiv R^{(k)}_{ij:ij}(r_1,r_2) \\
&= \int_0^\infty\!\!\int_0^\infty |R_{n_il_i}(r_1)|^2\,|R_{n_jl_j}(r_2)|^2\,\frac{(r_<)^k}{(r_>)^{k+1}}\,r_1^2r_2^2\,\mathrm{d}r_1\mathrm{d}r_2 \qquad \text{(D.22)}\\
G^{(k)}_{n_il_i:n_jl_j} &\equiv R^{(k)}_{ij:ji}(r_1,r_2) \\
&= \int_0^\infty\!\!\int_0^\infty R^*_{n_il_i}(r_1)\,R^*_{n_jl_j}(r_2)\,R_{n_jl_j}(r_1)\,R_{n_il_i}(r_2)\,\frac{(r_<)^k}{(r_>)^{k+1}}\,r_1^2r_2^2\,\mathrm{d}r_1\mathrm{d}r_2\ . \qquad \text{(D.23)}
\end{aligned}
$$

Having made these definitions, we can write the Coulomb and exchange integrals as:

$$J_{n_il_i:n_jl_j} = \sum_{k=0}^{\infty} c^{(k)}[l_im_{li}:l_im_{li}]\, c^{(k)}[l_jm_{lj}:l_jm_{lj}]\, F^{(k)}_{ij:ij} \tag{D.24}$$

$$K_{n_il_i:n_jl_j} = \sum_{k=0}^{\infty} \left\{c^{(k)}[l_im_{li}:l_jm_{lj}]\right\}^2 G^{(k)}_{ij:ji} \quad . \tag{D.25}$$

The integrals in (D.24) and (D.25) contain radial functions which typically are not known. However, the sums in the equations will always be limited to very few terms due to the triangular condition (D.10).

Even without explicit solutions for the *F*- and *G*-integrals, it is useful to express theoretical level energies in terms of these, and this can also serve as preparatory steps in numerical calculations of wave functions.

D.4 Coulomb Interaction Energies in *jj*-coupling

When computing perturbative energies for *jj*-coupling states (see chapter 8), the first perturbation to be applied to the CFA zero-order solutions is the spin–orbit interaction, H_{SO}. The diagonal representation is $|\gamma, n_1l_1j_1m_{j1}\ldots n_1l_1j_1m_{j1}\rangle$, where γ is shorthand for all electrons in filled orbitals, and the remaining quantum numbers are those of the valence electrons.

The next perturbation is the two-electron matrix elements of the Coulomb interaction Hamiltonian, H_{ee}. For this operator, the individual electron representation $|nlm_lm_s\rangle$ is the one leading to solutions in terms of Gaunt coefficients and Slater integrals (see previous sections of this appendix). This means that when fine-structure energies in *jj*-coupling need to be computed, it is necessary to perform a change of basis. The transformation relevant for this is that described by (8.13).

The change of representation is greatly facilitated by the fact that for a given m_l, m_j will always be restricted to the two values $m_j + 1/2$ and $m_j - 1/2$. Thus, the superposition never has more than two terms. This makes it practical to generate a table with energy contributions in terms of *F*- and *G*-integrals for different kinds of electron pairs. The procedure for this is detailed in section 8.2.2.

To be able to infer level energies in terms of Slater integrals, *jj*-coupling *T-polynomials* need to be calculated or tabulated, see (8.14). The definition of these coefficients is implicit in:

$$\begin{aligned}
&\sum_{\alpha>\beta}^{N}\sum_{\beta=1}^{N} T(l_\alpha j_\alpha m_{j\alpha}:l_\beta j_\beta m_{j\beta}) = \langle \mathbb{S}_{\sigma_i} | H_{\mathrm{ee}} | \mathbb{S}_{\sigma_i} \rangle \\
&= \sum_{\alpha>\beta}^{N}\sum_{\beta=1}^{N} \left[\left\langle \Psi_{l_\alpha j_\alpha m_{j\alpha}}(\mathbf{q}_a)\, \Psi_{l_\beta j_\beta m_{j\beta}}(\mathbf{q}_b) \left| \frac{1}{r_{12}} \right| \Psi_{l_\alpha j_\alpha m_{j\alpha}}(\mathbf{q}_a)\, \Psi_{l_\beta j_\beta m_{j\beta}}(\mathbf{q}_b) \right\rangle \right. \\
&\quad \left. - \left\langle \Psi_{l_\alpha j_\alpha m_{j\alpha}}(\mathbf{q}_a)\, \Psi_{l_\beta j_\beta m_{j\beta}}(\mathbf{q}_b) \left| \frac{1}{r_{12}} \right| \Psi_{l_\beta j_\beta m_{j\beta}}(\mathbf{q}_a)\, \Psi_{l_\alpha j_\alpha m_{j\alpha}}(\mathbf{q}_b) \right\rangle \right] ,
\end{aligned} \tag{D.26}$$

where *N* is the number of valence electrons.

In tables D.3, D.4, D.5 and D.6, we present values for $T(l_\alpha j_\alpha m_{j\alpha}{:}l_\beta j_\beta m_{j\beta})$ for all combinations of two electrons with orbital angular momenta $l \leq 2$. The tables were compiled by using (8.13), (D.1), (D.24), (D.25), (D.22), (D.23), and table D.1. They show T-polynomials for different configurations, which are needed in expansions of jj-coupling fine-structure energies in terms of Slater integrals.

Table D.3 Table of coefficients $T(l_\alpha j_\alpha m_{j\alpha}{:}l_\beta j_\beta m_{j\beta})$, used for expressing fine-structure energies in jj-coupling. The table is limited to atoms with two valence electrons, with orbital angular momentum quantum numbers $l \leq 2$ (l is expressed in terms of its spectroscopic symbols). The coefficients shown for each configuration are those corresponding to the Slater integral term given by the column headings. The subscripts $n_\alpha l_\alpha : n_\beta l_\beta$ to the F- and G-integrals have been omitted. This table covers the configurations ss, sp and sd.

l_α l_β	j_α	$m_{j\alpha}$	j_β	$m_{j\beta}$	$F^{(0)}$	$G^{(0)}$	$G^{(1)}$	$G^{(2)}$
s s	1/2	±1/2	1/2	±1/2	1	−1	0	0
				∓1/2	1	0	0	0
s p	1/2	±1/2	3/2	±3/2	1	0	−3	0
				±1/2	1	0	−2	0
				∓1/2	1	0	−1	0
				∓3/2	1	0	0	0
			1/2	±1/2	1	0	−1	0
				∓1/2	1	0	−2	0
s d	1/2	±1/2	5/2	±5/2	1	0	0	−5
				±3/2	1	0	0	−4
				±1/2	1	0	0	−3
				∓1/2	1	0	0	−2
				∓3/2	1	0	0	−1
				∓5/2	1	0	0	0
			3/2	±3/2	1	0	0	−1
				±1/2	1	0	0	−2
				∓1/2	1	0	0	−3
				∓3/2	1	0	0	−4

Table D.4 Table of coefficients $T(l_\alpha j_\alpha m_{j\alpha}\!:l_\beta j_\beta m_{j\beta})$. This is the same as table D.3, but for the configuration pp.

$l_\alpha\ l_\beta$	j_α	$m_{j\alpha}$	j_β	$m_{j\beta}$	$F^{(0)}$	$F^{(2)}$	$G^{(0)}$	$G^{(2)}$
p p	3/2	±3/2	3/2	±3/2	1	1	−1	−1
				±1/2	1	−1	0	−2
				∓1/2	1	−1	0	−2
				∓3/2	1	1	0	0
			1/2	±1/2	1	0	0	−1
				∓1/2	1	0	0	−4
		±1/2	3/2	±1/2	1	1	−1	−1
				∓1/2	1	1	0	0
			1/2	±1/2	1	0	0	−2
				∓1/2	1	0	0	−3
	1/2	±1/2	1/2	±1/2	1	0	−1	0
				∓1/2	1	0	0	0

Table D.5 Table of coefficients $T(l_\alpha j_\alpha m_{j\alpha}{:}l_\beta j_\beta m_{j\beta})$. This is the same as table D.3, but for the configuration pd.

l_α l_β	j_α	$m_{j\alpha}$	j_β	$m_{j\beta}$	$F^{(0)}$	$F^{(2)}$	$G^{(1)}$	$G^{(2)}$
p d	$^3/_2$	$\pm^3/_2$	$^5/_2$	$\pm^5/_2$	1	10	−90	−15
				$\pm^3/_2$	1	−2	−36	−36
				$\pm^1/_2$	1	−8	−9	−54
				$\mp^1/_2$	1	−8	0	−60
				$\mp^3/_2$	1	−2	0	−45
				$\mp^5/_2$	1	10	0	0
			$^3/_2$	$\pm^3/_2$	1	7	−9	−9
				$\pm^1/_2$	1	−7	−6	−36
				$\mp^1/_2$	1	−7	0	−90
				$\mp^3/_2$	1	7	0	−180
		$\pm^1/_2$	$^5/_2$	$\pm^5/_2$	1	−10	0	−50
				$\pm^3/_2$	1	2	−54	−49
				$\pm^1/_2$	1	8	−54	−24
				$\mp^1/_2$	1	8	−27	2
				$\mp^3/_2$	1	2	0	−10
				$\mp^5/_2$	1	−10	0	−75
			$^3/_2$	$\pm^3/_2$	1	−7	−6	−36
				$\pm^1/_2$	1	7	−1	−81
				$\mp^1/_2$	1	7	−8	−108
				$\mp^3/_2$	1	−7	0	−90
	$^1/_2$	$\pm^1/_2$	$^5/_2$	$\pm^5/_2$	1	0	0	−25
				$\pm^3/_2$	1	0	0	−50
				$\pm^1/_2$	1	0	0	−75
				$\mp^1/_2$	1	0	0	−100
				$\mp^3/_2$	1	0	0	−125
				$\mp^5/_2$	1	0	0	−150
			$^3/_2$	$\pm^3/_2$	1	0	−75	0
				$\pm^1/_2$	1	0	−50	0
				$\mp^1/_2$	1	0	−25	0
				$\mp^3/_2$	1	0	0	0

Table D.6 Table of coefficients $T(l_\alpha j_\alpha m_{j\alpha}{:}\, l_\beta j_\beta m_{j\beta})$. This is the same as table D.3, but for the configuration dd.

l_α l_β	j_α	$m_{j\alpha}$	j_β	$m_{j\beta}$	$F^{(0)}$	$F^{(2)}$	$F^{(4)}$	$G^{(0)}$	$G^{(2)}$	$G^{(4)}$
d d	5/2	$\pm 5/2$	5/2	$\pm 5/2$	1	100	1	−1	−100	−1
				$\pm 3/2$	1	−20	−3	0	−120	−4
				$\pm 1/2$	1	−80	2	0	−60	−9
				$\mp 1/2$	1	−80	2	0	0	−14
				$\mp 3/2$	1	−20	3	0	0	−14
				$\mp 5/2$	1	100	1	0	0	0
			3/2	$\pm 3/2$	1	70	0	0	−30	−1
				$\pm 1/2$	1	−70	0	0	−40	−6
				$\mp 1/2$	1	−70	0	0	0	−21
				$\mp 3/2$	1	70	0	0	0	−56
		$\pm 3/2$	5/2	$\pm 3/2$	1	4	9	−1	−4	−9
				$\pm 1/2$	1	16	−6	0	−48	−10
				$\mp 1/2$	1	16	−6	0	−108	−5
				$\mp 3/2$	1	4	9	0	0	0
			3/2	$\pm 3/2$	1	−14	0	0	−36	−4
				$\pm 1/2$	1	14	0	0	−2	−15
				$\mp 1/2$	1	14	0	0	−32	−30
				$\mp 3/2$	1	−14	0	0	0	−35
		$\pm 1/2$	5/2	$\pm 1/2$	1	64	4	−1	−64	−4
				$\mp 1/2$	1	64	4	0	0	0
			3/2	$\pm 3/2$	1	−56	0	0	−27	−10
				$\pm 1/2$	1	56	0	0	−6	−24
				$\mp 1/2$	1	56	0	0	−25	−30
				$\mp 3/2$	1	−56	0	0	−12	−20
	3/2	$\pm 3/2$	3/2	$\pm 3/2$	1	49	0	−1	−49	0
				$\pm 1/2$	1	−49	0	0	−98	0
				$\mp 1/2$	1	−49	0	0	−98	0
				$\mp 3/2$	1	49	0	0	0	0
		$\pm 1/2$	3/2	$\pm 1/2$	1	49	0	−1	−49	0
				$\mp 1/2$	1	49	0	0	0	0

Further Reading

The theory of atomic spectra, by Condon & Shortley [6]
Quantum theory of atomic structure, by Slater [2]
The theory of atomic structure and spectra, by Cowan [4]
Atomic Many-Body Theory, by Lindgren & Morrison [5]

References

1. J.A. Gaunt, Phil. Trans. Roy. Soc. (London) **A228**, 151 (1929)
2. J.C. Slater, *Quantum theory of atomic structure* (McGraw-Hill, New York, 1960)
3. G. Racah, Phys. Rev. **61**, 186 (1942)
4. R.D. Cowan, *The theory of atomic structure and spectra* (University of California press, Berkeley, 1981)
5. I. Lindgren, J. Morrison, *Atomic Many-Body Theory*, 2nd edn. (Springer Verlag, Berlin, 1986)
6. E.U. Condon, G.H. Shortley, *The theory of atomic spectra* (Cambridge University Press, Cambridge, 1935)

Appendix E
Electron Spin

The electron spin is a purely relativistic effect, and in contrast to the orbital angular momentum, which involves an evolution of the spatial coordinates of the centre of mass, spin is an *intrinsic* angular momentum. The electron is generally considered to have zero extension, but it does have a measurable magnetic dipole moment (and a predicted electric dipole moment). One way to classically model the magnetic properties of the electron is a magnetic moment related to a spin angular momentum of a charged particle, via a gyromagnetic ratio — or g-factor. The analogy with a spinning massive charged body is problematic for many reasons. Nevertheless, used with care, this simplified picture may be helpful for developing a basic understanding of an atom's magnetic interactions.

In this appendix, we will first discuss the relationship between angular momenta and magnetic moments from an essentially non-relativistic point of view. This facilitates the computation of spin interactions in vector model formalisms, used in several chapters of the book. In section E.2, we will provide a brief summary of Dirac's relativistic theory of the electron spin.

E.1 Classical Magnetic Moments

We consider an electromagnetic field characterised by its magnetic flux density $\mathbf{B}(\mathbf{r})$ and its electrostatic field $\mathcal{E}(\mathbf{r})$. In terms of the vector and scalar potentials of the electromagnetic field, $\mathbf{A}(\mathbf{r})$ and $\phi(\mathbf{r})$, the above vector fields can be expressed as:

$$\begin{aligned} \mathcal{E}(\mathbf{r}) &= -\nabla\phi(\mathbf{r}) - \frac{\partial \mathbf{A}(\mathbf{r})}{\partial t} \\ \mathbf{B}(\mathbf{r}) &= \nabla \times \mathbf{A}(\mathbf{r}) \quad . \end{aligned} \tag{E.1}$$

A. Kastberg, *Structure of Multielectron Atoms*, Springer Series on Atomic, Optical, and Plasma Physics 112, https://doi.org/10.1007/978-3-030-36420-5

A single classical spinless particle of mass m_i and charge q_i moving in such a field will see its motion governed by the Lagrangian function, in SI-units (see, for example, [1, 2] or [3]):

$$\mathscr{L}(\mathbf{r}_i,\mathbf{v}_i,t) = \frac{1}{2} m_i v_i^2 + q_i [\mathbf{v}_i \cdot \mathbf{A}(\mathbf{r}_i)] - q_i \phi(\mathbf{r}_i) \quad , \tag{E.2}$$

with $\mathbf{v}_i$ being the velocity, and $\mathbf{r}_i$ the position vector of the particle. The single particle Hamiltonian can then be formulated as:

$$H_i = \frac{[\mathbf{p}_i - q_i \mathbf{A}(\mathbf{r}_i)]^2}{2m_i} + q_i \phi(\mathbf{r}_i) \quad , \tag{E.3}$$

and the associated probability current:

$$\mathbf{j}_i = -\mathrm{i}\frac{\hbar}{2m_i}[\psi_i^*(\mathbf{r}_i)\nabla_i\psi_i(\mathbf{r}_i) - \psi_i(\mathbf{r}_i)\nabla_i\psi_i^*(\mathbf{r}_i)] - \frac{q_i}{m_i}\mathbf{A}(\mathbf{r}_i)|\psi_i(\mathbf{r}_i)|^2 \ . \tag{E.4}$$

If we consider an ensemble of N charged particles, the corresponding total Hamiltonian of the system will be a sum of terms such as (E.3), with the addition of the Coulomb interactions between the particles. If we introduce the potential energy $V_{ij}(\mathbf{r}_i,\mathbf{r}_j)$ as corresponding to the intercoupling between the particle pair i and j, the overall Hamiltonian becomes:

$$H = \sum_i^N \left\{ \frac{[\mathbf{p}_i - q_i \mathbf{A}(\mathbf{r}_i)]^2}{2m_i} + q_i \phi(\mathbf{r}_i) \right\} + \sum_i^N \sum_{j>i}^N V_{ij}(\mathbf{r}_i,\mathbf{r}_j) \ . \tag{E.5}$$

This leads to the following Schrödinger equation for the complete system:

$$\begin{aligned} &\left(\sum_i^N \left\{ -\frac{\hbar^2}{2m_i}\nabla_i^2 + \mathrm{i}\frac{\hbar q_i}{m_i}[\mathbf{A}(\mathbf{r}_i)\cdot\nabla_i] + \frac{[q_i \mathbf{A}(\mathbf{r}_i)]^2}{2m_i} + q_i\phi(\mathbf{r}_i) \right\} \right. \\ &\qquad \left. + \sum_i^N \sum_{j>i}^N V_{ij}(\mathbf{r}_i,\mathbf{r}_j) \right) \psi(\mathbf{r}_i \ldots \mathbf{r}_N) = E\,\psi(\mathbf{r}_i \ldots \mathbf{r}_N) \ . \end{aligned} \tag{E.6}$$

When expanding the square in the Hamiltonian (E.5), we have assumed that the two operators $\mathbf{p}$ and $\mathbf{A}(\mathbf{r}_i)$ commute. This is not always true, but it holds for $\nabla\cdot\mathbf{A} = 0$, which is, for example, the case for a homogeneous field. Note that in (E.6), we are still ignoring the spins of the particles, and we have assumed a static external field. If the time-dependent version of (E.6) is instead constructed, this formalism can also form the basis for a study of the interaction of an atom with a radiation field.

E.1.1 Magnetic Moment Due to the Orbital Angular Momentum

To see how a magnetic moment can arise from a moving charged particle, we take an atom as physical system and we first assume that there is no external field, that is $\mathbf{A}(\mathbf{r}_i)=\mathbf{0}$ and $\phi(\mathbf{r}_i)=0$. The Schrödinger equation in (E.6) will then be the standard atomic one (albeit with spin ignored), which is used in most early chapters in this book. If we simplify even further and consider just a one-electron atom, the solutions are those of (1.23), expressed as $\psi_{nlm_l}(\mathbf{r}_i)$.

If a single bound electron has $l>0$ and $m_l>0$, it implies that the negatively charged electron has a non-zero angular momentum projection along the chosen $\hat{e}_z$-axis (the quantisation axis). In classical terms, this corresponds to a current loop, and thus an associated classical magnetic moment along the $\hat{e}_z$-axis. With knowledge of the wave function, and by using (E.4) for the probability flux, this magnetic moment can be calculated.

In spherical coordinates, the azimuthal component of the current density for one electron becomes, in SI-units (the radial and zenith components are zero, at least on average):

$$\mathscr{J}_\varphi(\mathbf{r}_i) = -\frac{\hbar e m_l}{m_e\, r_i \sin\theta_i}\, \psi^*_{nlm_l}(\mathbf{r}_i)\, \psi_{nlm_l}(\mathbf{r}_i) \ . \tag{E.7}$$

See, for example, [3] for a detailed derivation. Note that m_e is the electron mass, whereas m_l is the quantum number for the projection of orbital angular momentum vector (also called the magnetic quantum number).

The azimuthal flux makes it natural to consider a circular filament located between r and $r+dr$ and between θ and $\theta+d\theta$, and to see this as a current loop generating a magnetic moment. The cross-sectional area and the volume of this 'ring' is:

$$\begin{aligned} \mathscr{A}_{\text{cs-ring}} &= r\,dr\,d\theta \\ \mathscr{V}_{\text{ring}} &= 2\pi r^2 \sin\theta\, dr\, d\theta \ . \end{aligned} \tag{E.8}$$

The area enclosed by the ring, in a plane perpendicular to $\hat{e}_z$, is:

$$\mathscr{A}_{\text{encl-ring}} = \pi r^2 \sin^2\theta \ . \tag{E.9}$$

The current flowing in the ring will be $\mathscr{J}_\varphi(\mathbf{r}_i)\,\mathscr{A}_{\text{cs-ring}}$ — with the current density from (E.7) — and this multiplied with $\mathscr{A}_{\text{encl-ring}}$ gives us the magnetic moment. It is directed along the negative $\hat{e}_z$-axis and has the magnitude:

$$\left|\mu_{l\text{-ring}}\right| = \left[\frac{e\hbar}{2m_e}\, \mathscr{V}_{\text{ring}}\, \psi^*_{nlm_l}(\mathbf{r}_i)\, \psi_{nlm_l}(\mathbf{r}_i)\right] m_l \ . \tag{E.10}$$

The final step is to integrate this over the whole volume, and with the assumption that the wave functions are normalised, the final expression for the magnetic moment arising due to the orbital angular momentum is:

$$\mu_l = -\frac{\hbar e}{2m_e} m_l = -\mu_B m_l \quad . \tag{E.11}$$

Here we have defined μ_l as the $\hat{e}_z$-component of the magnetic moment vector $\boldsymbol{\mu}_l$. The minus sign is a consequence of the electron's negative charge, and μ_B is the Bohr magneton (one half in a.u.).

E.1.2 The Electron Spin and Its Associated Magnetic Moment

The electron spin, **S**, is a quantum mechanical angular momentum, having all the general properties described in appendix C. This means that it can be represented by a wave function and that we have the eigenvalue relations (in SI-units):

$$\begin{aligned} \mathbf{S}^2\,\zeta(m_s) &= s(s+1)\hbar^2\,\zeta(m_s) \\ S_z\,\zeta(m_s) &= m_s\hbar\,\zeta(m_s) \quad . \end{aligned} \tag{E.12}$$

$\zeta(m_s)$ is the spin wave function, which depends on the discrete variable m_s.

We know empirically that the only possible value for s is one half (a mathematical proof of this will follow in section E.2), and the only two possible values for the eigenvalue m_s are $\pm 1/2$. Thus, we only have two eigenfunctions, which differ by the eigenvalues of the projections of their spins. We will refer to these functions as ζ_+ and ζ_- — *spin-up* and *spin-down*. ζ_+ and ζ_- are orthogonal, and they can be taken as normalised. Together they form a basis. From the commutation relations in C.3, and the ladder operators in (C.4), the action of the spin-component operators on ζ_+ and ζ_- can be derived as:

$$\begin{aligned} S_x\,\zeta_+ &= \frac{\hbar}{2}\,\zeta_- \qquad & S_y\,\zeta_+ &= \mathrm{i}\frac{\hbar}{2}\,\zeta_- \qquad & S_z\,\zeta_+ &= \frac{\hbar}{2}\,\zeta_+ \\ S_x\,\zeta_- &= \frac{\hbar}{2}\,\zeta_+ \qquad & S_y\,\zeta_- &= -\mathrm{i}\frac{\hbar}{2}\,\zeta_+ \qquad & S_z\,\zeta_- &= -\frac{\hbar}{2}\,\zeta_- \quad . \end{aligned} \tag{E.13}$$

In order to find the matrix form of the component operators in (E.13), we revert to the Dirac vector representation. For the basis functions ζ_+ and ζ_-, we introduce the *spinors*:

$$|+\rangle = \begin{pmatrix} 1 \\ 0 \end{pmatrix} \quad , \quad |-\rangle = \begin{pmatrix} 0 \\ 1 \end{pmatrix} \quad . \tag{E.14}$$

The matrix forms of the spin projection operators and $\mathbf{S}^2$ are then:

$$\begin{aligned} S_x &= \frac{\hbar}{2}\begin{pmatrix} 0 & 1 \\ 1 & 0 \end{pmatrix} \equiv \frac{\hbar}{2}\sigma_x \\ S_y &= \frac{\hbar}{2}\begin{pmatrix} 0 & -\mathrm{i} \\ \mathrm{i} & 0 \end{pmatrix} \equiv \frac{\hbar}{2}\sigma_y \end{aligned}$$

$$S_z = \frac{\hbar}{2}\begin{pmatrix}1 & 0\\ 0 & -1\end{pmatrix} \equiv \frac{\hbar}{2}\sigma_z$$

$$\mathbf{S}^2 = \frac{3\hbar^2}{4}\begin{pmatrix}1 & 0\\ 0 & 1\end{pmatrix} = S_x^2 + S_y^2 + S_z^2 \quad . \tag{E.15}$$

This includes a definition of the *Pauli spin-matrices*, σ_x, σ_y and σ_z. A consequence of the commutation relations of the spin matrices is that if an electron is specified to be in an eigenstate of J_z, for example, ζ_+, the components S_x and S_y will be completely undetermined, albeit their expectation values are zero. The total spin vector operator is:

$$\mathbf{S} = \frac{\hbar}{2}\boldsymbol{\sigma} \equiv \frac{\hbar}{2}\begin{pmatrix}\sigma_x\\ \sigma_y\\ \sigma_z\end{pmatrix} \quad . \tag{E.16}$$

Before the relativistic theory of the spin was developed, it was empirically known that the electron magnetic moment is μ_B, while the azimuthal projection of its spin is $\pm\hbar/2$ (regardless of choice of quantisation axis). This means that the relationship between the spin and the moment takes on a significant difference from that in (E.11) for the orbital angular momentum. The scalar spin magnetic moment is:

$$\mu_s = -\frac{\hbar e}{m_\mathrm{e}} m_s = -2\,\mu_\mathrm{B}\, m_s \ , \tag{E.17}$$

and the full vector is:

$$\boldsymbol{\mu}_s = -\frac{2\,\mu_\mathrm{B}}{\hbar}\mathbf{S} = -\mu_\mathrm{B}\,\boldsymbol{\sigma} \ . \tag{E.18}$$

For an interaction of the spin with a homogeneous magnetic field, parallel to $\hat{\mathrm{e}}_z$, the expectation values of this interaction term in the Hamiltonian for the spin states $\zeta_\pm$ are:

$$\left\langle \zeta_\pm \left| -\frac{2\,\mu_\mathrm{B}}{\hbar} B(\mathbf{r})\, S_z \right| \zeta_\pm \right\rangle = -\mu_\mathrm{B}\, B\, m_s = \mp\frac{\mu_\mathrm{B}}{2} B \, . \tag{E.19}$$

By introducing *g-factors* — also known as *gyromagnetic ratios* — (E.17) and (E.11) can be stated as:

$$\mu_l = -g_l \frac{\hbar e}{2 m_\mathrm{e}} m_l = -g_l\, \mu_\mathrm{B}\, m_l$$

$$\mu_s = -g_s \frac{\hbar e}{2 m_\mathrm{e}} m_s = -g_s\, \mu_\mathrm{B}\, m_s \quad . \tag{E.20}$$

The approximate values for the g-factors are $g_l \approx 1$ and $g_2 \approx 2$. There is no satisfactory classical explanation for why the spin g-factor is two. It is the result of a relativistic effect called Thomas precession. We shall not go through the details of this, but refer instead to general literature on the subject, for example, [4] or [3]. In section E.2, we show that a full relativistic calculation with the Dirac formalism directly gives the correct relation.

E.2 The Dirac Theory of the Electron Spin

In a relativistic treatment of a bound electron in an external field, a correct proportionality factor for the magnetic moment appears naturally, as does the spin–orbit interaction. For a more thorough treatise of this *Dirac theory*, we direct the reader to other volumes, such as [3, 5, 6] or [7]. Here, we will provide an overview and we will present some results, without going through all the details of the derivations. In relativistic quantum mechanics, it is no longer practical to use atomic units. An alternative is to use *natural units* [7], in which the speed of light is taken as unity. To avoid introducing another set of units in just one appendix section, we will here instead stick to SI-units.

E.2.1 The Dirac Equation

The relativistic version of the Schrödinger equation is formulated in a 4-vector notation. That is, an *event* is described by the vector:

$$\mathbf{x}^{\nu} = \begin{pmatrix} ct \\ x \\ y \\ x \end{pmatrix} \equiv \begin{pmatrix} x_0 \\ x_1 \\ x_2 \\ x_3 \end{pmatrix} . \tag{E.21}$$

Physical quantities and operators must be adapted accordingly. For example, the 4-vectors for energy-momentum and for the electromagnetic scalar-vector potentials are, respectively:

$$\mathbf{p}^{\nu} = \begin{pmatrix} E/c \\ p_x \\ p_y \\ p_x \end{pmatrix} , \quad \mathbf{A}^{\nu} = \begin{pmatrix} \phi/c \\ A_x \\ A_y \\ A_x \end{pmatrix} . \tag{E.22}$$

This formalism requires that the derivatives with regard to time and spatial parameters appear at the same order in equations of motion. The Dirac formulation of the Schrödinger equation fulfils this criterion by having a linear dependence on both energy and momentum. This is accomplished by introducing a set of four coefficient matrices, as will be shown below.

We write the field free Dirac equation as:

$$H\,\psi(\mathbf{x}^{\nu}) = \mathrm{i}\hbar \frac{\partial}{\partial t}\psi(\mathbf{x}^{\nu}) , \tag{E.23}$$

with the Hamiltonian:

$$H = c\boldsymbol{\alpha}\cdot\mathbf{p}^{\nu} = \alpha_0 mc^2 - \mathrm{i}\hbar c\left(\alpha_1\frac{\partial}{\partial x} + \alpha_2\frac{\partial}{\partial y} + \alpha_3\frac{\partial}{\partial z}\right) . \tag{E.24}$$

We have used the relativistic energy $E = mc^2$ and the matrix components of the 4-vector $\boldsymbol{\alpha}$ are:

$$\alpha_0 \equiv \begin{pmatrix} 1 & 0 & 0 & 0 \\ 0 & 1 & 0 & 0 \\ 0 & 0 & -1 & 0 \\ 0 & 0 & 0 & -1 \end{pmatrix} \quad , \quad \alpha_1 \equiv \begin{pmatrix} 0 & 0 & 0 & 1 \\ 0 & 0 & 1 & 0 \\ 0 & 1 & 0 & 0 \\ 1 & 0 & 0 & 0 \end{pmatrix}$$

$$\alpha_2 \equiv \begin{pmatrix} 0 & 0 & 0 & -\mathrm{i} \\ 0 & 0 & \mathrm{i} & 0 \\ 0 & -\mathrm{i} & 0 & 0 \\ \mathrm{i} & 0 & 0 & 0 \end{pmatrix} \quad , \quad \alpha_3 \equiv \begin{pmatrix} 0 & 0 & 1 & 0 \\ 0 & 0 & 0 & -1 \\ 1 & 0 & 0 & 0 \\ 0 & -1 & 0 & 0 \end{pmatrix} \quad . \tag{E.25}$$

Solutions to (E.23) will all be in the form of a vector, where each of the four components is a wave function, dependent on a 4-vector and on spin. For a free particle, a tentative solution for each of these component functions is a product of a spin function and a plane wave, as in:

$$\psi(\mathbf{x}^\nu) = \zeta(m_s) \exp\left[-\frac{\mathrm{i}}{\hbar}\mathbf{p}\cdot\mathbf{r} - Et\right] , \tag{E.26}$$

where $\zeta(m_s)$ is a spinor — see (E.14). If we substitute this in the Dirac equation, we get:

$$\begin{cases} \left[E - mc^2\right] \zeta_a(m_{sa}) = c\boldsymbol{\sigma}\cdot\mathbf{p}\, \zeta_b(m_{sb}) \\ \left[E + mc^2\right] \zeta_b(m_{sb}) = c\boldsymbol{\sigma}\cdot\mathbf{p}\, \zeta_a(m_{sa}) \end{cases} , \tag{E.27}$$

where $\boldsymbol{\sigma}$ is the Pauli matrix vector — see (E.15).

In the limit of zero momentum, this leaves us with two solutions with positive energy and two with negative:

$$E > 0 : \begin{cases} \psi_a(\mathbf{x}^\nu)\, \zeta_{a+} \\ \Psi_a(\mathbf{x}^\nu)\, \zeta_{a-} \end{cases} \quad \text{and} \quad E < 0 : \begin{cases} \psi_b(\mathbf{x}^\nu)\, \zeta_{b+} \\ \Psi_b(\mathbf{x}^\nu)\, \zeta_{b-} \end{cases} \quad . \tag{E.28}$$

The positive and negative values of the eigenenergies correspond to the electron and its anti-particle, the positron. If a positive energy is chosen, two of the solutions will dominate over the other two, and at a first approximation only these need to be retained. We choose the sign of the electron mass such that the two functions Ψ_{b+} and Ψ_{b-} can be discarded. We then make the assumption that $\Psi_a(\mathbf{x}^\nu)\zeta_{a+}$ and $\Psi_a(\mathbf{x}^\nu)\zeta_{a-}$ are very close to the true spin-up and spin-down eigenstates for the electron (see, for example, [7] or [3]).

E.2.2 A Relativistic Electron in an Electromagnetic Field

To describe an interacting electron, we need to transform the energy and momentum operator components in (E.23) according to:

$$
\begin{aligned}
E &\to E + e\phi(\mathbf{r}) = \mathrm{i}\hbar\frac{\partial}{\partial t} + e\phi(\mathbf{r}) \\
\mathbf{p} &\to \mathbf{p} + e\mathbf{A}(\mathbf{r}) = -\mathrm{i}\hbar\nabla + e\mathbf{A}(\mathbf{r}) \ .
\end{aligned} \tag{E.29}
$$

$\phi(\mathbf{r})$ is the spatially dependent scalar potential, and A_x, A_y and A_z are the Cartesian components of the vector potential $\mathbf{A}(\mathbf{r})$. The Dirac equation now becomes:

$$
\begin{aligned}
\Bigg\{ \left[\mathrm{i}\hbar\frac{\partial}{\partial t} + e\phi(\mathbf{r})\right] - m_\mathrm{e}c^2\alpha_0 + \left[\mathrm{i}\hbar c\frac{\partial}{\partial x} - ecA_x(\mathbf{r})\right]\alpha_1 + \left[\mathrm{i}\hbar c\frac{\partial}{\partial y} - ecA_y(\mathbf{r})\right]\alpha_2 \\
+ \left[\mathrm{i}\hbar c\frac{\partial}{\partial z} - ecA_z(\mathbf{r})\right]\alpha_3 \Bigg\}\psi(x^\nu) = 0 \ .
\end{aligned} \tag{E.30}
$$

For an electron in a single central potential, emanating from a massive positive charge at the coordinate origin, (E.30) represents a relativistic version of the hydrogen Schrödinger equation. With a somewhat lengthy derivation, see for example [3] or [7], one can from this derive an approximative form for the electron parts of the complete solution, where we assume that electron velocities are not too high and where only the first-order relativistic correction is kept:

$$
\begin{aligned}
\Bigg\{ \frac{[-\mathrm{i}\hbar\nabla + e\mathbf{A}(\mathbf{r})]^2}{2m_\mathrm{e}} - e\phi(\mathbf{r}) - \frac{(-\mathrm{i}\hbar\nabla)^4}{8c^2m_\mathrm{e}^3} + \frac{e\hbar}{2m_\mathrm{e}}\mathbf{B}(\mathbf{r})\cdot\boldsymbol{\sigma} \\
+ \frac{e\hbar}{4c^2m_\mathrm{e}^2}\boldsymbol{\sigma}\cdot[\boldsymbol{\mathcal{E}}(\mathbf{r})\times(-\mathrm{i}\hbar\nabla)] - \frac{e\hbar^2}{4c^2m_\mathrm{e}^2}\boldsymbol{\mathcal{E}}(\mathbf{r})\cdot\nabla \Bigg\}\psi(\mathbf{r}) = E\,\psi(\mathbf{r}) \ .
\end{aligned} \tag{E.31}
$$

The term in (E.31) with the factor $\mathbf{B}(\mathbf{r})\cdot\boldsymbol{\sigma}$ shows that there is an energy contribution from an interaction with an external magnetic field, and from a classical point of view, this can be seen as a consequence of the fact that the electron possesses a magnetic moment. This has not been artificially added to the equation but is an inherent quality emerging in the relativistic treatment. The thereafter following term corresponds to the spin-orbit interaction, and the last one is the Darwin term (see, for example, [5] or [6] for details about the relativistic hydrogen Schrödinger equation). Equation (E.31) will be used extensively in appendix F, where we further review magnetic interactions relevant for atomic systems.

Already here, we will take one closer look at the electron's interaction with a magnetic field. With the proportionality factor included, the relevant term in (E.31) is:

$$
\frac{e\hbar}{2m_\mathrm{e}}\mathbf{B}(\mathbf{r})\cdot\boldsymbol{\sigma} \quad . \tag{E.32}
$$

This we can compare with equations for the Zeeman effect (11.1), the definitions of the electron magnetic moment (4.2), the Pauli spin matrices (E.16), and the Bohr magneton, with the result:

$$
g_s = 2 \ . \tag{E.33}
$$

It should be noted that even though the Dirac theory gives the exact value of two for the electron spin g-factor, this is still an approximation. Taking into account quantum electrodynamics (see for example [5, 8] or [9]), other small corrections do appear. The currently (2019) best reported experimental value of g_s is [9, 10]:

$$g_s = 2(1 + a_e) \ , \tag{E.34}$$

where the electron magnetic moment anomaly, a_e, is:

$$a_e \approx 1.15965218091 \times 10^{-3} \ . \tag{E.35}$$

Further Reading

Quantum Mechanics of One- and Two-Electron Atoms, by Bethe [5]
Quantum theory of atomic structure, by Slater [3]
Quantum Electrodynamics, by Landau & Lifshitz [8]
Physics of Atoms and Molecules, by Bransden & Joachain [6]
Quantum Mechanics: An Experimentalists Approach, by Commins [7]

References

1. J.D. Jackson, *Classical Electrodynamics*, 3rd edn. (Wiley, New York, 1998)
2. L.D. Landau, E.M. Lifshitz, *Quantum Mechanics — Course of Theoretical Physics, volume 3*, 3rd edn. (Butterworth-Heinemann, Amsterdam, 1981)
3. J.C. Slater, *Quantum theory of atomic structure* (McGraw-Hill, New York, 1960)
4. L.D. Landau, E.M. Lifshitz, *The Classical Theory of Fields — Course of Theoretical Physics, volume 2*, 4th edn. (Butterworth-Heinemann, Waltham, 1987)
5. H.A. Bethe, E.E. Salpeter, *Quantum Mechanics of One- and Two-Electron Atoms* (Springer-Verlag, Berlin, 1957)
6. B.H. Bransden, C.J. Joachain, *Physics of Atoms and Molecules*, 2nd edn. (Prentice Hall, Harlow, England, 2003)
7. E.D. Commins, *Quantum mechanics: an experimentalist's approach* (Cambridge University Press, New York, 2014)
8. L.D. Landau, E.M. Lifshitz, *Quantum Electrodynamics — Course of Theoretical Physics, volume 4*, 3rd edn. (Butterworth-Heinemann, Waltham, 1982)
9. P.J. Mohr, B.N. Taylor, D.B. Newell, Rev. Mod. Phys. **84**, 1527 (2012)
10. D. Hanneke, S. Fogwell, G. Gabrielse, Phys. Rev. Lett. **100**, 120801 (2008)

Appendix F
Magnetic Interactions in Multielectron Atoms

Magnetic effects play central roles in the analyses for many chapters of this book. We often use a vector model approach for atoms with many electrons, with Hamiltonians that are somewhat approximative, and to an extent motivated in a handwaving fashion. Albeit the formalism presented in the above chapters reproduce empiric data well, a more robust physical theory is desirable for more profound studies.

In this appendix, we treat magnetic interactions with a more rigorous electromagnetic model and show that the Hamiltonians used in, for example, chapters 4, 9, 10, and 11 are indeed justified. In this endeavour, we will rely on the theory of the electron spin, presented in appendix E. An even more elaborate approach would be to treat the problem with many-body quantum electrodynamics (see, for example, [1] or [2]). However, such an approach falls beyond the scope of this volume.

F.1 General Formalism

The treatment of the electron spin in appendix E may serve as a starting point for a study of the spin–orbit and Zeeman effects for a hydrogenic atom. For a multielectron atom, we have to include *all* electrostatic and magnetic interactions of one electron with the nucleus, as well as with an external field, and not least also with every other electron. The approach here is to begin with the hydrogen atom Dirac equation in the form of (E.31). To that, we will then add the effects arising from the presence of many, mutually interacting, electrons.

To see how we must modify the interaction Hamiltonian included in (E.31), it is practical to divide it into one part that is a sum of all single-electron contributions, and another which is the sum of all cross-interaction terms:

$$H \equiv \sum_{i=1}^{N} H_i + \sum_{i=1}^{N} \sum_{j<i}^{N} H_{ij} \,. \tag{F.1}$$

A. Kastberg, *Structure of Multielectron Atoms*, Springer Series on Atomic, Optical, and Plasma Physics 112, https://doi.org/10.1007/978-3-030-36420-5

For the individual electron Hamiltonian, we need to take (E.31) and substitute the scalar potential and the electric field with the sums of the contributions from the nucleus and from a potential external field:

$$
\begin{aligned}
\mathcal{E}(\mathbf{r}) &\rightarrow \mathcal{E}_{\text{ext}}(\mathbf{r}_i) + \frac{Z}{r_i^3}\,\mathbf{r}_i \\
\phi(\mathbf{r}) &\rightarrow \phi(\mathbf{r}_i)_{\text{ext}} + \frac{Z}{r_i} \quad .
\end{aligned}
\tag{F.2}
$$

We have converted to atomic units. The vector fields $\mathbf{A}(\mathbf{r})$ and $\mathbf{B}(\mathbf{r})$ will be exclusively the contribution from external fields. With these modifications, H_i becomes:

$$
\begin{aligned}
H_i = {} & \frac{1}{2}\left[-\mathrm{i}\nabla_i + \mathbf{A}_{\text{ext}}(\mathbf{r}_i)\right]^2 - \phi_{\text{ext}}(\mathbf{r}_i) - \frac{Z}{r_i} - \frac{\alpha^2 \nabla_i^4}{8} + \frac{1}{2}\mathbf{B}_{\text{ext}}(\mathbf{r}_i)\cdot\boldsymbol{\sigma}_i \\
& - \frac{\mathrm{i}\alpha^2}{4}\boldsymbol{\sigma}_i \cdot \left\{\left[\mathcal{E}_{\text{ext}}(\mathbf{r}_i) + \frac{Z}{r_i^3}\mathbf{r}_i\right] \times \nabla_i\right\} - \frac{\alpha^2}{4}\left(\mathcal{E}_{\text{ext}}(\mathbf{r}_i) + \frac{Z}{r_i^3}\mathbf{r}_i\right)\cdot\nabla_i \quad .
\end{aligned}
\tag{F.3}
$$

In (F.3), $\boldsymbol{\sigma}_i$ is the Pauli spin operator for electron i, defined in (E.16). $\mathbf{A}_{\text{ext}}$, $\mathcal{E}_{\text{ext}}$, $\mathbf{B}_{\text{ext}}$ and ϕ_{ext} all refer to external fields, quantified in atomic units.

For the two-electron term, H_{ij}, we must consider (E.31) with all fields being ones appearing at $\mathbf{r}_i$ as a consequence of the presence of an electron at $\mathbf{r}_j$. On the other hand, we should not retain contributions from applied fields. Given the way the summation has been defined in (F.1), great care has to be taken to include magnetic effects not only from electron i on electron j, but also vice versa [3].

The relevant electric field expression that should go into (E.31) is:

$$
\mathcal{E}_{ji}(\mathbf{r}_i) \rightarrow -\frac{\mathbf{r}_{ij}}{r_{ij}^3} \quad .
\tag{F.4}
$$

This should be understood as the field caused by the electron at $\mathbf{r}_j$ felt by the one at $\mathbf{r}_i$. We use the definitions $\mathbf{r}_{ij} \equiv \mathbf{r}_i - \mathbf{r}_j$, $\mathbf{r}_{ji} = -\mathbf{r}_{ij}$, and $r_{ij} \equiv |\mathbf{r}_{ij}|$. The electric field is present in the last two terms on the left of (E.31), and we must here consider the field emanating from one electron, interacting with the momentum operator of the other. Since this interaction will occur in both directions, we must duplicate these two terms, but with reversed indices. The contributions to the pair interaction Hamiltonian H_{ij} from $\mathcal{E}_{ji}(\mathbf{r}_i)$ and $\mathcal{E}_{ij}(\mathbf{r}_j)$ will thus be:

$$
\frac{\alpha^2}{4r_{ij}^3}\left[\mathrm{i}\boldsymbol{\sigma}_i\cdot(\mathbf{r}_{ij}\times\nabla_i) + \mathrm{i}\boldsymbol{\sigma}_j\cdot(\mathbf{r}_{ji}\times\nabla_j) + \mathbf{r}_{ij}\cdot\nabla_i + \mathbf{r}_{ji}\cdot\nabla_j\right] \quad .
\tag{F.5}
$$

For the $\mathbf{A}$ and $\mathbf{B}$ fields, we think of each moving electron as representing a current density $\mathcal{J}$, which will contribute both to a vector potential $\mathbf{A}$ and to a magnetic flux density $\mathbf{B}$. Likewise, we can regard each electron spin as a manifestation of a magnetisation, $\mathcal{M}$, which will also contribute to both $\mathbf{A}$ and $\mathbf{B}$. These two contributions to the magnetic field felt by one electron could, respectively, be seen as those emanating from interactions with the other electron's orbital angular momentum and with its spin.

From classical electrodynamics, the relations between a current density and the fields **A** and **B** (in SI-units) are [4]:

$$\mathrm{d}\mathbf{A} = \frac{\mu_0}{4\pi}\frac{\mathcal{J}}{r}\,\mathrm{d}V$$
$$\mathrm{d}\mathbf{B} = \frac{\mu_0}{4\pi}\frac{\mathcal{J}\times\mathbf{r}}{r^3}\,\mathrm{d}V \ . \tag{F.6}$$

This should be understood as the contributions to **A** and **B** from $\mathcal{J}$ in the volume element dV, and the meaning of **r** is the distance from the current element to the point where the field should be measured. The corresponding equations for the influence of a magnetisation are:

$$\mathrm{d}\mathbf{A} = \frac{\mu_0}{4\pi}\frac{\mathcal{M}\times\mathbf{r}}{r^3}\,\mathrm{d}V$$
$$\mathrm{d}\mathbf{B} = -\frac{\mu_0}{4\pi}\left[\frac{\mathcal{M}}{r^3} - \frac{3(\mathcal{M}\cdot\mathbf{r})\mathbf{r}}{r^5}\right]\mathrm{d}V \ , \tag{F.7}$$

with **r** now the distance from the spin giving rise to the magnetisation to the observation point.

The vector potential $\mathbf{A}_{ji}$ and the magnetic field $\mathbf{B}_{ji}$ that we need are those felt by the electron at $\mathbf{r}_i$ due to the interaction with that at $\mathbf{r}_j$. This we can now take from (F.6) and (F.7), and if we at the same time revert to atomic units, the relevant fields become:

$$\mathbf{A}_{ji}(\mathbf{r}_i) = \alpha^2\frac{\mathcal{J}_j(\mathbf{r}_i)}{r_{ij}} + \alpha^2\frac{\mathcal{M}_{ji}(\mathbf{r}_i)\times\mathbf{r}_{ij}}{r_{ij}^3}$$
$$\mathbf{B}_{ji}(\mathbf{r}_i) = \alpha^2\frac{\mathcal{J}_j(\mathbf{r}_i)\times\mathbf{r}_{ij}}{r_{ij}^3} - \alpha^2\left\{\frac{\mathcal{M}_{ji}(\mathbf{r}_i)}{r_{ij}^3} - \frac{3\left[\mathcal{M}_{ji}(\mathbf{r}_i)\cdot\mathbf{r}_{ij}\right]\mathbf{r}_{ij}}{r_{ij}^5}\right\} . \tag{F.8}$$

The equations for $\mathbf{A}_{ji}$ and $\mathbf{B}_{ji}$ in (F.8) need to be converted to a more convenient operator form. This should be expressions that yield correct expectation values when operating on the product wave function $\psi_i\psi_j$, then multiplied from the left by $\psi_i^*\psi_j^*$ and integrated over all spatial parameters. Taking that into account, we can use, for the magnetisation vector, magnetic moment per volume. The density part of that will be provided when the expectation values are taken, and the magnetic moment can be related to the Pauli spin matrix vector for electron j, as given in (E.18) — with the Bohr magneton being one half in a.u. and keeping in mind that the negative charge means a negative magnetic moment. The relevant current density is the charge density for electron j multiplied by the velocity. The former will naturally occur when we take the expectation value and multiply with the charge (which is minus one in a.u.), and the latter is the momentum vector ($-\mathrm{i}\nabla$) divided by the electron mass (unity in a.u.). The equations that emerge are:

$$\mathbf{A}_{ji}(\mathbf{r}_i) = \alpha^2 \left(\frac{\mathrm{i}\nabla_j}{r_{ij}} - \frac{\boldsymbol{\sigma}_j \times \mathbf{r}_{ij}}{2r_{ij}^3} \right)$$

$$\mathbf{B}_{ji}(\mathbf{r}_i) = \alpha^2 \left[\frac{\mathrm{i}\nabla_j \times \mathbf{r}_{ij}}{r_{ij}^3} + \frac{\boldsymbol{\sigma}_j}{2r_{ij}^3} - \frac{3(\boldsymbol{\sigma}_j \cdot \mathbf{r}_{ij})\mathbf{r}_{ij}}{2r_{ij}^5} \right] . \tag{F.9}$$

The vector potential $\mathbf{A}$ is present in (E.31) in the square in the first term on the left. The square of the vector potential itself will be a very small contribution, and at this level of approximation, we chose to ignore it. The remaining contribution from $\mathbf{A}$ is the scalar product $-\mathrm{i}\nabla\cdot\mathbf{A}(\mathbf{r})$, where the vector potential should be taken from $\mathbf{A}_{ji}(\mathbf{r}_i)$ in (F.9), and the gradient as ∇_i. This results in the following terms for the pair Hamiltonian:

$$\alpha^2 \left[\frac{\nabla_i \cdot \nabla_j}{r_{ij}} + \mathrm{i}\frac{\boldsymbol{\sigma}_j \cdot (\mathbf{r}_{ij} \times \nabla_i)}{2r_{ij}^3} \right] . \tag{F.10}$$

In the second term, we have rearranged the vectors in the triple product.

The magnetic field $\mathbf{B}$ in (E.31) appears in just one term, as a scalar product with the Pauli spin vector. Using the expression in (F.9), the three contributions to H_{ij} become:

$$\frac{\alpha^2}{2} \left[\mathrm{i}\frac{\boldsymbol{\sigma}_i \cdot (\nabla_j \times \mathbf{r}_{ij})}{r_{ij}^3} + \frac{\boldsymbol{\sigma}_i \cdot \boldsymbol{\sigma}_j}{2r_{ij}^3} - \frac{3(\boldsymbol{\sigma}_j \cdot \mathbf{r}_{ij})(\boldsymbol{\sigma}_i \cdot \mathbf{r}_{ij})}{2r_{ij}^5} \right] . \tag{F.11}$$

The last two terms in (F.11) represent the spin–spin interaction. They involve both spins and thus we do not need to add terms with inversed indices. The same is true for the first term in (F.10). It covers the mutual interaction between the two orbital angular momenta. The second term in (F.10) with reversed indices would yield exactly the first term in (F.11). Thus, the equations cover all mutual interactions between the two electrons.

A more complicated electromagnetic contribution to H_{ij} is the scalar potential. The scalar potential resulting from a moving charge should be a so-called Lienard-Weichert type [4]:

$$-\phi_j(\mathbf{r}_i) = -\frac{1}{r_{ij} - \frac{\mathbf{v}_j \times \mathbf{r}_{ij}}{c}} . \tag{F.12}$$

$\mathbf{v}_j$ is the velocity of electron j. Also the vector potential $\mathbf{A}$ should be modified in a corresponding way. In addition, since both electrons are moving, the interaction from one of them on the other must be a retarded potential [3].

The detailed derivation of the necessary modifications due to the above is long and complicated, and instead of including it, we refer to the original work on the issue [5]. Using the results from there, we find that we will get correct expressions if we substitute the scalar potential in (E.31) according to:

$$-\phi(\mathbf{r}) \rightarrow \frac{1}{r_{ij}} + \alpha^2 \frac{\nabla_i \cdot \nabla_j}{2r_{ij}} + \alpha^2 \frac{(\mathbf{r}_{ij} \cdot \nabla_i)(\mathbf{r}_{ij} \cdot \nabla_j)}{2r_{ij}^3} \quad . \tag{F.13}$$

We take into account the expressions in (F.5), (F.10), (F.11) and (F.13), and we can then write the pair interaction Hamiltonian as:

$$\begin{aligned} H_{ij} = & \frac{1}{r_{ij}} + \alpha^2 \frac{\nabla_i \cdot \nabla_j}{2r_{ij}} \\ & + \frac{\alpha^2}{4r_{ij}^3} \{2 (\mathbf{r}_{ij} \cdot \nabla_i) (\mathbf{r}_{ij} \cdot \nabla_j) - 2\mathrm{i} [\boldsymbol{\sigma}_i \cdot (\mathbf{r}_{ij} \times \nabla_j) - \boldsymbol{\sigma}_j \cdot (\mathbf{r}_{ij} \times \nabla_i)] \\ & + \boldsymbol{\sigma}_i \cdot \boldsymbol{\sigma}_j + \mathrm{i}\boldsymbol{\sigma}_i \cdot (\mathbf{r}_{ij} \times \nabla_i) - \mathrm{i}\boldsymbol{\sigma}_j \cdot (\mathbf{r}_{ij} \times \nabla_j) + \mathbf{r}_{ij} \cdot \nabla_i - \mathbf{r}_{ij} \cdot \nabla_j\} \\ & - \frac{3\alpha^2 (\boldsymbol{\sigma}_i \cdot \mathbf{r}_{ij}) (\boldsymbol{\sigma}_j \cdot \mathbf{r}_{ij})}{4r_{ij}^5} \quad . \end{aligned} \tag{F.14}$$

Equations (F.3) and (F.14) will enable us to calculate the most important magnetic interactions relevant for atomic structure, at the level of approximation used in this book. In the upcoming section we will look at specific magnetic interactions, and see which terms in (F.3) and (F.14) that are, respectively, responsible for them.

It should be noted that more rigorous methods can be applied for the analysis of magnetic effects in atoms. This is, for example, reviewed in [6]. One thing that has been left out is a so-called *Fermi contact term*, which appears when two electrons occupy the same point in space. More generally, we will use the equations herein derived only for the express purposes of validating assumptions made for more qualitative arguments deployed in various chapters, when analysing the spin-orbit effect, the Zeeman effect and hyperfine structure.

F.2 The Terms of the Hamiltonian

The Hamiltonians in (F.3) and (F.14) need to be dissected. We shall associate physical effects to the various components, and we will focus on those related to magnetic interactions. For electron orbital angular momenta, we use the identity:

$$\mathbf{L}_i = \mathbf{r}_i \times (-\mathrm{i}\nabla_i) \quad . \tag{F.15}$$

We also introduce the following:

$$\begin{aligned} \mathbf{L}_{ij} &\equiv \mathbf{r}_{ij} \times (-\mathrm{i}\nabla_i) \\ \mathbf{L}_{ji} &\equiv \mathbf{r}_{ij} \times (\mathrm{i}\nabla_j) \quad . \end{aligned} \tag{F.16}$$

The top definition represents the angular momentum of electron i around the position of electron j, and the bottom one the opposite.

Before identifying the magnetic interaction terms, we begin by discriminating other contributions. In the first term of the single-electron Hamiltonian (F.3), we can develop the square. This first yields $-\nabla^2/2$, which is just the kinetic energy. The other two parts of the square are $-\mathrm{i}\nabla_i \cdot \mathbf{A}_{\mathrm{ext}}$ and $\mathbf{A}^2_{\mathrm{ext}}/2$.

The atomic unit for magnetic flux density is $\hbar/ea_0^2$. In SI-units, that corresponds to a field of the order of 10^5 T, which means that a laboratory magnetic field expressed in a.u. will be represented by a very small number. The a.u. for the magnitude of the vector potential differs from the one for $\mathbf{B}$ by a factor a_0, which means that also that will be very small, and as a consequence we neglect the term $\mathbf{A}^2_{\mathrm{ext}}/2$ (it is the diamagnetic Zeeman effect — see section 11.5). The cross term, $-\mathrm{i}\nabla_i \cdot \mathbf{A}_{\mathrm{ext}}$, is an interaction between the orbital angular momentum of one electron and the external field, and we will come back to that shortly.

The contribution $-Z/r_i$ in (F.3) is the usual Coulomb potential for the interaction between an electron and a nucleus, and in (F.14), $1/r_{ij}$ is the interelectronic repulsion. The term proportional to ∇^4 in (F.3) is a relativistic correction. It will affect the overall energy, but it will not contribute to the magnetic interaction level splittings. The same is true for the very last term in (F.3), $-Z\alpha^2 \mathbf{r}_i \cdot \nabla_i/(4r_i^3)$. Also this is related to relativistic effects.

F.2.1 The Zeeman Effect

The Zeeman effect is the action on atomic charges by an external magnetic field. We neglect the magnetic moment of the nucleus, and we take the total Zeeman Hamiltonian as the sum of all terms in (F.3) that include $\mathbf{B}_{\mathrm{ext}}$ or $\mathbf{A}_{\mathrm{ext}}$. Every electron will experience a contribution to its Zeeman effect interaction, both from its orbital angular momentum and from its spin.

In the previous section, we identified the term responsible for the interaction between the orbital angular momentum and the external field as $-\mathrm{i}\nabla_i \cdot \mathbf{A}_{\mathrm{ext}}$. The general relation between the vector potential and the magnetic flux density, in the Coulomb gauge, is:

$$\mathbf{A} = (\mathbf{B} \times \mathbf{r})/2 \ , \tag{F.17}$$

and, using vector identities, we can rewrite the interaction term as:

$$-\mathrm{i}\nabla_i \cdot \mathbf{A}_{\mathrm{ext}} = -\frac{\mathrm{i}}{2} \nabla_i \cdot (\mathbf{B}_{\mathrm{ext}} \times \mathbf{r}_i) = \frac{\mathbf{B}_{\mathrm{ext}}}{2} \cdot [\mathbf{r}_i \times (-\mathrm{i}\nabla_i)] = \frac{1}{2} \mathbf{B}_{\mathrm{ext}} \cdot \mathbf{L}_i \ . \tag{F.18}$$

We have here taken the field $\mathbf{B}_{\mathrm{ext}}$ to be homogeneous.

The term responsible for the spin part of the Zeeman effect in (F.3) is:

$$\frac{1}{2} \mathbf{B}_{\mathrm{ext}} \cdot \sigma_i = \mathbf{B}_{\mathrm{ext}} \cdot \mathbf{S}_i \ . \tag{F.19}$$

If we sum (F.18) and (F.19), we find that the total Zeeman effect Hamiltonian (to the first order) is:

$$H_Z = \frac{1}{2}\sum_{i=1}^{N}(\mathbf{L}_i + 2\mathbf{S}_i)\cdot\mathbf{B}_{\text{ext}} \ . \tag{F.20}$$

This is in agreement with (11.6) and (11.13), which are used for the Zeeman effect in chapter 11.

F.2.2 Interelectronic Magnetic Interactions

Before analysing the intrinsic spin–orbit interaction, we look at smaller magnetic effects that involve angular momenta from more than one electron. In the cross-interaction Hamiltonian (F.14), there are several terms that fall into this category. There are two that together represent the *spin–spin interaction* — the magnetic interaction between two spins. These are the ones containing products of $\boldsymbol{\sigma}_i$ and $\boldsymbol{\sigma}_j$. Together, these terms make up the spin–spin interaction Hamiltonian for one electron pair:

$$\begin{aligned} H_{\text{ss}}(\mathbf{r}_{ij}) &= \alpha^2\,\frac{\boldsymbol{\sigma}_i\cdot\boldsymbol{\sigma}_j}{4\,r_{ij}^3} - \frac{3\,\alpha^2\,(\boldsymbol{\sigma}_i\cdot\mathbf{r}_{ij})\,(\boldsymbol{\sigma}_j\cdot\mathbf{r}_{ij})}{4\,r_{ij}^5} \\ &= \alpha^2\left[\frac{\mathbf{S}_i\cdot\mathbf{S}_j}{r_{ij}^3} - \frac{3\,(\mathbf{S}_i\cdot\mathbf{r}_{ij})(\mathbf{S}_j\cdot\mathbf{r}_{ij})}{r_{ij}^5}\right] \ . \end{aligned} \tag{F.21}$$

This is the expression used in (9.6) in section 9.1.

Also appearing in (F.14) are the terms proportional to $\boldsymbol{\sigma}_i(\mathbf{r}_{ij}\times\nabla_j)$ and $\boldsymbol{\sigma}_j(\mathbf{r}_{ij}\times\nabla_i)$. These constitute the coupling between the spin of electron i with the orbital angular momentum of electron j, and vice versa. This *spin-other-orbit interaction* can, for one pair of electrons, be written as:

$$H_{\text{soo}} = -\frac{\mathrm{i}\,\alpha^2}{2\,r_{ij}^3}\left[\boldsymbol{\sigma}_i\,(\mathbf{r}_{ij}\times\nabla_j) - \boldsymbol{\sigma}_j\,(\mathbf{r}_{ij}\times\nabla_i)\right] = \frac{\alpha^2}{r_{ij}^3}\left[\mathbf{L}_{ij}\cdot\mathbf{S}_j + \mathbf{L}_{ji}\cdot\mathbf{S}_i\right] \ , \tag{F.22}$$

which is the form used in (9.7).

Remaining in (F.14) are three terms that describe the *magnetic orbit–orbit interaction*. The corresponding Hamiltonian, used also in (9.5), is:

$$H_{\text{moo}} = \alpha^2\left[\frac{\nabla_i\cdot\nabla_j}{2\,r_{ij}} + \frac{(\mathbf{r}_{ij}\cdot\nabla_i)\,(\mathbf{r}_{ij}\cdot\nabla_j)}{2\,r_{ij}^3} + \frac{\mathbf{r}_{ij}\cdot(\nabla_i - \nabla_j)}{4\,r_{ij}^3}\right] \ . \tag{F.23}$$

The interactions discussed in this section are typically much smaller than the intrinsic spin-orbit coupling, at least for medium and heavy atoms. This comes about due to the fact that the intrinsic spin–orbit interaction scales as the fourth power of Z — see (4.15). For the lightest of the multielectron atoms, the situation is less clear cut. A prime example is helium, for which the contributions above are substantial enough to cause a breakdown of the Landé interval rule (see the discussion in section 9.1.4).

F.2.3 The Intrinsic Spin–Orbit Interaction

For the interaction between an electron's orbital angular momentum and its own spin, we take the external electric field to be zero ($\mathscr{E}_{\text{ext}}=0$ and $\phi_{\text{ext}}=0$), and we look for contributions of the form $\mathbf{L}_i \cdot \mathbf{S}_i$.

In the single-electron Hamiltonian (F.3), the following spin-orbit term can be identified:

$$-\frac{\mathrm{i}\,\alpha^2}{4}\boldsymbol{\sigma}_i \cdot \left(\frac{Z}{r_i^3}\mathbf{r}_i \times \nabla_i\right) = \frac{\alpha^2 Z}{2r_i^3}\mathbf{L}_i \cdot \mathbf{S}_i \ . \tag{F.24}$$

This is the spin–orbit interaction for a single electron, and for a multielectron atom this should be summed over all electrons.

Additionally, there are terms related to the spin–orbit interaction present also in the cross-term Hamiltonian (F.14), namely:

$$\frac{\mathrm{i}\,\alpha^2}{4r_{ij}^3}\left[\boldsymbol{\sigma}_i \cdot (\mathbf{r}_{ij} \times \nabla_i) - \boldsymbol{\sigma}_j \cdot (\mathbf{r}_{ij} \times \nabla_j)\right] = -\frac{\alpha^2}{2r_{ij}^3}\left[\mathbf{L}_{ij} \cdot \mathbf{S}_i + \mathbf{L}_{ji} \cdot \mathbf{S}_j\right] . \tag{F.25}$$

Note that this is *not* the interaction between the spin of one electron and the orbital angular momentum of another, which was dealt with in the previous section. Also this is an intrinsic spin–orbit interaction, but in this case it is one due to the orbital angular momentum contribution to an electron from the torque caused by its electrostatic interaction with another electron. Note the subtle difference from (F.22).

Adding the two contributions and summing for all N valence electrons must be done with some care. Equation (F.25) involves a pair of electrons, while (F.24) is for a single electron. Furthermore, the factor Z in (F.24) is for a single electron interacting with an unscreened nucleus. One way of writing the complete intrinsic spin-orbit Hamiltonian is:

$$H_{\text{SO}} = \frac{\alpha^2}{2}\left\{\sum_{i=1}^{N}\left[\frac{Z\,\mathbf{L}_i \cdot \mathbf{S}_i}{r_i^3} - \sum_{j\neq i}^{N}\frac{1}{r_{ij}^3}\left(\mathbf{L}_{ij} \cdot \mathbf{S}_i + \mathbf{L}_{ji} \cdot \mathbf{S}_j\right)\right]\right\} . \tag{F.26}$$

We can see in (F.26) that for a valence electron, inner orbital electrons will screen the nuclear charge also with regard to the spin-orbit effect. We also know from section 6.2 that, to a good approximation, contributions to the overall $\mathbf{L}$ and $\mathbf{S}$ from closed inner orbitals can be neglected, since the time average of their combined contributions is spherically symmetric.

It is a natural extension from this to look for an approximation to (F.26), which includes an effective nuclear charge — in the same spirit as in (5.4) — and where we only include the valence electrons. The motivation for this is strengthened by the practicality of obtaining a spin–orbit interaction Hamiltonian for multielectron atoms roughly analogous to the hydrogenic version in (4.8), yielding energy expectation values that can be clearly expressed in terms of quantum numbers, or sums thereof. This leads to a simplification of (F.26) in the form of:

$$H_{\mathrm{SO}} = \frac{\alpha^2}{2} \sum_{i=1}^{N_{\mathrm{v}}} \frac{Z_{\mathrm{eff}-i}}{r_i^3} \mathbf{L}_i \cdot \mathbf{S}_i = \sum_{i=1}^{N_{\mathrm{v}}} \xi(r_i)\, \mathbf{L}_i \cdot \mathbf{S}_i \quad . \tag{F.27}$$

The summation is now only over the valence electrons, and in the last line we define the radial fine-structure function $\xi(r_i)$.

The equations (F.26) and (F.27) are consistent with (4.18) and (4.19). Note that, for a neutral alkali atom, $Z_{\mathrm{eff}-i}$ will be of the order of unity. The summation made in (4.19) is a considerable approximation, but within the CFA and with the simplifications of the vector model, this is warranted.

F.2.4 Hyperfine Structure

From the single electron Hamiltonian (F.3), we can also derive an expression for the magnetic dipole hyperfine structure. The approach here is to consider the external field in (F.3) as one occurring due to the finite magnetic dipole moment of the nucleus. We will derive the basic formalism for an example with a one-electron atom.

The terms in (F.3) associated with an external magnetic field are the ones containing $\mathbf{A}_{\mathrm{ext}}(\mathbf{r})$ and $\mathbf{B}_{\mathrm{ext}}(\mathbf{r})$. The vector potential appears in a square, and after developing the latter, we can neglect the contribution proportional to $\mathbf{A}^2_{\mathrm{ext}}$, with the rationale that we are dealing with a small perturbation (the same argument was made earlier). Writing the two fields as $\mathbf{A}_I(\mathbf{r})$ and $\mathbf{B}_I(\mathbf{r})$, to make the point that their source is the nuclear magnetic moment (with the nuclear spin described by the quantum numbers I and M_I), the hfs Hamiltonian can be taken as:

$$H_{\mathrm{hfs-MD}} = -\mathrm{i}\,\nabla_i \cdot \mathbf{A}_I(\mathbf{r}_i) + \frac{1}{2}\mathbf{B}_I(\mathbf{r}_i) \cdot \boldsymbol{\sigma}_i \quad . \tag{F.28}$$

For consistency, we retain the index i, even though we are for the moment only concerned with a one-electron atom.

We have again a magnetic field generated by a spinning charge. This means that we can follow an analysis analogous to that beginning with (F.7) for the magnetic interaction between two spinning electrons. We can take the relation between the nuclear magnetisation and the two fields in (F.28 (expressed in a.u.) as:

$$\mathrm{d}\mathbf{A}_I = \alpha^2 \frac{\boldsymbol{\mathcal{M}}_I \times \mathbf{r}_i}{r_i^3}\, \mathrm{d}V_i$$

$$\mathrm{d}\mathbf{B}_I = -\alpha^2 \left[\frac{\boldsymbol{\mathcal{M}}_I}{r_i^3} - \frac{3(\boldsymbol{\mathcal{M}}_I \cdot \mathbf{r}_i)\,\mathbf{r}_i}{r_i^5} \right] \mathrm{d}V_i \quad . \tag{F.29}$$

As usual we have chosen the nucleus as the CoM, and thus dV_i is the volume element around the position vector of the electron, $\mathbf{r}_i$.

When the magnetisation field appears in an expectation value integral, it will yield the magnetic moment, but in contrast to the discussion following (F.8), we must here take the nuclear magnetic moment. If we write the latter in atomic units — see (10.18) — it is:

$$\boldsymbol{\mu}_I = \frac{g_I}{2m_\mathrm{p}}\mathbf{I} \quad , \tag{F.30}$$

with the proton mass expressed in units of m_e. All this taken into account, we can rewrite (F.28) as:

$$H_{\mathrm{hfs-MD}} = -\frac{\mathrm{i}\alpha^2 g_I}{2m_\mathrm{p}}\frac{\nabla_i\cdot(\mathbf{I}\times\mathbf{r}_i)}{r_i^3} - \frac{\alpha^2 g_I}{4m_\mathrm{p}}\left[\frac{\mathbf{I}\cdot\boldsymbol{\sigma}_i}{r_i^3} - \frac{3(\mathbf{r}_i\cdot\boldsymbol{\sigma}_i)(\mathbf{I}\cdot\mathbf{r}_i)}{r_i^5}\right] \quad . \tag{F.31}$$

Following the same procedures as in the preceding section, this can be converted to:

$$H_{\mathrm{hfs-MD}} = \frac{\alpha^2 g_I}{2m_\mathrm{p}}\mathbf{I}\cdot\left[\frac{\mathbf{L}_i}{r_i^3} - \frac{\mathbf{S}_i}{r_i^3} + \frac{3\mathbf{r}_i(\mathbf{S}_i\cdot\mathbf{r}_i)}{r_i^5}\right] \quad . \tag{F.32}$$

This closely resembles the operator expression inside the expectation value bracket in (10.41) for the hfs a-constant. A reasonable way to proceed from here is to assume that the perturbation is small, and that a good approximation to the expectation value of the Hamiltonian can be obtained by projecting $\mathbf{L}_i$ and $\mathbf{S}_i$ in (F.32), using, for example, the Wigner–Eckart theorem. This follows precisely the philosophy proscribed in section 10.2.3, and the end result will be the same.

Further Reading

Quantum theory of atomic structure, by Slater [3]
Atomic Many-Body Theory, by Lindgren & Morrison [1]
The Classical Theory of Fields, by Landau & Lifshitz [4]

References

1. I. Lindgren, J. Morrison, *Atomic Many-Body Theory*, 2nd edn. (Springer Verlag, Berlin, 1986)
2. I. Lindgren, *Relativistic Many-Body Theory, A New Field-Theoretical Approach*, 2nd edn. (Springer-Verlag, Switzerland, 2016)
3. J.C. Slater, *Quantum theory of atomic structure* (McGraw-Hill, New York, 1960)
4. L.D. Landau, E.M. Lifshitz, *The Classical Theory of Fields — Course of Theoretical Physics, volume 2*, 4th edn. (Butterworth-Heinemann, Waltham, 1987)
5. C.G. Darwin, The London, Edinburgh, and Dublin Philosophical Magazine and Journal of Science **39**, 537 (1920)
6. H.A. Bethe, E.E. Salpeter, *Quantum Mechanics of One- and Two-Electron Atoms* (Springer-Verlag, Berlin, 1957)

Index

A. Kastberg, *Structure of Multielectron Atoms*, Springer Series on Atomic, Optical, and Plasma Physics 112, https://doi.org/10.1007/978-3-030-36420-5

The manufacturer's authorised representative in the EU is Springer Nature Customer Service Centre GmbH, Europaplatz 3, 69115 Heidelberg, Germany. If you have any concerns regarding our products, please contact ProductSafety@springernature.com

Printed and bound by CPI Group (UK) Ltd, Croydon, CR0 4YY
30/09/2026
02225648-0015